Cooperating Embedded Systems and Wireless Sensor Networks

Cooperating Embedded Systems and Wireless Sensor Networks

Edited by
Michel Banâtre
Pedro José Marrón
Anibal Ollero
Adam Wolisz

First published in Great Britain and the United States in 2008 by ISTE Ltd and John Wiley & Sons, Inc.

ISTE Ltd
6 Fitzroy Square
London W1T 5DX
UK

John Wiley & Sons, Inc.
111 River Street
Hoboken, NJ 07030
USA

www.iste.co.uk

www.wiley.com

Library of Congress Cataloging-in-Publication Data

Cooperating embedded systems and wireless sensor networks / edited by Michel Banâtre... [et al.].
    p. cm.
  Includes bibliographical references and index.
  ISBN 978-1-84821-000-4
  1. Embedded computer systems. 2. Sensor networks. I. Banâtre, Michel,
1950-
  TK7895.E42C68 2008
  681'.2--dc22

2007043927

British Library Cataloguing-in-Publication Data
A CIP record for this book is available from the British Library
ISBN: 978-1-84821-000-4

Printed and bound in Great Britain by Antony Rowe Ltd, Chippenham, Wiltshire.

FSC
Mixed Sources
Product group from well-managed
forests and other controlled sources
Cert no. SGS-COC-2953
www.fsc.org
© 1996 Forest Stewardship Council

Table of Contents

Chapter 3. Paradigms for Algorithms and Interactions 115
Andrea ZANELLA, Michele ZORZI, Elena FASOLO, Anibal OLLERO,
Ivan MAZA, Antidio VIGURIA, Marcelo PIAS, George COULOURIS
and Chiara PETRIOLI

Chapter 4. Vertical System Functions 259
Marcelo PIAS, George COULOURIS, Pedro José MARRÓN, Daniel MINDER,
Nirvana MERATNIA, Maria LIJDING, Paul HAVINGA, Şebnem BAYDERE,
Erdal ÇAYIRCI and Chiara PETRIOLI

Chapter 1

An Introduction to the Concept of Cooperating Objects and Sensor Networks

This chapter introduces concepts on cooperating objects and wireless sensor networks that will be used in the following chapters. It also introduces the Embedded WiSeNts Coordination Action. Finally, it presents an overview of the book and the relations between the following chapters.

1.1. Cooperating objects and wireless sensor networks

The evolution of embedded systems, together with the developments of technologies for integration and miniaturization, has led to the emergence of diverse devices, machines and physical objects of everyday use with embedded capabilities for computation, communication and interaction with the environment. These entities or objects range from millimeter-scale devices containing sensors, computing resources, bi-directional wireless communications and power supply to portable devices, home appliances, consumer electronic products or even machines and vehicles with on-board embedded controllers. The performance of all the above objects is limited by technology constraints and by the cost of performing particular functions and tasks. However, the cooperation of these objects may provide unprecedented capabilities and support for applications in many different fields. This cooperation requires the development of suitable methods and tools to support distributed interactions with the environment, adaptation to dynamic evolving working conditions, efficient use of resources, dependability and security constraints, amongst others.

Chapter written by Anibal OLLERO, Adam WOLISZ and Michel BANÂTRE.

Today's technical systems are becoming more and more complex. While, so far, individual entities have frequently been sufficient for efficient control of individual parts of the system, the growing complexity of the system necessitates the cooperation of individual entities. This is particularly true for embedded systems. Embedded systems are themselves characterized by an intrinsic need to interact with the environment. This interaction can take place in the form of sensing as well as actuation. Because of system complexity, isolated entities can no longer perform this interaction efficiently and achieve the required control objectives. Hence, in the interaction with, exploration of and control of the environment, cooperation between individual entities becomes a necessity.

A fundamental notion in the above context is the concept of a *Cooperating Object* (CO). A CO can be defined as a collection of sensors, actuators, controllers or other COs that communicate with each other and are able to achieve, more or less autonomously, a common goal. The sensors retrieve information from the physical environment. The actuators modify the environment in response to appropriate commands. Thus, sensors and actuators form the hardware interfaces with the physical world. The controllers process the information gathered by sensors and issue the appropriate commands to the actuators, in order to achieve control objectives. The inclusion of other cooperating objects as part of a CO itself in a recursive manner indicates that these objects can combine their sensors, controllers and actuators in a hierarchical way and are therefore able to create arbitrarily complex structures.

Wireless Sensor Networks, or more generally *Wireless Sensor and Actuator Networks*, are typical examples of the above-mentioned cooperation. Such networks consist of objects that are individually capable of simple sensing, actuation, communication and computation, but the full capabilities of such networks are reached only by the cooperation of all these objects. These networks can, in turn, cooperate themselves with other individual, intelligent objects, other networks, other controllers, or even users via proper interfaces. While these "cooperating objects" represent a potentially disruptive technology, the concrete realization of this vision is still unclear.

A number of different system concepts have become apparent in the broader context of embedded systems over the past years. First, there is the classic concept of embedded systems as being mainly a control system for a physical process (machinery, automobiles, etc.). Second, more recently, the notion of pervasive and ubiquitous computing has evolved, where objects of everyday use can be endowed with some form of computational capacity, and perhaps with some simple sensing and communication facilities. Third, and most recently, the idea of "wireless sensor networks" has arisen, where entities that sense their environment are not operating individually, but collaborate together to achieve a well-defined purpose of supervision of a particular area, process, etc. We claim that these three types of actually quite diverse state of the art systems on one hand share some principal common features and, on the other hand, have some complementary aspects that make their combination very promising.

In particular, the important notions of control, heterogenity, wireless communication, dynamic/ad hoc nature and cost are prevalent to various degrees in each of these concepts.

The main vision is the conception of a future-proof system that combines the strong points of all three system concepts at least in the following functional aspects:

• support the control of physical processes in a similar way that embedded systems are able to do today;

• have as good support for device heterogeneity and spontaneity of usage as pervasive and ubiquitous computing approaches have today;

• be as cost-efficient and versatile in terms of the use of wireless technology as wireless sensor networks are.

Figure 1.1 illustrates the above concepts.

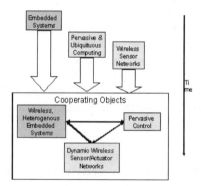

Figure 1.1. *Cooperating objects*

The commercial application of some concepts and technologies presented in this book is heavily dependent on the development of several supporting technologies. The first group of supporting technologies deals with miniaturization. It includes micro and nanotechnologies for new sensors and actuators and also the progress on micro and nano systems for computing and communications, including, for example, the "nets-on-chip" technology. The second group is the communication technologies themselves including new developments to support increasing mobility and bandwidth. The third group deals with power sources and includes not only the primary technologies to generate power but also the so-called energy harvesting technologies. The last group includes all aspects related to the intelligent management of cooperating objects interacting autonomously with the environment, and thus involving technologies such as networked control and networked robotics. All these technologies are not the subject of this book but they will influence the practical implementation of the presented concepts and techniques.

1.2. Embedded WiSeNts

This book is the result of the work developed in the Coordination Action (CA) entitled *Cooperating Embedded Systems for Exploration and Control featuring Wireless Sensor Networks – Embedded WiSeNts*, funded by the European Commission under the Information Society Technology (IST) priority within the 6th Framework Programme (FP6).

The project supports the establishment of a new research domain that integrates the broad context of embedded systems with ubiquitous computing and wireless sensor networks in support of COs. The key actions are concerned with integration of existing research in the field and related fields, supporting teaching and training in the area of COs and developing a technology roadmap to drive the vision forward.

Embedded WiSeNts is a joint effort between 12 partners from 10 different European countries: Technische Universität Berlin (coordinator, Germany), University of Cambridge (United Kingdom), University of Copenhagen (Denmark), Swedish Insitute of Computer Science (Sweden), University of Twente (The Netherlands), Yeditepe University (Turkey), Consorzio Interuniversitario Nazionale per l'Informatica (Italy), University of Padua (Italy), Swiss Federal Institute of Technology Zurich (Switzerland), Asociación de Investigación y Cooperación Industrial de Andalucía (Spain), Institut National de Recherche en Informatique et en Automatique (France) and Universität Stuttgart (Germany).

These partners are among the top research institutions in wireless communication and distributed computing as well as in cooperating objects in general, and are at the forefront of ubiquitous communication and wireless sensor networks in particular. The project took place from September 1st 2004 to December 31st 2006.

The three main goals of this CA have been as follows.

1. Supporting the integration of existing research

This goal included the development of a survey of Platforms and Tools with a critical evaluation of Wireless Sensor Network platforms, operating systems, programming and simulation environments and testbeds.

2. Road mapping for technology adoption

This goal included:

• State of the art studies: a survey of the current state of the art and open research issues.

• Visions for innovative applications: exploring application areas that could potentially be realized in a 10-year horizon once CO-technology becomes widely available.

• Research roadmap: describing emerging trends and technological opportunities and proposing a research agenda. Distinguishing features of the roadmap are:
      - list of important gaps in the field and current trends, including an estimation of the time when each gap will be solved;
      - market analysis rating the importance of different application areas; suggestions for predominant work areas that should be tackled in the future;
      - organization of research describing needed interactions between main research groups;
      - potential roadblocks or major inhibitors that hinder the acceptance of CO technology in society.

3. Promoting excellence in teaching and training on systems of cooperating objects

This goal included the improvement of teaching material (Teachware) as well as the dissemination and development of a repository for teachers and students listing courses and teaching materials on COs.

An intended important result has been to increase the awareness of cooperating object technology within the academic community and, most importantly, within the industrial producer and user community. The resulting strategic impact will be that the options and the potential for the use of Embedded WiSeNts technologies will be available to selected decision makers in academia and industry.

The practical impact could be double on the manufacturers of such devices, providing them with required information on the type of devices that will be required in the future as well as incorporating their feedback on technical feasibility, enabling these manufacturers to lead the market in the production of the required technology.

The impact could also be on the adopters of such devices and the concept of the cooperating object system more generally, allowing them to form an understanding of the possibilities, market chances, product options, possible services surrounding this concept, etc., again contributing to and ensuring decisive competitiveness advantages. As a consequence, the introduction of cooperating objects in actual products will be hastened and the overall efficiency could be improved.

1.3. Overview of the book

The four following chapters of this book correspond to the studies carried out in the framework of the Embedded WiSeNts CA. The objectives of these studies are:
   • an in-depth analysis of the current state of the art in Cooperating Embedded Systems and Wireless Sensor Networks;
   • to identify open issues and trends in the field.

The book aims to provide a comprehensive and detailed overview of the scenarios, paradigms, functions and system architectures dealing with COs.

Chapter 2: Applications and application scenarios

Chapter 2 provides an initial overview of CO applications and application scenarios. The main objective is to identify relevant state of the art, projects and activities in the CO domain. Some application scenarios enable us to better understand the area of CO in the broad sense of the term from two different perspectives: socio-economic and application-type aspects have been identified and developed. First, general application characteristics for all domains are identified. Then, a survey of state of the art CO projects is given.

In the next section of the chapter, CO applications have been classified in sectors which have a social and economic impact on society. The classification is then used as a basis for the analysis of common characteristics in the field of action. Some application scenarios from different domains have been chosen and they are given along with their typical parameters, requirements, roles, traffic, threads, legal/economic issues that best characterize the research performed in the field of cooperating objects in the broad sense of the term. Considering the current trends, the following areas which can benefit from cooperating objects are defined:

- control and automation
- home and office
- logistics
- transportation
- environmental monitoring for emergency services
- healthcare
- security and surveillance
- tourism
- education and training

In this chapter you can also find a summary of the outcome of the study which will later be used as a measurement for the importance of the application domain and will act as a means of weighting conflicting requirements.

Chapter 3: Paradigms for algorithms and interactions

This chapter provides an up-to-date overview of the fundamental design paradigms, algorithms and interaction patterns that enable the realization of systems based on COs. In order to cope with the large heterogenity of CO systems, the literature has been divided into four thematic areas, namely:

- Wireless Sensor Networks for Environmental Monitoring characterized by a large number of stationary sensor nodes, disseminated in a wide area and one (or a few) sink nodes, designated to collect information from the sensors and act accordingly. Depending on the user scenario, nodes can be either accurately placed in the area according to a pre-planned topology or randomly scattered over the area, except for a limited number of nodes that are placed in specific positions, for instance to guarantee connectivity or to act as beacons for the other nodes. Finally, a random topology is obtained when nodes are scattered in the area without any plan. Sensor nodes are often inaccessible, battery powered, or prone to failure due to energy depletion or crashes. Furthermore, network topology can vary over time due to the power on/off cycles that nodes go through to save energy. Generally, traffic in Wireless Sensor Networks for Environmental Monitoring flows mostly from sensors to one or more sinks and vice versa, and data may show a strong spatial correlation. Since nodes are usually battery powered and not (easily) rechargeable, power consumption is a primary issue.

• Wireless Sensor Networks with Mobile Nodes ranging from a network with only mobile nodes to a network with a trade-off between static and mobile nodes. The use of mobile nodes in sensor networks increases the capabilities of the network and allows dynamic adaptation to the environmental changes. Some applications of mobile nodes could be: collecting and storing sensor data in sensor networks reducing the power consumption due to multi-hop data forwarding, sensor calibration using mobile nodes with different and possibly more accurate sensors, reprogramming nodes "by air" for a particular application and network repairing when the static nodes are failing to sense and/or to communicate. The mobile nodes can collect the data from the sensors and send it to a central station by using a long-range radio technology, thus acting as mobile sinks. Usually, the traffic between static and mobile nodes is low, whereas the traffic among mobile nodes and the central station is high. The mobile nodes can also process the information gathered from the static nodes in order to reduce the data traffic. Power consumption is still an issue but mobile nodes can reduce the power consumed in multi-hop data forwarding. Furthermore, it is possible to have energy stations where mobile nodes can recharge their batteries.

• Autonomous Robotic Teams that make it possible to build more robust and reliable systems by combining redundant components. The topology is often pre-planned because the motion of the robots is always to some extent controlled and predictable, and the data traffic among the robots is usually higher (images, telemetry, etc.) than the traffic among the nodes of a WSN. Moreover, traffic patterns are more similar to the classic all-to-all paradigm considered in ad hoc networks. Furthermore, the cooperation and/or coordination of the robots require very heavy data processing. Nevertheless, Autonomous Robotic Teams usually do not have severe energy constraints, since robots can autonomously recharge their batteries when a low level of energy is detected. Also, mixed solutions including solar panels are possible. In opposition to the other distributed systems here considered, these teams can require real-time features, depending on the particular application.

• Inter-vehicular Networks, which is a promising approach to address critical road safety and efficiency, for example, coordinated collision avoidance systems. These networks exhibit characteristics that are specific to scenarios of high node mobility and, despite the constraints on the movement of vehicles, the network will tend to experience very rapid changes in topology. Some applications require communication with destinations that are groups of vehicle and thus the traffic pattern is predominantly multicast, while the data rate is generally rather high (car-to-car voice/video connections, web surfing, etc.). Unlike typical WSN deployment scenarios, vehicles can be instrumented with powerful sensors and radio to achieve long transmission ranges and high quality sensor data, since low-power consumption and small physical size are not an issue in this context.

As such, the above thematic areas differ in characteristics and requirements and, we believe, are representative of a number of possible application scenarios.

For each thematic area, the chapter provides a rather comprehensive survey of the most interesting algorithms proposed in the literature for the following aspects:

- MAC, routing, localization,
- data aggregation and data fusion,
- time synchronization,
- navigation,
- object coordination and cooperation.

The algorithms and paradigms are thus classified according to the requirements that have been identified in Chapter 2. In this way, the study makes it possible to clearly identify common features and differences among the different design paradigms adopted in each specific system, thus revealing the most promising research trends and research gaps that should be covered in the near future.

Therefore, the study collects and compares the different approaches and solutions proposed in the literature for the realization of the basic functionalities of CO-system, and a summary of the research gaps and the promising approaches regarding the algorithms and paradigms for the realization of CO-systems.

Chapter 4: Vertical system functions

This chapter complements the previous three chapters in identifying the relevant state of the art in the context of distributed COs and WSNs.

The set of characteristics exhibited in CO applications are more diverse than those found in applications of traditional wireless and wired networks. The set of requirements includes among other things the location and context of COs, security, privacy and trust, system scalability and reliability, to mention a few.

Critical factors impact on the architectural and protocol design of such applications. Organizing software and hardware components of COs into a framework that can cope with the inherent complexity of the overall system will be an important exercise for application developers.

The current operating systems proposed for WSNs and COs cannot offer all the required functionality to these applications. The main goal of this chapter is to discuss the roles and effects of vertical system functions (VFs), which are defined in this chapter as the functionalities that address the needs of applications in specific domains and in some cases a VF also offers minimal essential functionality that is missing from available real-time operating systems.

The chapter revisits and reviews the most relevant application requirements and discusses the most suitable VFs to address these needs. It is organized as follows:

• Definition of vertical system functions in the context of cooperating objects.

• Discussion of the characteristics and requirements of the CO application studied in Chapter 2.

• Different types of VFs to address the application needs (section 4.4). The VFs discussed are:

      - Context and location management.
      - Data consistency.
      - Communication functionality.
      - Security, privacy and trust.
      - Distributed storage and data search.
      - Data aggregation.
      - Resource management.
      - Time synchronization.

The chapter concludes with a summary of trends and open issues identified. Briefly, the most urgent issues that deserve the attention of the research community are the distributed accurate location of COs and system support for node mobility and sensor heterogenity. The software framework requires consolidation using standardized APIs and high-level descriptive language. In practical terms, there is a need for deployment and configuration of sensors in real application scenarios.

Chapter 5: System architectures and programming models

Key to the successful and widespread deployment of cooperating objects and sensor network technologies is the provision of appropriate programming abstractions and the establishment of efficient system architectures able to deal with the complexity of such systems. This chapter provides a survey about the current state of the art of programming models and system architecture for COs and motivates their importance for a successful development of these technologies. The second section of this chapter provides a brief introduction to the topics and motivates the need to design suitable programming abstractions for COs.

In the third section, the most relevant existing programming abstractions are surveyed and classified. A programming model is considered as "a set of abstractions and paradigms designed to support the use of computing, communication and sensing resources in an application" and to a system architecture as "the structure and organization of a computing system, as a set of functional modules and their interactions". The main reason for the development of these abstractions is to allow a programmer to design applications in terms of global goals and to specify interactions between high level entities (such as agents or roles), instead of explicitly managing the cooperation between individual sensors, devices or services. For example, the database abstraction makes it possible to consider a whole sensor network as a logical database, and performs network-wide queries over the set of sensors. The various paradigms are surveyed and a set of criteria making it possible to easily review their strengths and weaknesses are presented.

The next section presents the existing system architectures for COs at two different levels: first, system architectures of individual nodes, which includes the structure of the operating system running at node level and its facilities; second, system architectures supporting the cooperation of different nodes, such as communication models.

Finally, the chapter points out some of the limitations of current approaches, and proposes some research perspectives. In particular, programming paradigms should provide more support to facilitate programming and heterogenity, as well as scalability issues. Regarding system architectures, real-time aspects, which currently are not well addressed, will become increasingly important for COs. Dynamic maintenance (such as code deployment and runtime update support) is another important issue to address in future systems. Finally, effort is required to better integrate the various paradigms and systems into a unified framework.

Although the four chapters are focused on different aspects concerning COs, they are strictly inter-related. Figure 1.2 points out some of these relations.

Chapter 6: Cooperating objects roadmap and conclusions

The last chapter of this book provides a brief summary of the roadmap developed as part of the Embedded WiSeNts work with input from associated industrial and academic partners. A full version of this roadmap can be obtained as a separate book from Logos Verlag, Berlin and is also available online from the project website (http://www.embedded-wisents.org).

The roadmap, being the final result of the project, takes into account the information contained in the previous chapters to define the research trends and gaps, therefore identifying the potential research directions that the CO community will take in the

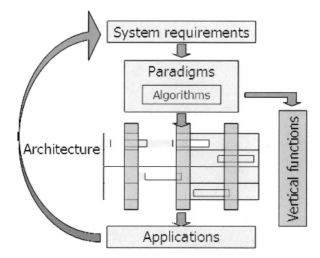

Figure 1.2. *Relations between chapters*

next 10 to 15 years. The executive summary provided as part of this chapter concludes with a series of recommendations and specific actions that can be taken into account by the research community, industry and end-users.

Chapter 2

Applications and Application Scenarios

2.1. Summary

This chapter provides a summary of the relevant state of the art, projects and activities in the cooperating object domain. It is the first of four studies, in Chapters 2-5 of this book, which are intended to give an in-depth analysis of current state of the art research in the domain. Results of the studies will be used as an input to the road mapping task identifying essential open issues critical for the development of future cooperating objects. The first step in these studies is the identification of relevant state of the art activities in the cooperating object domain, which is given here.

2.2. Introduction

The newly emerging micro-sensors and actuators open revolutionary ways for new applications in wireless communications area. By using low-power, low-bandwidth, low-cost tiny sensor nodes and pervasive computing phenomenon, it will be possible to change the way people live and their habits. Recent developments in wireless sensor technology have made people aware of environmental changes. The AmI (Ambient Intelligence) paradigm explains the case where the user is surrounded by intelligent and intuitive interfaces able to recognize and respond to his/her needs. It is possible to integrate AmI and wireless sensor network technology in order to monitor different environments and act according to sudden changes.

In the scope of this book, a cooperating object (CO) is defined as a collection of sensors, actuators, controllers or other COs that communicate with each other and are

Chapter written by Şebnem BAYDERE, Erdal ÇAYIRCI, İsa HACIOĞLU, Onur ERGIN, Anibal OLLERO, Ivan MAZA, Antidio VIGURIA, Phillipe BONNET and Maria LIJDING.

able to achieve, fairly autonomously, a common goal. The inclusion of other COs as part of a CO itself indicates that these objects can combine their sensors, controllers and actuators in a hierarchical way and are therefore able to create arbitrarily complex structures.

This chapter provides an overview of CO applications and application scenarios that can be readily understood today. The main objective of this study is to identify relevant state of the art projects and activities in the CO domain. For this purpose, both European and other projects outside Europe are considered. Some application scenarios that enable us to better understand the area of CO from socio-economic and application-type points of view have been identified and analyzed.

The applications and scenarios take into account the state of the art of current service-centric (control applications, pervasive or ubiquitous computing) as well as data-centric approaches (wireless sensor networks). In data-centric approaches, efficient management of data is the major concern whereas service-centric approaches are mostly concerned with the definition of the interface or API in order to provide functionality for the user. Hybrid scenarios where service-centric and data-centric technologies must be combined are also considered. Hybrid scenarios and applications have the potential to provide valuable clues for the identification of distinguishing and overlapping features of service and data-centric approaches. The wide spectrum of potential applications indicate that the constraints for one CO application domain may be much different from another CO application domain. CO applications can be classified in many different ways as each application has common features with others. In this chapter, sectors that can benefit from the CO paradigm and have social and economic impact in society are used as the basis of classification. Sectoral classification helped us to analyze the common characteristics and requirements of a specific field of action. Considering the current trends, sectoral areas which can benefit from cooperating objects are defined as follows: control and automation, healthcare, environmental monitoring, security and surveillance, logistics, home and office, transportation, tourism and education and training.

The rest of the document is organized as follows: section 2.3 contains general characteristics and requirements of data-centric, service-centric and hybrid CO applications. In section 2.4 a survey of state of the art CO projects are given in the order of evolvement. Section 2.5 contains classification of CO applications and projects into sectoral areas. Section 2.6 illustrates the object symbol set used for the functional description of CO scenarios. Section 2.7 contains scenarios from control, surveillance, monitoring and transportation domains along with their typical parameters, requirements, roles, traffic, threads, legal/economic issues that best characterize the research performed in the field. Finally, section 2.8 summarizes the outcome of the study to be used as a measurement for application domain's importance and acting as a means of weighting conflicting requirements.

2.3. Characteristics and requirements of applications

The characteristics of CO applications are quite different from traditional wireless and wired networks. There are critical factors influencing the architectural and protocol design of these applications, and these factors introduce some stringent constraints. Moreover, the constraints for one CO application domain may be much different from another CO application domain. For example, security requirements in health and security applications can be more critical than in a home application. Therefore, the "one size fits all" approach does not work for cooperating objects. In this section we enumerate and briefly explain the typical architectural trends that best characterize the CO applications and system requirements that influence the protocol and algorithm design in the wide sense of the term. These characteristics are used as the basis of analysis of open research issues in different application domains. State of the art research on common characteristics are given in Chapters 3 and 4. Also, in [32], a good categorization of the requirements and characteristics of wireless sensor networks is presented. The following two properties of CO applications are general *characteristics* that they can have.

Data traffic flow: the amount of data traveling inside the network determines the traffic characteristics of an application. In one application the data transferred among nodes can be limited to a few bytes for simple measurements, whereas heavy video-audio traffic can be conveyed in another application. Potentially, wireless sensor network traffic does not follow any known traffic patterns. It is non-stationary and highly correlated because of the event-driven characteristic of the WSN. When an event is detected, there are sporadic outbursts of high traffic; otherwise, most sensor nodes will remain asleep for long durations. The traffic characterization of the WSNs is very complex and difficult. All layers of the protocol stack affect the traffic pattern of the network.

Multipath phenomenon, human activities, background noise, node orientation, and interference from other nodes cause severe changes in the traffic pattern of a WSN. The protocols running on the network layer have significant effects on the traffic pattern of the sensor network as well. For example, the traffic characteristics change if two packets with the same destination are combined into one packet. This is a kind of data aggregation.

Network topology: in a CO application, sensor nodes may directly communicate with an actuator, or a sensor node sends its data to the actuator through several sensors. The first case implies a *single-hop* topology whereas the second requires a *multi-hop* sensor network. With current technology, the single-hop communication model is more trivial to establish than multi-hop communication networks. Multi-hop topologies have significant challenges such as routing, support for mobility and scalability, etc. Substantial research is still needed on the requirements of multi-hop set-ups in real life applications.

Indoor or outdoor: generally speaking, operating environments for CO applications are categorized as *indoor* and *outdoor*. Indoor applications are mostly implemented in home and office environments whereas roads, railways and forests may be some examples of outdoor environments for CO applications. Most of the factors listed in this section, such as localization, security and mobility, introduce more challenging requirements for outdoor applications.

The other properties, given below, are the general system *requirements* of CO applications. CO applications require some of these requirements, depending on the task on which they are focused. Different applications require different levels of importance of the below properties.

Automation: nodes can be remotely controlled or fully unattended and autonomous. In the applications of the latter class, nodes make autonomous decisions according to the collected information. Particularly, applications operating without human intervention and including robots and automated machinery require a high degree of autonomy.

Context awareness: in CO applications, some devices or objects may need to have information about the circumstances under which they operate and can react accordingly. These context aware objects may also try to make assumptions about themselves or the objects' (which they monitor or control) current situations.

Fault tolerance: it is highly possible that some sensor nodes will be lost during the operation of the network due to their limited power capacity and challenging operation environment conditions. In many outdoor applications, it is impossible to change batteries of sensor nodes. As a result, a CO network must be able to sustain its operations although it faces node failures.

Localization: there are several CO applications for target tracking and event detection, e.g. intrusion, forest fire, etc., that necessitate node and/or target localization. For this reason satellite based positioning systems are used. The most popular positioning system is GPS (global positioning systems) and it can be used in applications where scalability and cost per node requirements can be satisfied. However, the cost of equipping every node with a GPS unit cannot be tolerated in many applications. Also, mounting a GPS receiver to sensor nodes increases their size and their power consumption which is not wanted. Furthermore, in some environments such as indoors and underwater, GPS do not work. There are also some proposed GPS-free localization schemes for wireless sensor networks; however, it is still a significant challenging issue.

Mobility: in some applications, all physical components of the system may be static, whereas in others, the architecture may contain mobile nodes, especially applications which can benefit from autonomous robots in the field of action may require special support for mobility. Mobility support for multi-hop routing in infrastructureless networks is still a challenging issue. High mobility requirement of the application also affects the design for other characteristics such as localization and synchronization.

Networking infrastructure: CO networks can be *infrastructured* or *infrastructure-less (ad hoc)*. Even in some applications the data can be collected by some mobile nodes when passing by the source nodes. Having an infrastructure or not mostly depends on the operational area of the CO application. For example, some environmental monitoring and surveillance applications established in remote regions require infrastructureless operation, whereas others may benefit from other wireless and/or wired systems in the environment.

Node heterogenity: most CO applications include different types of nodes that have distinct hardware and software characteristics. For instance, in a precision agriculture application, there may be various sensor types like biological, chemical, temperature and humidity sensors.

Packaging for robustness: in many CO applications, the low-power, low-bandwidth, tiny sensors will be used in challenging operational environments. For instance, in a desert, sensor nodes must be covered with some housing structure in order to prevent them from high temperatures or other harsh desert conditions.

Power awareness: it is obvious that power consumption is one of the most crucial performance metrics and limiting factors in almost every CO application. In order to make a system practical for real-life scenarios which require long lifetime, efficient power consumption strategies must be developed. Long-lifetime is particularly expected from an application established for environmental monitoring. For instance, a CO application set up in a remote site such as a desert must have a long lifetime, since it is not easy to access that environment and replace the network with a new one. There is much research in the literature focused on developing more power efficient strategies.

Production and maintenance cost: depending on the application type, CO applications containing a large number of nodes and aiming at operating for a long time require low production or low maintenance cost. This need can be determined by some characteristics of the applications. For example, for the networks which are expected to stay alive and operate for a long time, low maintenance cost becomes more important, which can be achieved through higher production costs. Also, these cost constraints have a great influence on the capabilities of the nodes. Cheaper nodes have higher capacity limitations and lower fault tolerance.

Scalability: the number of entities in the application may vary depending on the environment where it is implemented and on its task. Consider an application used for early detection of forest fires which is implemented in a huge forest such as the Amazon. Due to the fact that sensors should have a small transmission range of the order of a few meters, the network must have in the order of thousands of nodes in order to cover the whole area. In such a case, the algorithms running inside the network should scale well in parallel to the increasing number of nodes in a region maintaining the given task of operation properly. In other words, the network should adapt itself to changing node density without affecting the application performance. Scalability is a significant issue, especially in outdoor applications.

Security: ad hoc networking and wireless mediums introduce many security flaws which make CO networks open for various types of malicious attacks. The system may be threatened by unauthorized users trying to access the network. Also, there are security risks on the physical layer of the network. For example, a jamming signal may corrupt the radio communication between the entities in the mission-critical networks.

Real-time/low end-to-end delay: end-to-end delay requirements are very stringent in real-time applications. For instance, in a manufacturing automation application, an actuation signal is required in real-time. Additionally, low end-to-end delay may be an essential requirement in some delay-sensitive applications such as target tracking.

Reliability: end-to-end reliability guarantees that the transmitted data is properly received by the receiving-end. In some applications end-to-end reliability may be a dominating performance metric, whereas it may not be important for others. In particular, in security and surveillance applications, end-to-end delivery has high importance.

Time synchronization: associating the data coming from multiple objects with the same event, data aggregation, data fusion, target or event tracking tasks and cooperation make time synchronization among communicating entities a key issue in many applications.

2.4. State of the art projects

Several projects are underway in the CO domain. In this section, a survey of activities in the order of starting period is given. It is worth noting that only a few readily available application scenarios featuring wireless sensor networks exist today. Current projects are mostly at the stage of understanding and analyzing some application-specific requirements. This is a natural outcome of the fact that WSN technology currently provides solutions for very basic requirements. Therefore, in the survey, projects in design phases are also given in order to illustrate the wide spectrum of the potential.

The criteria used in selecting the projects are twofold: studying various deployment environments such as home, office, factory, desert and forest to identify the requirements of different physical set-ups and social and economic impact of the application in improving people's life.

CyberGuide (1995): this project, developed by Georgia Tech, aims at providing people with navigating physical and cyber spaces by using portable computers. The visitors carry CyberGuide intelligent tour guide and obtain information about the surrounding environment, such as the positions and features of the objects in the area, their locations, etc. It is also possible for them to communicate with each other and with central computers. Furthermore, e-mail and web connectivity are provided. Some potential applications of the guide are translating Japanese signs and menus for a visitor to Japan, providing the map of a space, a tour guide for an art museum, recognizing faces and people at cocktail parties, etc. In the late 1990s, a series of CyberGuide indoor and outdoor prototypes were produced by students at Georgia Tech. These used GPS navigation system for positioning. The project supported both indoor and outdoor communication. Mobility had to be provided as well. Furthermore, it had to be cost-effective in order to attract visitors' interest and provide accurate location information of the visitor and/or surrounding places, streets, etc. For further information, see [43].

ActiveBat (1998): this system, developed by Sentient Project Group in ATT Labs, Cambridge, provides the location information of objects in indoor environments. It is an ultrasonic location finding system using the trilateration principle where position finding is done by measuring the distances between the reference points and objects with unknown locations. In the ActiveBat network, transceivers mounted on walls and ceilings are the reference points and they communicate with transmitters (Bats) carried by users in the environment. Also, there are RF controllers using radio signals for control and time synchronization purposes. First of all, the Bats send ultrasound signals, and transceivers mounted at known points on the ceilings receive those signals. In the second step, they estimate the time-of-arrival of ultrasonic signals, and by using the speed of sound, the distance from the reference points to the object is calculated and the position of the object is estimated using trilateration [22]. ActiveBat products have been deployed in some pilot sites since 1998. The location estimation accuracy of ActiveBat is about 9 cm (95%). The tags and transceivers are cheap, however, the system administration is expensive. The main limitation of ActiveBat is the requirement of a ceiling sensor grid [33]. The Computer Laboratory, Digital Technology Group, another research group in Cambridge University, also works on indoor tracking systems, reliable location systems, and location privacy issues.

Smart Dust Inventory Control (1998): the Smart Dust project, developed by Robotics and Intelligent Machines Laboratory, Department of Electrical Engineering and Computer Sciences, University of California at Berkeley, aims at building a self-contained, millimeter-scale sensing and communication platform for a massively distributed sensor network. The projection is that this device will be about the size of a grain of sand and will contain sensors, computational ability, bi-directional wireless communications, and a power supply, whereas the cost of deploying hundreds of those devices stays feasible. Inventory control is an application area targeted by Smart Dust. A product is monitored from the manufacturing step until it is delivered to the end-user. The operation is as follows: the carton talks to the box, the box talks to the palette, the palette talks to the truck, and the truck talks to the warehouse, and the truck and the warehouse talk to the Internet. It will be possible to know where your products are and what shape they are in, any time, anywhere. FedEx tracking on steroids for all products in the whole production stream from raw materials to delivered goods is planned to be an implementation area of this project [31]. The nodes in the network build a single-hop network that monitors a product from the manufacturing step until arriving to the end-user. Maintenance cost and power consumption must be taken into consideration because it is a long-lasting application and it is impossible to change the battery of a node mounted on the product.

Smart Dust Product Quality Monitoring (1998): another application area of the Smart Dust project is product quality monitoring. In a factory environment, the products can be monitored for their quality assessment. For instance, the temperature and humidity monitoring of meat and dairy products give information about their freshness [31].

Smart Sight (1999): this is a tourist assistant system equipped with a unique combination of sensors and software and developed by Carnegie Mellon University. The aim is to translate from and to local language, handle queries and answer in spoken language, and provide the positions of the users and the surrounding objects. The assistant is a wearable computer (consisting of a Xybernaut MAIV and a Thinkpad 600) including a microphone, earphone, video camera, and GPS to determine user location. This combination enables a multimodal interface to take advantage of speech and gesture input to provide assistance for a tourist. The system would have better knowledge of the environment than the tourist with accessing local database and the Internet. On the other hand, the software supports natural language processing, speech recognition, machine translation, handwriting recognition and multimodal fusion. As an example, a tourist in a foreign country may stand in front of an information sign, circle the text and ask "what does it mean?" – for which the language translation module can then offer an informative interpretation. Two applications were in development in 1999, requesting information of close by landmarks and a possibility for tourists to store information about places they find interesting [52]. Smart Sight users carrying wearable computers directly communicate with the data center, that is, the network is single-hop. Devices in the network are carried by people, therefore it must support mobility. Furthermore, the product and maintenance cost must be low in order to get people's interest. Power consumption is another issue, because it would not be easy for tourists to replace an exhausted battery with a new one.

EasyLiving (1999): the adaptation of cooperating object and wireless sensor networks to indoor environments has attracted much interest in recent years.

Research has focused especially on creating autonomous, intelligent environments. EasyLiving, developed by Vision Group at Microsoft Research, aims to adapt cooperating objects and wireless sensor networks to indoor environments for creating autonomous, intelligent environments which are spaces that contain myriad devices that work together to provide users with access to information and services. Both stationary (acoustic speakers, ceiling lights) and mobile (laptops, portable phones) devices may participate in the cooperation. The project will allow the dynamic aggregation of diverse I/O devices into a single coherent user experience [34]. In an intelligent environment, devices such as mice, keyboards, active badge systems, cameras or wall switches, providing the input to the system, and output of the system is given on the devices such as home entertainment systems, wall-mounted displays, speakers or lighting. Apart from I/O devices, there will be devices dedicated to provide computational capacity to the system. The main components of the system are middleware (to facilitate distributed computing), world modeling (to provide location-based context), perception (to collect information about world state), and service description (to support decomposition of device control, internal logic, and user interface). Media Control is one of the EasyLiving demo systems. When a user is authenticated to the system, his preferences are loaded that direct automatic behaviors. Users can have behaviors that direct various media types, such as a CD, MP3, DVD or VCR, that plays based on their location context. It is still an open challenge to define automatic behaviors and preferences for an intelligent environment in a consistent user-friendly manner.

RoboCup (2000): this is an international joint project aiming at promoting Artificial Intelligence (AI), robotics and related fields. It chose to use a soccer game as a central topic of research where a wide range of technologies can be integrated and examined. The ultimate goal of the project is to develop a team of fully autonomous humanoid robots that can beat the human world soccer champion team. It is apparent that innovations obtained from the RoboCup can be applied for socially significant problems and industries. Design principles of autonomous agents, multi-agent collaboration, strategy acquisition, real-time reasoning, robotics and sensor-fusion are some of the most significant topics that must be investigated for developing a robot team actually playing a soccer game [50].

Probeware (2000): this project refers to educational hardware and software used for real-time data acquisition, display and analysis with a computer or calculator. It is a ubiquitous computing application. It is also known as Microcomputer-Based Labs (MBL). When it is used with a calculator, it is known as Calculator Based Labs (CBL). The Probeware hardware consists of probes that use sensors to convert some physical property into an electrical signal. Temperature, light and distance probes are the most common, with over 40 kinds of probes used in education. By connecting probes to a computer running suitable software, students can observe data displayed in a variety of formats as it is being collected. Probeware has been widely used in science, mathematics and technology education. Software running on probes can usually represent the data from the probe as a number, dial or graph. By allowing students to see the display change in real-time, that is, as soon as the physical input changes, learners quickly catch the physical change with the way the representation changes. For example, an increase of temperature at the sensor causes the line to go up on a graph on the display. Also, some software may be needed to analyze the data as soon as it has been collected. For example, the user may want to fit data to a function or filter out noise. The main role of the probeware is to decrease the drudgery, allow students to focus more on the experiment and increase the amount and range of experimentation students can undertake. While it is becoming cheaper, the probeware continues to be refined, and is gaining increased flexibility. Current developments in the probeware search for the ways data can get from sensors into computers. Smart probes include a microprocessor that converts the sensor signal directly into a computer-readable format that can be plugged into a computer. They may communicate with computers via standard serial inputs, USB, or computer-specific ports. Also, there are future plans that investigate the possibility of wireless probes that communicate over a infrared or microwaves as well as sensors that connect directly to the Internet and can be read anywhere with an Internet connection [44].

Traffic Pulse (2000): the Traffic Pulse network is the foundation for all mobility technology applications [12]. It uses a process of data collection, data processing and data distribution to generate the necessary information for travellers. The data is collected through a sensor network, processed and stored in the data center, and distributed through a wide range of applications. The sensor network installed along major highways collects information such as travel speeds, lane occupancy and vehicle counts. The data is then transmitted to the data center for reformatting. The roadway conditions are continuously monitored on a 24/7 basis and updates are provided to the data center in real-time by the sensor network. In each city, the data center, called a traffic operations center, is established. It collects and reports on real-time events, construction and incident data. Each center produces the information through a wide range of methods: video, aircraft, mobile units, and monitoring of emergency and maintenance services frequencies. The sensors established along the roads communicate with the data center via multi-hop networks. They are static and therefore mobility is not an issue for this application. Algorithms run on the network must maintain their operations when the number of sensors is increased, that is, scalability is a need. Besides, the maintenance cost must be as low as possible, because the network will be in operation for a long time. The state of the monitored roads must be distributed among cars as quickly as possible in order to properly arrange traffic that dictates low end-to-end delay. The data center hosts both real-time applications and a database of archived traffic information collected by the sensors and the traffic operations center. A multi-tiered architecture is used to provide a highly scalable, flexible and secure platform on which products and services are based. Today, Traffic Pulse networks are in operation and broadcast benefits such as highest quality digital and analog local traffic content, web-based product delivery, unique broadcast applications system, 2D mapping solutions, 3D animated fly-over, for radio and TV broadcasters, cable operators, and advertisers who sponsor local programming.

Smart Kindergarten (2000): according to the Smart Kindergarten project proposed by the University of California, Los Angeles, children learn by exploring and interacting with objects such as toys in their environments, and the experience of having the environment respond (casually) to their actions is one key aspect of their development. The project aims at implementing wireless sensor network technology in the early childhood education environment, thus providing parents and teachers with the comprehensive investigation of students' learning processes [16]. Some questions that can be answered with the help of such a project are: "How well is student X reading book Y?", "Is student A usually isolated?", "Does student A tend to confront other students?", etc. A group of wirelessly networked toys capable of processing wireless communication and sensing the environment would be used as the application platform, together with a background computing and data management infrastructure. Aural, visual, motion, tactile and other types of feedback may be provided by a networked toy, and it may be able to sense speech, physical manipulation and absolute and relative location. As a result, Smart Kindergarten enhances the education process by providing a childhood learning environment which is tailored to each child, adapts to the context, coordinates activities of multiple children and allows unobtrusive evaluation of the learning process by the teacher [39].

WaterNet (2000): the WaterNet project, funded by the EU, uses wireless sensors in order to design a monitoring system for the sources of drinking water, which are mainly rivers in Europe. The goal of the WaterNet project is to provide the users, especially drinking water authorities, with a suitable monitoring technology. It has demonstrated the usability and appropriateness of newly deployed methods in real-life applications. Operators and authorities responsible for monitoring large-scale processes such as drinking water production and waste-water treatment plants have to manage large amounts of data coming from online sensors and laboratory analysis. The WaterNet project team has developed and validated, in real applications, a tool capable of providing real-time environmental and process monitoring system information processing. It provides a number of useful methods which allow operators to effectively deal with large quantities of real-time data by processing high level assessment information in a preferred form.

First of all, sensor readings are aggregated. In the second step, the river state and its evolution are assessed and interpreted. In the last step, information of different levels of abstraction are presented to specific but widely varying user classes.

The sensors communicate in a multi-hop network fashion. The system must support scalability in order to maintain its operation when the number of nodes is increased. It must also use power-efficient algorithms in order to attain the longest possible lifetime. Furthermore, some level of robustness is required, because it is an outdoor environmental monitoring application that may suffer from harsh operational conditions. Even when robustness is provided, there will be some node failures. In such a case, fault tolerance should be provided for accurate monitoring. The WaterNet requires a system architecture based on distributed, self-contained computational units enabling multiple users with different needs to share and exchange data. The tool developed by the WaterNet team was validated at two application sites (Paris and Barcelona) over a period of 4 months. A tool designed to operate on a Windows NT platform and SQL Server was developed during the project programme. Specific methods for validation, assessment and classification were implemented, tested and validated at the application sites [49].

Zebra RFID Product Tracking (early 2000s): this project is based on the RFID technology. It is a method of remotely storing and retrieving data using devices called RFID tags. An RFID tag is a small object, such as an adhesive sticker, that can be attached to or incorporated into a product. RFID tags contain antennae to enable them to receive and respond to radio frequency queries from an RFID transceiver. They are on-the-shelf products making automatically tracking and identifying products possible, no matter how they are produced and processed, with RFID technology [51].

CarTALK 2000 (2001): this EU Project works on driver assistance systems based on interaction among vehicles. The development of cooperative driver assistance systems and a self-organizing ad hoc network as a communication basis with the aim of preparing a future standard is the main target of CarTALK 2000 [42]. In terms of assistance systems, the main points are the assessment of today's and future applications for cooperative driver assistance systems, development of software structures and algorithms such as new fusion techniques, and testing and demonstrating assistance functions in probe vehicles in real or reconstructed traffic scenarios. On the other hand, in order to attain a suitable communication system, algorithms providing ad hoc networks with extremely high dynamic network topologies are developed and prototypes are tested in the vehicles. Some applications of the project are information and warning functions (a vehicle breakdown, traffic density), communication-based longitudinal control (preventing accidents that occur because of the inattention of the vehicle in front) and cooperative assistance systems (controlling highway entry and merging).

CORTEX Car Control (2001): transportation is one of the most suitable areas for CO applications. Future car systems will potentially be able to transport people without human intervention. CORTEX Car Control is an EU project researching the possibility of such a system. In this application, each car has several sensors with different tasks and an actuator that makes decisions according to sensor readings. Cars cooperate with each other to arrange traffic and to share information about road conditions [5]. The control car system automatically selects the optimal route according to desired time of arrival, distance, current and predicted traffic, weather conditions, etc. The cooperation of cars provides people with moving safely on the road and reduces traffic conditions. Also, traffic lights are able to communicate with cars so that each car senses the traffic lights and moves according to their state. The network inside a car is single-hop, that is, sensors directly communicate with an actuator, whereas there is a multi-hop network among cars. Event localization, such as obstacle detection and determination of jammed roads, is a significant characteristic of the project and therefore must be considered. Furthermore, the number of cars participating in the communication changes randomly which requires ad hoc (infrastructureless) communication support. Also, when the number of cars increases, scalability is needed in order to provide proper work of algorithms. When a car encounters an obstacle on the road, it must broadcast that news to other cars as quickly as possible, that is, minimum end-to-end delay must be provided. Currently, the project is implemented in laboratory environments. There are some developed robot cars equipped with a GPS receiver, 802.11b communication module and sensors. However, it seems that the realization of the project in the real world still needs some time, because problems faced in the requirements of the project have not been solved yet [37].

CORTEX Smart Room (2001): recent advances in mobile and pervasive computing and in wireless sensor networks have led to the idea of integrating them into our daily lives. The CORTEX project works on the concept of a sentient object, an intelligent software component that is also able to sense its environment via sensors and react to sensed information via actuators. There are mainly two application categories, namely personalized intelligent services and multi-person triggered actions. An example scenario of the first category is as follows: when Alice enters the sentient room (assume that it is a semi-public living room), her identity is captured by some device in the room and it starts to behave intelligently in her preferred way. The room realizes that it is too dim for Alice to do a given task under the current light density and automatically turns on extra lights. It adjusts the room temperature to the level that Alice feels comfortable. When she sits in her armchair, the room knows that she is relaxing and does the following according to her preferences: switches on the TV and the Hi-Fi system, tunes to her favourite music channel and starts playing it. By collecting raw sensory data from embedded sensors, the intelligent room can then process and analyze them to deduce high-level context and personal preferences that change with various situations, such as time of the day, location, working or relaxing, etc. On the other hand, consider a scenario where more than two people are involved in the room simultaneously. Assume that a group of people are in the room and talking to each other. In this case, the room might infer that there is a meeting and start to react according to the meeting scenario, for instance, switching on the plasma screen, displaying one of the attendee's slides [5]. The main requirements of the CORTEX Smart Room application are sentience, autonomy, cooperation and distribution. The sentient room consists of a number of sensors and actuators, which are actually abstract wrapper objects around driver software for particular hardware devices. Sensors collect data about the environment, while actuators attempt to change the state of the real world in some pre-defined way. A wrapper object around the air-conditioner is an actuator example, which processes CORTEX related software events and changes the room temperature through this conventional room device. Furthermore, every sentient object has its own internal control logic and is able to make autonomous decisions by itself, thus requiring some degree of autonomy. The sentient objects should be capable of cooperating with other sentient objects in order to fulfill their task. Distribution scale is another important issue, because the sentient room application should accommodate a reasonable number of sensors and actuators.

Habitat Monitoring on Great Duck Island (2002): the Intel Research Laboratory at Berkeley initiated a collaboration with the College of the Atlantic in Bar Harbor and the University of California at Berkeley to deploy wireless sensor networks on Great Duck Island, Maine, in 2002. The task of the network is to monitor the microclimates in and around nesting burrows used by the Leach's Storm Petrel. The objective of the project is to develop a habitat monitoring kit that enables researchers worldwide to participate in the non-intrusive and non-disruptive monitoring of sensitive wildlife and habitats. For this project, temperature, humidity, barometric pressure and mid-range infrared sensors are used. Motes periodically sample and relay their sensor readings to computer base stations on the island. This information is fed into a satellite link in order to allow researchers to access real-time environmental data over the Internet. Firstly, 32 motes were deployed on the island and, at the end of the field season in November 2002, over 1 million readings had been logged from those sensors. In June 2003, a second generation network with 56 nodes were deployed. The network was extended in July 2003 with 49 additional nodes and again in August 2003 with over 60 more burrow nodes and 25 new weather station nodes. These nodes form a multi-hop network transferring their data back "bucket brigade" style through dense forest. Some nodes are put more than 1,000 feet deep in the forest, providing data through a low power wireless transceiver [47].

BioWatch (2002): in recent years, bioterrorism has become a significant threat for humanity. The developments in micro and nanotechnology have brought about the idea of early detection and response to a bioterrorism event with wireless sensor networks. The BioWatch project has been developed by the US Department of Homeland Security for this purpose. It is an early warning system rapidly detecting biological materials in the air, possibly stemming from an intentional release. The system helps public health experts to determine the presence and the geographical extent of the biological substance by collecting air samples by a series of sensors. It allows federal, state and local officials to more quickly determine emergency response, medical care and consequence management needs.

During the operation of the system, aerosol samplers mounted on pre-existing EPA (Environmental Protection Agency) air quality monitoring stations collect air, passing it through filters. These filters are manually collected at regular intervals and are analyzed for potential biological weapon pathogens using polymerase chain reaction (PCR) techniques [36]. It is foreseen that this system will provide an early warning of a pathogen release, thus alerting authorities before victims begin to show symptoms. Today, BioWatch is a nationwide operation focusing on major urban centers. It became one of the most important tools used by public health agencies to warn citizens against the presence of biological agents.

COMETS (2002): COMETS Real-time Coordination and Control of Multiple Heterogenous Unmanned Aerial Vehicles (EU Project, IST-2001-34304) is a system that integrates multiple heterogenous Unmanned Aerial Vehicles (UAVs) for missions such as surveillance, monitoring, mapping and search [13, 30]. Currently both helicopters and airships have been integrated into COMETS. The COMETS consortium includes seven partners from five countries, and the scientific and technical coordinator is AICIA.

The COMETS system is based on the cooperation of heterogenous objects with different properties and characteristics. It exploits the complementarities of different UAVs in missions where the cooperation of several autonomous vehicles is very valuable due to the requirements on the required coverage, redundancies and flexibility when comparing the use of a single UAV with long endurance flight and important on-board capabilities.

The COMETS architecture comprises a ground segment and the UAVs. The ground control center has a mission planning system, a monitoring and control system, and the ground part of the perception system. Also, a teleoperation station for the guidance of remotely piloted vehicles is present. The cooperative perception system has a set of basic functions for aerial image processing, and integrates detection, monitoring and terrain mapping functions.

Each UAV in the flying segment is endowed with: a) its onboard proprietary components which gather the various functions specific to the UAV (flight control, data acquisition, possible data process), b) a generic supervisor which interfaces the UAV with the other COMETS sub-systems (ground segment and other UAVs) and controls its activities, and c) a deliberative layer which provides autonomous decisional capabilities to the UAV. The communications system used in COMETS is realized via a distributed shared memory, the blackboard.

The first mission application of COMETS is fire alarm detection, confirmation, localization and monitoring. Forest fire experiments were conducted in Portugal in May 2003 and May 2004. The final demonstration was also in Portugal in May 2005.

Scalable Coordination of Wireless Robots (SCOWR) (early 2000s): this is a wireless sensor network application developed by the University of Southern California [17]. The aim of the project is to merge artificial intelligence in robotic applications. It is a dynamic adaptive wireless network with autonomous robot nodes. The main target is to develop, test and characterize algorithms for scalable, application-driven, wireless network services using a heterogenous collection of communicating mobile nodes. Some of the nodes will be autonomous (robots) in that their movements will not be human controlled. The others will be portable, thus making them dependent on humans for transportation. There will be not only mobile nodes but also static computers on the network, that is, the system must support heterogenity. Due to the fact that the system will have numerous wireless mobile robots, it must support mobility and ad hoc communication. Also, the system will mainly consists of robots and since it will make autonomous decisions, it requires a high degree of autonomy. The mobility leads to frequent changes in the number of nodes and the topology of the network. In terms of power efficiency, it would not be a good strategy to run centralized algorithms for this case. Also, the nature of wireless sensor networks requires decentralized (distributed) algorithms. Therefore, the degree of distribution has importance for SCOWR. One application area of the project may be forest-fire detection. In such a case, the location information of the event or node may be significant. Also, the end-to-end delay may be required to be as low as possible for such event-detection applications. To sum up, the SCOWR team has already designed prototypes of autonomous robots and aerial vehicles. In the near future, it may be possible to use them in various wireless sensor network applications.

Oxygen (early 2000s): the aim of the project, developed by the MIT Computer Science and Artificial Intelligence Laboratory, is to provide people with direct interaction with devices either handheld or embedded in the environment. As a result of this interaction, they will learn and adopt our needs and wishes. Instead of traditional interaction means such as typing and clicking, users will use speech and gestures that describe their commands. For example, when somebody enters the room, the conditions in the room will be adjusted with respect to his/her intent and mood. With the help of speech and vision technologies, people will be able to communicate with Oxygen devices as they interact with other people.

The operational environment of the Oxygen will be highly dynamic and include various human activities. Therefore, it must have some properties such as pervasiveness, being embedded, nomadic, easily adaptable, etc. Microphone and antenna arrays, and different handheld devices will be used. The main difference of those components from other such devices is the fact that they must be power efficient. For further information, see [18].

Smart Mesh Weather Forecasting (early 2000s): this project was deployed for meteorology and hydrology monitoring of Yosemite National Park. It was developed by the Scripps Institution of Oceanography and United States Geological Survey. Over half of California's water supply comes from high elevations in the snowmelt-dominated Sierra Nevada. Natural climate fluctuations, global warming and the growing needs of water by consumers demand intelligent management of this water resource. This requires a comprehensive monitoring system across and within the Sierra Nevada. A prototype network of meteorological and hydrological sensors has been deployed in Yosemite National Park, traversing elevation zones from 1,200 to 3,700 m. Communication techniques are tailored to suit each location, resulting in a hybrid network of radio, cell-phone, land-line, and satellite transmissions. Results are showing how, in some years, snowmelt may occur quite uniformly over the Sierra, while in others it varies with elevation [15].

FloodNet (early 2000s): FloodNet, developed by Envisense consortium, aims at monitoring rivers that have flood threats by deploying a wireless sensor network across the river or floodplain. By processing and synthesizing collected information over a river and functional floodplain, FloodNet obtains an environmental self-awareness and resilience to ensure robust transmission of data in adverse conditions and environments. The coordinated efforts between sensor nodes which are fixed at specific points within a river or floodplain provide the spatio-temporal monitoring of an environment. The sensor readings are collected at a data center where FloodNet management provides a platform by which expert decisions can be made regarding flood warning and mitigating activities. In an emergency case, operational instructions indicating that operation rooms should become active and take necessary measurements can be issued through various media such as pagers, telephones, computer screens. Also, the design of the data management should provide responsible authorities with direct activation of flood warning road signs, flood warden notification, emergency services data provision, etc [46]. The FloodNet requires efficient power management strategy and a long lifetime. Sensor nodes must also be robust against harsh river and floodplain conditions. The number of nodes in the network may be in the order of thousands and therefore scalability is another issue that needs to be solved. Also, the mission of the network requires the lowest possible end-to-end delay. Today, there is a FloodNet application deployed at the River Crouch, Essex. It was deployed and activated on 30th April, 2004 and presently consists of six nodes, each equipped with a pressure sensor. The sensor is mounted underwater close to the riverbed and has a thin breath pipe that connects it to the surface in order to allow it to compare pressure at its location to that in the atmosphere. The difference in pressure can then be used to estimate water depth. The topology was determined with respect to the topology of a hydraulic model of the study site and was not limited by technology considerations such as radio range [46].

GlacsWeb (early 2000s): this is Envisense's monitoring system project for a glacial environment that will be transferable to other remote locales both on Earth and in space. The aim is to obtain information from glaciers about global warming and climate change. The network consists of probes embedded in the ice, a base station on the ice surface, a reference station relatively far from the glacier (in the order of kilometers), and the sensor network server. Subglacial probes containing various sensors beneath the glacier are used. Sensor readings are collected by a base station on the surface close to the glaciers. The information is then sent to the data center via a reference station. In order to accurately research this environment, the system must autonomously record glaciers over a reasonable geographic area and a relatively long lifetime. Due to the harsh environmental conditions of glaciers, robustness, fault tolerance and network reliability are the other main needs of the project. It is infrastructure-based, therefore, all nodes are only one hop away from the base station. Scalability is also a significant issue together with power consumption and production cost. It also must be as non-invasive as possible to let the sensor nodes, or probes, mimic the movement of stones and sediment under the ice [46]. The GlacsWeb was installed in the summer of 2004 at Briksdalsbreen, Norway. The Sensor Network Server (SNS) is based in Southampton. The low-powered probes are placed near the bottom of the glaciers and move with the ice, recording temperature, pressure, speed and the makeup of the glacier's sediment. They send back their data to the surface by radio, and these are picked up on a surface base station, which also records temperature and velocity. It has a webcam and snow meter, and it is able to track the position of these probes put in the glacier. The base station then sends the information by radio to the monitoring team's campsite (reference station). That data is then fed into a computer and put online to make it instantly available to glaciologists around the world.

PinPtr (early 2000s): this is developed by the Electrical Engineering and Computer Science Department at Vanderbilt University. Currently, most of the communication systems based on acoustic signals face severe performance degradations when used in urban terrain because multipath effects typically corrupt the available sensor readings. However, PinPtr claims that they have developed an acoustic system that works well even in complex urban environments. It provides an accuracy of 1 m and latency smaller than 2 seconds. The PinPtr system uses a wireless network of many low-cost sensors to determine both a shooter's location and the bullet's trajectory by measuring both the muzzle blast and the shock wave. It estimates the source location based on the time-of-arrival (TOA) of acoustic events, reference sensor nodes with known locations, and the speed of sound. The PinPtr sensor-fusion algorithm running on a base station performs a search on a hyper-surface defined by a consistency function that provides the number of sensor measurements. Those measurements are consistent with hypothetical shooter positions and shot times. The algorithm automatically categorizes measurements and eliminates those resulting from multipath effects. A fast search algorithm finds the global maximum of the surface that gives the shooter position. The number of sensors in the network may be in the region of hundreds. They can be deployed manually and placed in predetermined locations, or disseminated randomly by some other means. After the deployment phase, the sensors automatically set up an ad hoc communication network to define their locations and establish a time base. When an event is detected, the TOA is measured and by means of a specially tailored data aggregation and routing service the measurement is sent to the base station via the network [27]. The sensors communicate with the base station via a multi-hop network. It seems that the most challenging issues of the PinPtr are time synchronization and the random nature of the radio channel. The accuracy of the measurement strongly depends on those points.

Vehicle Tracking and Autonomous Interception (early 2000s): this project investigates a networked system of distributed sensor nodes, called PEG, that detects an uncooperative agent called the *evader* and helps an autonomous robot called the *pursuer* in capturing the evader [35]. Unlike environmental monitoring, it not only obtains measurements of the physical disturbance caused by evader, but also takes action as quickly as possible. The *pursuer*, a cooperative mobile agent, tries to intercept the evader using information collected by the sensor network and its own autonomous control capabilities. There is a clustering structure in the network and each cluster of nodes sensing sufficiently strong events can compute an estimate of the location of disruptive vehicle. Evader localization and mobility are the main characteristics of this multi-hop network. Furthermore, the number of nodes employed in the network may vary according to the application area. In such a case, algorithms must allow the network to maintain its operation. Also, the pursuer will make autonomous decisions about the evader; thus autonomy must be supported. This network may be established in outdoor environments where conditions may be harsh, which requires robustness and fault tolerance. A network of 100 motes was built in a 400 m^2 field in 2003. The evader was a four-wheeled robot remotely controlled by a person. The pursuer was an identical robot with laptop-class computing resources. As a result, it was demonstrated that PEG successfully intercepted the evader in all runs.

Ubiquitous Computing Support for Medical Work in Hospitals (early 2000s): this is developed by the Center for Pervasive Healthcare Department of Computer Science, University of Aarhus, Denmark. This project aims at creating new computer systems using ubiquitous computing instead of traditional computer technology designed for office use. The vision of the project for the hospital of the future is a highly *interactive hospital*, where clinicians can access relevant information and can collaborate with colleagues and patients independent of factors like time or place. Suppose that a patient is lying on an interactive bed. There is a public wall-display in another room, clinic or hospital, and a nurse is having a real-time conference with a radiologist. This project must support localization of patients. Scalability is a must due to the fact that it may comprise many hospitals and clinics distributed over a large geographical area. Mobility is another significant characteristic of this medical work. The work so far has been concentrating on creating a basic infrastructure to be used in hospitals, and on creating some example of clinical applications running on top of this framework. The most significant challenges for such an infrastructure are collection of services, security (in terms of authentication and authorization), context-awareness, and collaboration. In the framework of creating interactive hospitals, interactive hospital beds will play an important role. The bed has an integrated computer and a touch sensitive display. What is more, various sensors are mounted on the bed which can identify the patient lying on it, the clinician standing beside it, and various medical stuff embedding RFID tags. For example, when the nurse arrives with the patient's medicine, the bed logs the nurse, checks if the nurse is carrying the right medicine for the right patient, and displays the relevant information on the screen [2].

Monitoring Volcanic Eruptions with WSN (early 2000s): this project aims to implement a wireless sensor network composed of low-frequency acoustic sensors to monitor volcanic eruptions. The network, based on Mica2 sensor mote platform and consisting of three infrasonic (low-frequency acoustic) microphone nodes, was deployed on Volcanò Tungurahua, an active volcano in central Ecuador, in July 2004. Those microphone nodes transmit data to an aggregation node, which relays the data over a 9 km wireless link to a laptop at the volcano observatory. In order to provide infrasonic sensors with synchronization, a separate GPS receiver was used. Deployment on a small scale proved a proof-of-concept as well as a wealth of real acoustic signals. For providing scalability, a distributed signal correlation scheme, in which individual infrasonic motes capture signals locally and communicate only to determine whether an "interesting" event has occurred, was developed. One significant benefit of this scheme is that by only transmitting well-correlated signals to the base station, effective radio-bandwidth usage is provided [41]. Apart from scalability and synchronization, localization and robustness are the other main characteristics of the project. The location information of the volcanic explosions is necessary for determining the places that can be affected by the volcano. On the other hand, it is apparent that operational conditions are not good for sensor networks. They must be robust against harsh environmental conditions.

Wireless Sensors for Wildfire Monitoring (early 2000s): national parks, wilderness areas, environmentally and economically sensitive urban-wildland interfaces are prone to wildfires that may lead to significant disasters in terms of both the risk to lives and property. Environmental monitoring applications with wireless sensors have attracted significant interest in recent years. Wireless, low-power, low-bandwidth and low-cost sensors can be used to collect environmental data such as temperature, humidity and pressure in order to monitor the wildfire-prone regions. Those sensors can be equipped with GPS receivers which provide location information for collected data. The aim of the project is to develop some part of a set of real-time database management and wireless data acqusition tools for rapid and adaptive assessment of the impact of catastrophic events such as earthquakes, fires, hurricanes or floods [7]. For this purpose, the project mainly focuses on developing and field testing an asset tracking system for location and environmental monitoring of firefighting personnel, and investigating possible "spin-off" applications in other domains, such as monitoring of structural health, geologic hazards and/or environment [7]. Currently, the results from a testbed implemented near San Fransisco, California are available. The system includes environmental sensors measuring temperature, humidity and barometric pressure with an on-board GPS unit attached to a wireless, networked mote, a base station and a database server. The sensor readings are stored into a MySQL database, which is queried by a browser-based client interacting with a web server database bridge. The system can be operated using any web browser. The sensors are aggregated on a printed circuit board plugged into the Crossbow's Mica2 mote running the TinyOS. Performance of the monitoring system during two prescribed burns at Pinole Point Regional Park (Contra Costa County, California, near San Fransisco) is promising [7]. The wildfire monitoring system with wireless sensors provides a proof-of-concept implementation for wireless instrumentation in destructive, environmentally hostile wildfires. Results from testbed implementation reveal that the commercial development of the system is not far off. Furthermore, the protection of the motes should be better in heavier fuels.

Ubisense (early 2000s): this is a location-finding system developed by Ubisense Limited. The main difference between Ubisense and other location-finding systems such as ActiveBat is that its sensor products uses ultrawideband (UWB) technology to communicate with and locate the objects in real-time up to an accuracy of 15 cm in 3D. The main advantage of UWB technology is the fact that it provides a high level of accuracy with low infrastructure requirements. It is claimed that Ubisense is the first commercially available platform that provides both high accuracy and high scalability requirements together with cost-effectiveness [21].

Networked Ubisensors are installed and connected to the existing infrastructure of the building. They use UWB communication in order to detect the position of Ubitags carried by the objects in the environment. The positioning is done by using speed of light and trilateration. In the second phase, the Ubisensors send the Ubitag location information to the Ubisense software platform that creates a real-time view of the environment which can be used by any number of simultaneous programs used for different purposes, such as responding immediately to the changes in the environment.

Sustainable Bridges (2003): this project, funded by the EU, examines the high-speed train railway bridges which are expected to meet the needs of the future. These needs are basically expected to be an increase on the capacities, heavier loads to be carried or increments in the speeds of the trains concerning the increased traffic on the railways. All types of bridges are to be considered in this scenario. The railway network in Europe will be primarily inspected. Meeting the increasing capacities and loads can basically be realized by assessing the bridge structure, determining the true behavior of the structure, strengthening of certain portions of the bridge or by monitoring of critical properties. Any irregularities during the design and construction of bridges and railways or while they are in service can be detected with existing structures. However, it is very obscure if they could carry much greater loads or handle much faster trains. That is why a new monitoring and alerting mechanism should be used. The overall goal is to enable the delivery of improved capacity without compromising the safety and economy of the working railway. The most recent technique to determine the irregularities in the bridges is the installation of wired monitoring systems to analyze the structural behavior. However, such monitoring systems use standard sensors and several other devices which are time-consuming to install and very expensive to spread around. Also, kilometers of wires are needed for their communications, which means a waste of money and a high risk of communication failure in case of an irregularity in the medium. An alternative way to monitor such railways and bridges is equipping them with a wireless monitoring system, using micro-electromechanical systems. Hence, the set-up cost of a communication system will disappear (getting rid of wires) and using small and intelligent sensor-carrying devices will dramatically reduce the production and maintenance costs [9, 28].

AUBADE – A wearable EMG augmentation system for robust emotional understanding (2003): this project, funded by the EU, aims to improve the monitoring and assessment of the psychological state of patients, people feeling high stress or people under extraordinary conditions. In other words, it will create a wearable platform, also called a wearable mask, to ubiquitously monitor and recognize the emotional state of people in real-time. Several kinds of biosensors construct the wearable mask. They will be on the user's face and collect physiological data such as skin conductivity, heart rate variability, respiration rate, etc. These raw data will be sent to the centralized system for decision-making via the data acquisition module where pre-processing of the data is done. Also, the intelligent emotion recognition module will be used to extract the psychological state of the user from sensor readings.

The main topics of the project work will be biosensors, wearable systems, signal processing, decision support systems, communications standards, security mechanisms and facial muscle movement representation. Currently, laboratory work has been continuing and some prototypes have been developed [20].

Sun RFID Industry Solution Architecture (2003): this project was developed by Sun Microsystems and uses RFID technology for pharmaceutical supply chain management. The main aim is to develop a global RFID architecture to prevent drug counterfeiting within the pharmaceutical industry, thus improving the health and safety of the people. The scenario is as follows: at the bottom of the supply chain, RFID tags are mounted on vials or bottles, boxes and cases. They are read by RFID readers which can be handheld devices or located at fixed positions. As a result, drugs are traced along the supply chain.

The main components of the system are Sun Java System RFID software, RFID readers, RFID printers and applicators, integration middleware, electronic pedigree application, and product authentication services. Due to the fact that RFID technology is still in the development phase, a few years are needed in order to fully benefit from RFID [19].

Rapid Intelligent Sensing and Control of Forest Fires-RiscOFF (2003): this project studies and tests a sensor network system for fire detection and alarm signaling. Its two main targets are to raise an alarm when the dimensions of the fire are still small (about tens of meters in size) and to locate the fire as best as possible to minimize spreading (the spatial resolution in localization of the detected fire should be about 200 meters). The cells of the wireless sensor network communicate via a wireless data transmission system with a central control or data collection point connected to the warning system of the emergency services, thus immediately raising a response from the fire alarm for a wide area of forest or grass [1]. Currently, the project has been focusing on the type of sensors to be used. Considering the application area, it must be multi-hop and support scalability. Also, it is desired that responsible authorities take necessary measurements against fire as quick as possible, therefore, end-to-end delay must be low. Mobility has not been mentioned in the project yet, however, there may be some mobile extinguisher robots in the monitored forest. Furthermore, the need for robustness and fault tolerance are compulsory due to the harsh environmental conditions. Apart from those characteristics, efficient power management strategies are necessary for prolonging the network lifetime.

UbiBus (2003): bus networks work in a similar way in different countries. To take a bus we just have to go to the closest bus stop and make a sign to the bus driver when the bus arrives. However, this simple task is complicated for a blind or partially blind person. The objective of the project, developed by INRIA, is to design an application that helps blind people to benefit from public transportation. The UbiBus user can say in advance to the bus that he needs to get on and be warned when his/her bus has arrived. This application should be easy to adopt and not disturb the service for other users and bus drivers.

UbiBus scenario is demonstrated as follows.

Three types of entities will interact: the bus rider (say Peter), the bus stop, and the bus. Peter has a mobile phone equipped with a short range communication interface such as Bluetooth or WiFi. Peter interacts with UbiBus via speech recognition. The only thing he has to do is to say the line number. Once Peter has said which bus he wants to take, he walks towards the bus stop. When he is close enough to the bus stop, his phone notifies him with the estimated waiting time (6 minutes for instance), which is received from the bus stop. Peter walks slowly to the bus stop, having plenty of time. At the bus stop, a non-blind girl is also waiting for a bus. Her attention is captivated by an ad on the bus stop for a new movie which interests her. To know more about it, she gets her device: the screen already shows the same poster as on the bus stop. Touching the screen makes her "dive" into the poster as the device plays the movie trailer. Two minutes later, a bus arrives and the girl gets inside. Peter stays sat down at the stop. A man gets out of the bus and looks around; he seems to be lost. He looks at his device, just as he would look at his watch: the device spontaneously displays a map of the area. Peter is still waiting for his bus. Another bus is approaching, but Peter cannot see it. However, inside the bus the driver notices a flashing "stop request" message displayed on the screen of device installed on the dashboard. The driver stops the bus, opens the door and Peter is notified by a voice message that his bus has arrived.

MyHeart (2003): cardio-vascular diseases (CVD) are the main cause of death in the western world. MyHeart is an EU project aiming to empower citizens to fight cardio-vascular diseases by preventive lifestyle and early diagnosis. The first step of the project is to obtain knowledge on a citizen's actual health status. In order to gain this information, continuous monitoring of vital signs is a must. The approach is therefore to integrate system solutions into functional clothes with integrated textile sensors. With the combination of functional clothes and integrated electronics and processing them on-body, intelligent biomedical clothes will be used in order to fight CVD. MyHeart will bring about the solutions that will continuously monitor vital signs and context information, diagnose and analyze the health status and acute events, provide user feedback and seamlessly provide access to clinical and professional expertise if required. The main technical challenges of the project are continuous monitoring, continuous personalized diagnosis, continuous therapy, feedback to user, and remote access and professional interaction. There are no public results available yet for real-world applications of the project [48].

CROMAT (2003): this is a project funded by the Spanish National Research and Technology Development Programme. The main objective is the development of new methods and techniques for the cooperation of aerial and ground mobile robots [14]. The intention is to develop technologies that could be used in applications such as the inspection of utilities, infrastructure and large buildings, disaster detection and monitoring (fires, floods, volcano eruptions and earthquakes), exploration, surveillance, urban safety and humanitarian demining. CROMAT is a project coordinated by the University of Seville with three subprojects that share the design and development of a new control architecture for the coordination of aerial and ground mobile robots. The first subproject is devoted to the development of an autonomous helicopter and its cooperation with a ground mobile robot. The second deals with teleoperation and cooperation of mobile robots, and the third is devoted to the development of helicopter control techniques.

Smart Surroundings (2004): the overall mission of the Smart Surroundings project, which is funded by the EU, is to investigate, define, develop and demonstrate the core architectures and frameworks for future ambient systems [11] which are networked embedded systems integrated with everyday environments which support people in their activities. As a result of ambient systems, a Smart Surrounding will be created for people to facilitate and enrich their daily life, and increase productivity at work. They will be quite different from current computer systems as they will be based on an unbounded set of hardware artifacts and software entities, embedded in everyday objects or realized as new types of device. Over the last 10 years, the ubiquitous computing vision has inspired research into computing systems and applications that become pervasively embedded in our everyday environments, and that bring the unique flexibility of digital technology to the activities around which our lives evolve. Caused by rapid progress in technology, early research tended to focus on experimental prototypes of infrastructure, devices and applications. As the field is progressing, the most important research challenge, and the focus of this project is to develop the fundamental architecture of ubiquitous computing environments. The project's ambition is to move beyond prototypes toward sustainable systems for implementation of the ubiquitous computing vision. The ubiquitous computing research community at large has been very successful in advancing the infrastructure components for pervasive systems and in exploring the design opportunities for novel applications. This work is compelling but has mostly remained centered around single devices as opposed to distributed systems composed of many devices. It is still observed that there is a very wide gap between the new design materials at hand (e.g. smart artifacts, ad hoc networks, location technologies) and the potential applications. For example, microprocessors and wireless radio can now be built into practically everything to create smart networked objects, but we lack the technology to integrate these in an open platform for a wide range of applications. Likewise, the components are in place for prototyping and exploring application ideas, but we lack the foundations for the principled study of ubiquitous systems and application designs.

CoBIs (2004): this EU project aims at using smart sensor technology in industrial/ supply chain/sensitive settings. It will make it possible to implement networked embedded systems technologies in large-scale business processes and enterprise systems by developing the technologies for directly handling processes at the relevant point of action rather than in a centralized back-end system. Modeling embedded business services, developing the collaborative and technology frameworks for CoBIs with necessary management support, and investigating and evaluating CoBIs in real-world application trials in the oil and gas industry are the main objectives of the project. The collaborative items are physical entities in business environments with embedded sensing, computing and wireless short-range communication. Items have unique digital IDs, embody sensors to monitor their state and environmental conditions, are able to communicate peer-to-peer, collaborate in order to fulfill collective services such as observation of conditions that no single item could obtain independently, and interface back-end systems to make their service integral with overarching business processes [45].

GoodFood (2004): agriculture is a challenging area where newly emerged wireless sensors based on micro and nanotechnologies can be applied in order to increase the safety and quality of food products. The implementation of micro and nanotechnology into agriculture is called precision agriculture, where a number of distributed elementary sensors communicating wirelessly are used to monitor the processes of food production and to detect the remainder chemical substances in products. Wireless sensor networks can monitor every step of the food chain from farmer to consumer and introduce smart agro-food processing. Precision agriculture is a multidisciplinary application that includes the cooperation of microelectronics, biochemistry, physics, computer science and telecommunications [10, 25]. The Integrated Project GoodFood is a precision agriculture project that aims at assuring the quality and safety of the food chain. Principally, it is based on the massive use of tiny detection systems capable of being close to the foodstuff and merged in AmI paradigm. In GoodFood, high density sensors are deployed in the agricultural area to be monitored and are wirelessly interconnected, as well as being capable of locally implementing computations based on predetermined mathematical models and forwarding information to a remote site. The first GoodFood application scenario was implemented in spring 2005 on a pilot site in a vineyard at Montepaldi farm, a property of the University of Florence. Based on MICA motes and TinyOS, a WSN is disseminated in the area and is primarily used for detection of insurgence toxins in the vineyard. The main characteristics of the GoodFood scenario are network lifetime and power efficiency, synchronization, security, localization, address and data-centric communications, scalability, fault tolerance and node heterogenity. Each characteristic is a challenging area for the scenario and requires research in order to provide proper operation.

Hogthrob (2004): this project is developed by the Technical University of Denmark and aims at monitoring sows. Consider sows wearing sensor nodes incorporating movement detectors as well as a micro-controller and a radio. A sensor node with an integrated radio placed on each sow could transmit the sow's identification to the farmer's hand-held PC, thus alleviating the need for a tag reader. The use of sensor nodes on the animals could facilitate other monitoring activities: detecting the heat period (missing the day where a sow can become pregnant has a major impact on pig production), possibly detecting illness (such as a broken leg) or detecting the start of farrowing (turning on the heating system for newborns when farrowing starts). Sows carrying sensor nodes directly communicate with the farmer's hand-held PC, hence the network is single-hop. Additionally, the system must support mobility. Locating the pigs may also be an issue and furthermore, farmers cannot afford to buy an expensive technology; therefore, the product and maintenance costs must be feasible.

Safe Traffic (2004): this project, developed in Sweden, aims to build a novel communication system which will provide all vehicles, persons and other objects close to the road with the necessary information for establishing safer traffic. Furthermore, all road-users must be provided with an accurate positioning device. As a result, the plan of Safe Traffic is that each road-user is equipped with a small unit including a transceiver and a positioning device. Also, wireless tiny low-power, low-cost sensors will be used for traffic monitoring, traffic light coordination, emergency services and highway surveillance. The proposed system could also offer Internet access and multimedia services such as video-on-demand and real-time games for commercial and personal benefits. The project is currently focusing on the major components of the physical and network layers of the wireless communications system that will become the platform for all related traffic safety applications [40].

2.5. Taxonomy of CO applications

The CO projects surveyed in section 2.4 can be classified in many different ways as all applications have some common features with others. However, sectoral classification clearly highlights the application domains which can benefit from the CO paradigm to improve social and economic life and is therefore preferred. Considering current research trends, areas which can benefit from COs are determined as follows: control and automation, healthcare, environmental monitoring, security and surveillance, logistics, home and office, transportation, tourism, and education and training. Other sectors such as entertainment are not included in the categorization since there is hardly any ongoing project in this domain yet that best characterize the requirements. Common characteristics and requirements of each sectoral area are summarized below. In the following tables, requirements written in italics represent challenging issues.

2.5.1. *Control and Automation (CA)*

The applications that fall into this category may be used in indoor or outdoor environments and they should provide the ability to enable distributed process control with ad hoc and robust networking in challenging environments. They include robotics, control and automation technologies. The benefits of those applications may be explained by giving examples in the process manufacturing area where continuous research and implementation of new production technologies must be done in order to reduce waste, reduce time of operations and improve quality and volumes. Through the use of networked sensors, actuators, robots and process control algorithms, the manufacturing process performance is readily increased.

In many control and automation applications, apart from basic sensor nodes, there are transportation systems and other entities which are capable of making autonomous decisions. These applications are mostly event-driven.

The applications in this category are required to work in real-time. Hence, having support for fault tolerance, end-to-end delay and synchronization of the components is very important. These applications are expected to decrease the need for human interference as much as possible, and therefore the degree of automation is important. Also, the components in such a system need not be identical. Other issues are very dependent on the applications and their scenarios. So, some characteristics, like security, can be very important for some applications, while they are not an issue for other applications. Table 2.1 summarizes the common and application-dependent characteristics of control and automation applications.

Common characteristics and requirements	Application-dependent characteristics and requirements
• Degree of automation • Fault tolerance • Heterogenity • Production and maintenance cost • Real-time • Time synchronization	• Context awareness • Localization • Mobility • Packaging • Power awareness • Scalability • Security

Table 2.1. *Characteristics of control and automation applications*

2.5.2. *Home and Office (HO)*

Traditional lifestyles and habits have been changing with the emerging applications of embedded systems in home or office environments. Networked sensors with different tasks may arrange room conditions such as temperature and light according to the needs of the person. Furthermore, carbon monoxide sensors may be used to detect unsafe levels emitted from the heating systems. The security of an indoor environment may be increased as well by linking the home to private security companies. Those applications change the way that people live. In home and office applications, the components of the system behave depending on the state of the environment. Since almost all of the applications will be very personalized, security and privacy is highly important. Since the main input of the system will come from the person, localization is also important to state the location of the input to determine what to do. Such systems are inevitably designed to be context aware. The data to be processed or the feedback data of the system are expected to require a high bandwidth. Also, the system must be affordable. A multi-hop network may not be necessary, and the topology and the routing will probably not be changed, hence mobility does not seem to be a challenging issue for this category. Table 2.2 summarizes the common and application-dependent characteristics of home and office applications.

Common characteristics and requirements	Application-dependent characteristics and requirements
• Context awareness • High bandwidth • Indoor • Localization • Production and maintenance cost • Security	• Degree of automation • Fault tolerance • Heterogenity • Mobility • Power awareness • Real-time • Time synchronization

Table 2.2. *Characteristics of home and office applications*

The following are some state-of-art projects in this category: *Oxygen* [18], an integrated vision and speech system, uses cameras and microphone arrays to respond to a combination of pointing gestures and verbal commands and is developed by MIT Computer Science and Artificial Intelligence Lab. When you arrive home, you say "I'm home", and the space comes alive. Lights turn on and music starts up on the stereo. *ActiveBat* [22] is a context-aware ultrasonic indoor positioning system, developed by ATT Labs, Cambridge. ActiveBat transceivers are mounted on walls. They communicate with ActiveBat tags carried by objects in the environment and determine their positions. The location-finding system uses trilateration and speed-of-sound to estimate the distance between the objects and reference points from TOA measurements. *Ubisense* is another location-finding systems used for creating smart spaces. Ubisensors mounted on walls use UWB signals to communicate with Ubitags carried by objects in the environment. As a result, the Ubisensors locate those objects [21]. *CORTEX's Smart room* [5] is an EU project. When someone enters the sentient room, her identity is captured by a device in the room and it starts to behave intelligently in her preferred way. *EasyLiving* [34] is a ubiquitous computing project developed by Microsoft Research. It aims at developing architecture and technologies for building intelligent environments. *Smart Surroundings* [11, 26] is an EU project and the aim is to investigate, define, develop and demonstrate the core architectures and frameworks for future ambient systems. It projects that people will be surrounded by embedded and flexibly networked systems that provide easily accessible yet unobtrusive support for an open-ended range of activities, enriching daily life and increasing productivity at work. This idea requires ubiquitous computing. Although ActiveBat and Ubisense only provide location information to the users and do not promise to create intelligent information, they are mentioned in this category in order to give some examples of the first home/office ubiquitous computing applications. Characteristics and requirements of the given scenarios are summarized in Table 2.3. There is no conflict in the common features dictated by the application domain. Some application-dependent features such as fault tolerance also seem to be common. Additionally, varying requirements such as heterogenity exist in the scenarios.

2.5.3. *Logistics (L)*

Logistics is a hot research field for new micro and nanotechnologies. With the use of distributed networked sensors, a product can readily be followed from production until it is delivered to the end-user. For this purpose, it may cover a large geographical area and many diverse entities in order to establish communication between entities involved in the application. Therefore, a high degree of distribution may be required. Also, a manufacturer or company would like to minimize its expenses in order to increase its profit. For a corporation, the maintenance cost of a logistics application is a significant parameter in order to decide to establish it or not.

In logistics applications, it is not easy to let humans interfere with individual components of the system at any time for maintenance purposes. Therefore providing a fault tolerant system is important for this class. The system must also be able to serve to a scale of hundreds or thousands of inventories as well as to tens of them. Mobility and localization of the components are basic requirements of the applications in this category. Table 2.4 summarizes the common and application-dependent characteristics of logistics applications.

Due to the fact that many logistics applications benefit from the RFID technology, there are only a few scenarios mentioned in this category: *CoBIs-Collaborative*

	Oxygen	Smart Surroundings	ActiveBat	Easyliving	CORTEX Smart Room	Ubisense
Network topology	Single	Single	Single	Single	Single	Single
Indoor/outdoor	Both	In	In	In	In	In
Scalability	Low	Low	Low	Low	Low	Medium
Packaging	Low	Low	Low	Low	Low	Low
Fault tolerance	Low	Low	Low	Low	Low	Low
Localization	High	High	High	High	High	High
Time synch.	*Medium*	*Medium*	Medium	*Medium*	*Medium*	Medium
Security	*Medium*	*Medium*	*Medium*	*Medium*	*Medium*	Medium
Infrastructure	Ad hoc	Yes	Yes	Yes	Yes	Yes
Prod-maint.cost	*High*	*High*	*High*	*High*	*High*	High
Mobility	Low	Low	Low	Low	Low	Low
Heterogenity	High	High	Low	High	High	Low
Data traffic flow	High	High	Low	High	High	Medium
Automation	*High*	*High*	Low	*High*	*High*	Low
Power awareness	Medium	Medium	Medium	Medium	Medium	Medium
Real-time	*Medium*	*Medium*	Medium	*Medium*	*Medium*	Medium
Context-awareness	High	High	High	High	High	High
Reliability	*Medium*	*Medium*	*Medium*	*Medium*	*Medium*	Medium
Start date	Early 2000s	Early 2000s	Late 1990s	Late 1990s	Early 2000s	Early 2000s
Current stage	Testbed	Design step	Testbed	Testbed	Testbed	Available

Table 2.3. *Home and office scenarios and their characteristics*

Common characteristics and requirements	Application-dependent characteristics and requirements
• Degree of automation • Fault tolerance • Localization • Mobility • Power awareness • Production and maintenance cost • Reliability • Scalability	• Heterogenity • Packaging • Real-time • Security • Time synchronization

Table 2.4. *Characteristics of logistics applications*

Business Items [45] aims to use smart sensor technology in industrial/supply chain/ sensitive settings. It will make it possible to implement networked embedded system technologies in large-scale business processes and enterprise systems by developing the technologies for directly handling processes at the relevant point of action rather than in a centralized back-end system. Modeling embedded business services, developing the collaborative and technology frameworks for CoBIs with necessary management support, and investigating and evaluating CoBIs in real-world application trials in the oil and gas industry are the main objectives of the project. *Zebra RFID (Radio Frequency Identification) Product Tracking* [51] is a method of remotely storing and retrieving data using devices called RFID tags. An RFID tag is a small object, such as an adhesive sticker, that can be attached to or incorporated into a product. RFID tags contain antennae to enable them to receive and respond to radio frequency queries from an RFID transceiver. The *Smart Dust Inventory Control (SDIC)* [31] project can be described with a simple scenario, based on a communication sequence. The carton talks to the box, the box talks to the palette, the palette talks to the truck, the truck talks to the warehouse, and the truck and the warehouse talk to the Internet. You know where your products are and what shape they are in any time, anywhere. This is similar to FedEx tracking on steroids for all products in the production stream, from raw materials to delivered goods. In *Sun RFID Industry Solution Architecture* [19], the aim is to monitor the pharmaceutical supply chain in order to prevent drug counterfeiting and improve the health and safety of the people. It is based on RFID technology.

In this category, several requirements such as localization, scalability, end-to-end reliability are still challenging research issues. Characteristics and requirements of the readily available scenarios are given in Table 2.5.

2.5.4. *Transportation (TA)*

Applications in this class aim at providing people with more comfortable and safer transportation conditions. They offer valuable real-time data for a variety of governmental or commercial services. It is possible to design different scenarios of transportation applications. For instance, in one scenario cars may communicate with each other in order to organize traffic, whereas in another traffic organization may be done by installing static entities along highways and monitoring traffic conditions of roads. The desired result of these kinds of applications is to attain human-independent transportation.

The nature of transportation applications starts from being ad hoc and hence infrastructureless and mobile. One vital issue for those mobile components is localization. Systems like car control or traffic applications are highly required to run in real-time. Thus, the synchronization of the components and the end-to-end delay of the whole system are quite critical for such systems. Another issue that affects all these requirements is scalability. All the characteristic requirements stated here should be handled by taking care of different scales of the network. Table 2.6 summarizes the common and application-dependent characteristics of transportation applications.

	Zebra RFID Product Tracking	CoBIs	Smart Dust Inventory Control	Sun RFID Industry Solution Architecture
Network topology	Single	Multi	Single	Single
Indoor/outdoor	Both	Both	Both	Both
Scalability	*High*	*High*	*High*	*High*
Robustness	Medium	Medium	Medium	Medium
Fault tolerance	Medium	Medium	Medium	Medium
Localization	High	*High*	*High*	High
Time synch.	Medium	*Medium*	Medium	Medium
Security	Medium	Medium	Medium	Medium
Infrastructure	Yes	Ad hoc	Ad hoc	Ad hoc
Prod-maint.cost	*High*	*High*	*High*	*High*
Mobility	High	*High*	*High*	High
Heterogenity	Low	High	Low	Low
Data traffic flow	Medium	Medium	Medium	Medium
Automation	Low	Low	Low	Low
Power awareness	High	High	High	High
Real-time	Medium	Medium	Medium	Medium
Context-awareness	Low	Low	Low	Low
Reliability	High	*High*	*High*	High
Start date	Early 2000s	Early 2000s	Late 1990s	2003
Current stage	Available	Design step	Testbed	Testbed

Table 2.5. *Logistics scenarios and their characteristics*

Common characteristics and requirements	Application-dependent characteristics and requirements
• Heterogenity • Infrastructureless • Localization • Mobility • Real-time • Scalability • Time synchronization	• Degree of automation • Fault tolerance • Packaging • Power awareness • Security

Table 2.6. *Characteristics of transportation applications*

There are many projects that aim at providing safer public transport. Only the following applications are covered since the field of action does not necessarily involve a sensor network in the scenarios. *Traffic Pulse* is developed by Mobile Technologies,

USA. This project is the foundation for all Mobility Technologies applications. It collects data through a sensor network, processes and stores the data in a data center and distributes it using a wide range of applications. The Digital Traffic Pulse Sensor Network has been installed along major highways, and the digital sensor network gathers lane-by-lane data on travel speeds, lane occupancy and vehicle counts. These basic elements make it possible to calculate average speeds and travel times. The data will then be transmitted to the data center. In *CORTEX's Car Control project* [37], the implemented system will automatically select the optimal route according to desired arrival time, distance, current and predicted traffic, weather conditions, and any other necessary information. Cars cooperate with each other to move safely on the road, reduce traffic conditions and reach their destinations. Cars automatically slow down if there are obstacles or they are approaching other cars and speed up if there are no cars or obstacles. Cars automatically obey traffic lights. The principal target of this application scenario is to present the sentient object paradigm for real-time and ad hoc environments. It needs decentralized (distributed) algorithms. *Safe Traffic* [40] aims at the implementation of an intelligent communication infrastructure. This communication system would provide all vehicles, persons and other objects located on or near a road with the necessary information needed to make traffic safer. In addition, all road-users should be provided with an accurate positioning device. The purpose is simple: to reduce the number of traffic-related deaths and injuries. *CarTALK 2000* [42] is a European project focusing on new driver assistance systems which are based upon inter-vehicle communication. The main objectives are the development of cooperative driver assistance systems and the development of a self-organizing ad hoc radio network as a communication basis with the aim of preparing a future standard. To achieve a suitable communication system, algorithms for radio ad hoc networks with extremely high dynamic network topologies are developed and prototypes will be tested in probe vehicles.

These are mostly ad hoc applications with high mobility requirements. The characteristics of scenarios are illustrated in Table 2.7.

2.5.5. *Environmental monitoring for emergency services (EM)*

Environmental monitoring applications is of crucial importance to scientific communities and society as a whole. Those applications may monitor indoor or outdoor environments. The supervised area may be thousands of square kilometers and the duration of the supervision may last years. Networked microsensors make it possible to obtain localized measurements and detailed information about natural spaces where it is not possible to do this through known methods. Not only communication but also cooperation such as statistical sampling and data aggregation are possible between nodes. An environmental monitoring application may be used in either a small or a wide area for the same purposes.

One of the first ideas of a wireless sensor network concept is to design it to use the system to monitor environments. The main issue is to determine the location of

	Traffic Pulse	CORTEX Car Control	Safe Traffic	CarTALK 2000
Network topology	Multi	Multi	Multi	Multi
Indoor/outdoor	Out	Out	Out	Out
Scalability	High	*High*	*High*	*High*
Robustness	High	Low	High	Low
Fault tolerance	High	Medium	Medium	Medium
Localization	High	High	High	High
Time synch.	High	*High*	*High*	*High*
Security	Medium	*Medium*	*Medium*	*Medium*
Infrastructure	Yes	Ad hoc	Ad hoc	Ad hoc
Prod-maint.cost	Low	*Low*	*Low*	*Low*
Mobility	Low	*High*	*High*	*High*
Heterogenity	Low	*High*	*High*	*High*
Data traffic flow	Low	High	High	High
Automation	low	*High*	*High*	*High*
Power awareness	High	Low	Low	Low
Real-time	*High*	*High*	*High*	*High*
Context-awareness	High	High	High	High
Reliability	Low	High	High	High
Start date	Early 2000s	Early 2000s	Early 2000s	Early 2000s
Current stage	Available	Testbed	Design step	Testbed

Table 2.7. *Transportation scenarios and their characteristics*

the events. Such systems are to be infrastructureless and very robust, because of the inevitable challenges in nature, like living things or atmospheric events. Since the nodes are untethered and unattended in this class of applications, the system must be power-efficient and fault tolerant. The long lifetime of the network must be preserved when the scale increases in the order of tens or hundreds. Table 2.8 summarizes the common and application-dependent characteristics of environmental monitoring for emergency services applications.

Environmental monitoring for emergency services is a typical domain which can benefit from networked tiny sensors. Several projects are underway: *GoodFood* [10, 25] project aims to develop the new generation of analytical methods based on micro and nanotechnologies (MST and MNT) for safety and quality assurance along the food chain in the agro-food industry. *Hogthrob* aims at monitoring sensor wearing sows. The use of sensor nodes on the animals could facilitate other monitoring activities: detecting the heat period (missing the day where a sow can become pregnant has

Common characteristics and requirements	Application-dependent characteristics and requirements
• Fault tolerance	• Degree of automation
• Infrastructureless	• Heterogenity
• Localization	• Mobility
• Packaging	• Real-time
• Power awareness	• Security
• Production and maintenance cost	• Time synchronization
• Scalability	

Table 2.8. *Characteristics of environmental monitoring for emergency services applications*

a major impact on pig production), possibly detecting illness (such as a broken leg) or detecting the start of farrowing (turning on the heating system for newborns when farrowing starts). The objective of *WaterNet* [49] is to provide the users, namely drinking water authorities, with a suitable technology and it has demonstrated the usability and appropriateness of newly developed methods in real-life applications. The *GlacsWeb* [46] project is being developed in order to be able to monitor glacier behavior via different sensors and link them together into an intelligent web of resources. Probes are placed on and under glaciers and data collected from them by a base station on the surface. Measurements include temperature, pressure and subglacial movement. The main goal of *Sustainable Bridges* [9, 28] is to develop a cost effective solution for detecting structural defects and to better predict the remaining lifetime of the bridges by providing the necessary infrastructure and algorithms. Meteorology and hydrology in Yosemite National Park is monitored by *Smart Mesh Weather Forecasting (SMWF)* system. Results show how, in some years, snowmelt may occur quite uniformly over the Sierra, while in others it varies with elevation.

The number of commercial applications will increase when algorithms and paradigms satisfy challenging requirements such as scalability, robustness or power efficiency. The characteristics and the requirements of the above scenarios are summarized in Table 2.9.

2.5.6. Healthcare (H)

Applications in this category include telemonitoring of human physiological data, tracking and monitoring of doctors and patients inside a hospital, drug administrators in hospitals, etc. Merging wireless sensor technology into health and medicine applications will make life much easier for doctors, disabled people and patients. They will also make diagnosis and consultancy processes faster by patient monitoring

	GoodFood	Hogthrob	WaterNet	GlacsWeb	Smart Mesh Weather Forecasting	Sustinable Bridges	Wildfire Monitoring
Net. top.	Multi	Single	Multi	Multi	Multi	Multi	Multi
In/out	Out	Both	Out	Out	Out	Out	Out
Scal.	*High*	*Medium*	*High*	*High*	*High*	*High*	*High*
Pack.	*High*	High	High	High	High	High	High
F. Tol.	*High*	*High*	*High*	*High*	*High*	*High*	High
Loc.	Medium	High	Low	High	Medium	High	High
T. synch.	Medium	Low	Low	Medium	Medium	High	High
Sec.	Medium	Low	Medium	Medium	Low	Medium	Medium
Infr.	Ad hoc	Ad hoc	Ad hoc	Ad hoc	Ad hoc	Ad hoc	Ad hoc
Pr.-ma.cost	*High*	*High*	*High*	*High*	*High*	*High*	*High*
Mob.	Low	High	Low	Low	Low	Low	Low
Heter.	*High*	Low	Medium	Medium	Low	Medium	High
D. tr. flow.	Medium	Low	Low	Medium	Medium	Medium	Medium
Aut.	Low	Low	Low	Low	Low	Low	Low
P. aw.	*High*	Medium	*High*	*High*	*High*	*High*	*High*
R. time	Medium	High	Medium	Medium	Medium	High	High
Cont.-aw.	Low	Low	Low	Low	Low	Low	Low
Reli.	Medium	Medium	Medium	Medium	Medium	High	Medium
St. date	Early 2000s	Early 2000s	Early 2000s	Early 2000s	Early 2000s	Early 2000s	Early 2000s
Cur. stage	Pilot application	Available	Pilot application	Pilot application	Pilot application	Design step	Testbed

Table 2.9. *Environmental monitoring for emergency services scenarios and their characteristics*

entities consisting of sensors which provide the same information regardless of location, automatically transferring data from one network in a clinic to the other installed in patient's home. As a result, high quality healthcare services will be more accessible to patients.

Health applications are critical, since vital events of humans will be monitored and automatically interfered. Heterogenity is an issue, because the sensed materials will be various. Localization is important because it is critical to determine where exactly the person is; if he carries a heart rate control device and it detects a sudden heart attack, there must be no mistake or problem in finding his location. However, since in most cases single-hop networks will be used and neither topology nor routing will be changed, mobility is not considered to be a challenging issue for this kind of application. The delay between the source of the event and the other end-point of the system is also important. The data has to be conserved as original, which points to reliability of transmission. Also, the context is important. A sensing node must take into account that a person is doing sports at that moment in order to tolerate higher heart rate differences. Although the idea of embedding wireless biomedical sensors inside the human body is promising, many additional challenges exist: system safety and reliability; minimal system maintenance; and energy-harnessing from body heat. With more research and progress in this field, better quality of life can be achieved and medical cost can be reduced. Table 2.10 summarizes the common and application-dependent characteristics of healthcare applications.

Common characteristics and requirements	Application-dependent characteristics and requirements
• Context awareness • Heterogenity • Localization • Real-time • Reliability	• Degree of automation • Fault tolerance • Infrastructureless • Mobility • Power awareness • Security • Time synchronization

Table 2.10. *Characteristics of healthcare applications*

The projects examined in this category are as follows: *MyHeart* [48] aims at empowering a person to fight cardio-vascular diseases by lifestyle changes and early diagnosis. The first step is to obtain knowledge on a person's actual health status. In order to gain this information, continuous monitoring of vital signs is a must. The approach is therefore to integrate system solutions into functional clothes with integrated textile sensors. The combination of functional clothes and integrated electronics and processing them on-body, we define as intelligent biomedical clothes. The processing consists of making diagnoses, detecting trends and reacting to them. The MyHeart system is formed together with feedback devices able to interact with the user as well as with professional services. In the *Ubiquitous Support for Medical Work in Hospitals (US-MW)* [2] project, clinics can access relevant medical information and collaborate with colleagues and patient independent of time, place and activity. Suppose a patient is lying on an interactive bed and there is a public wall-display in another room, clinic or hospital; a nurse can then have a real-time conference with a radiologist. In *AUBADE*, the aim is to create a wearable platform consisting of biosensors to ubiquitously monitor and recognize the emotional state of patients in real-time.

Healthcare projects are summarized in Table 2.11 regarding their characteristics and requirements.

	MyHeart	Ubiquitous Support for Medical Work in Hospitals	AUBADE
Network topology	Single	Single	Single
Indoor/outdoor	Both	Indoor	Indoor/Outdoor
Scalability	Low	*Medium*	Low
Packaging	Low	High	Low
Fault tolerance	Low	Low	Low
Localization	High	High	High
Time synch.	Medium	High	Medium
Security	Medium	Medium	Medium
Infrastructure	Yes	Yes	Yes
Prod-maint.cost	Low	Low	Low
Mobility	Low	Low	Low
Heterogenity	High	High	High
Data traffic flow	Low	High	Low
Automation	Low	Low	Low
Power awareness	Medium	Medium	Medium
Real-time	High	High	High
Context-awareness	High	High	High
Reliability	Medium	Medium	Medium
Start date	Early 2000s	Early 2000s	2003
Current stage	Design step	Design step	Design step

Table 2.11. *Healthcare scenarios and their characteristics*

2.5.7. *Security and Surveillance (SS)*

Sensors and embedded systems provide solutions for security and surveillance concerns. These kinds of applications may be established in varying environments such as deserts, forests, urban areas, etc. Communication and cooperation among networked devices increase the security of the concerned environment without human intervention. Natural disasters such as floods and earthquakes may be detected earlier by installing networked embedded systems closer to places where these phenomena may occur. The system should respond to the changes to the environment as quickly as possible.

Security and surveillance applications have the most number of challenging requirements. Almost all issues must be covered to develop such systems. These applications require real-time monitoring technologies with high security cautions. The mediums to be observed will mostly be inaccessible to humans all the time. Hence, robustness takes an important place. Also, maintenance may not be possible. Thus, power efficiency and fault tolerance must be satisfied. Table 2.12 summarizes the common and application-dependent characteristics of security and surveillance applications.

Common characteristics and requirements	Application-dependent characteristics and requirements
• Fault tolerance • Infrastructureless • Localization • Packaging • Power awareness • Real-time • Scalability • Security • Time synchronization	• Degree of automation • Heterogenity • Mobility • Production and maintenance cost

Table 2.12. *Characteristics of security and surveillance applications*

Sample projects covered in this category are: *Bio Watch* [36], which aims at providing early warning of a mass pathogen release, including anthrax, smallpox and plague. The primary goal of *FloodNet* [46] is to demonstrate a methodology whereby a set of sensors monitoring the river and functional floodplain environment at a particular location are connected by wireless links to other nodes to provide an "intelligent" sensor network. *Vehicle Tracking and Autonomous Interception (VT-AI)* [35] is a networked system of distributed sensor nodes that detects an uncooperative agent called the evader and assists an autonomous robot called pursuer in capturing evader, developed by the University of California at Berkeley. *Monitoring Volcanic Eruptions with a Wireless Sensor Network (MVE-WSN)* [41] is a wireless sensor network to monitor volcanic eruptions with low-frequency acoustic sensors developed by Harvard University and the University of North Carolina. The *PinPtr* [27] system uses a wireless network of many low-cost sensors to determine both a shooter's location and the bullet's trajectory by measuring both the muzzle blast and the shock wave. The *COMETS project* [13, 30] (EU funded, IST-2001-34304) on real-time coordination and control of multiple heterogenous UAVs includes the experimentation and demonstration of the system on forest fire detection and monitoring. The *CROMAT project* [14] on the cooperation of aerial and ground robots also considers applications in disaster scenarios. Lastly, the *RISCOFF* project studies and tests a sensor network system for forest fire detection and alarm signaling.

The results of requirement analysis are shown in Table 2.13.

	PinPtr	Vehicle Tracking	COMETS and CROMAT	Flood Net	Bio Watch	RISCOFF	Monitoring Volcanic Eruptions
Network topology	Multi	Multi	Multi	Multi	Multi	Multi	Multi
Indoor or outdoor	Out	Out	Out	Out	Out	Out	Out
Scalability	*High*	*High*	*High*	*High*	*High*	*High*	*High*
Packaging	Low	*High*	High	High	High	High	High
Fault tolerance	High	High	High	High	High	High	High
Localization	High	High	High	High	High	*High*	High
Time synch.	High	High	High	Medium	High	High	High
Security	High	High	Medium	Medium	High	Medium	Medium
Infrastructure	Ad hoc	Ad hoc	Ad hoc	Ad hoc	Ad hoc	Ad hoc	Ad hoc
Prod-maint.cost	*High*	*High*	*High*	*High*	High	*High*	*High*
Mobility	Low	High	High	Low	Low	Low	Low
Heterogenity	Low	Low	High	Low	High	High	Low
D. tr. flow	Low	High	High	Low	Low	Medium	Low
Automation	Low	*High*	*High*	Low	Medium	Low	Low
Power awareness	High	*High*	*High*	*High*	High	High	*High*
Real-time	*High*	*High*	*High*	*High*	*High*	*High*	*High*
Context-awareness	Low	Low	Low	Low	Low	Low	Low
Reliability	Medium	Medium	Medium	Medium	Medium	Medium	Medium
Start date	Early 2000s	Early 2000s	Early 2000s	Early 2000s	Early 2000s	Early 2000s	Early 2000s
Current stage	Testbed	Testbed	Pilot appl	Pilot appl	Available	Design step	Testbed

Table 2.13. *Security and surveillance scenarios and their characteristics*

2.5.8. *Tourism (T)*

Everybody wants to feel safe and comfortable when they are in a new environment. When you visit a new country you want to find your way without much effort. New micro and nanotechnologies may help tourists in a foreign environment. For example, sensors and hand-held devices may act as a city guide, or may help people in an art museum. The location of museums, restaurants and information about the weather can be provided to tourists.

Tourism oriented applications do not have high dependencies on the characteristics we mention here. However, they must be personalized. These applications must be service and context-aware and cost-effective. They must also support mobility of the user. Table 2.14 summarizes the common and application-dependent characteristics of tourism applications.

Common characteristics and requirements	Application-dependent characteristics and requirements
• Context awareness	• Degree of automation
• Mobility	• Localization
• Production and maintenance cost	• Fault tolerance
	• Packaging
	• Power awareness
	• Real-time
	• Scalability
	• Security
	• Time synchronization

Table 2.14. *Characteristics of tourism applications*

We present two sample studies: *Smart Sight* [52] is intended to translate from and to the local language, handle queries posed and answer in spoken language, and to be a navigational aid. The assistant is a "wearable computer" (consisting of a Xybernaut MAIV and a Thinkpad 600) with microphone, earphone, video camera, and GPS to determine the user's location. The system would have better knowledge of the environment than the tourist with access to local databases and the Internet. The goal of *Cyber-Guide* [43] is to provide information to a tourist based on her position and orientation. Initial prototypes of the Cyberguide were designed to assist visitors on a tour of the Graphics, Visualization and Usability Center during monthly open house sessions. The user will be able to see her current location and the demonstrators in her surroundings on a map. The characteristics and challenging requirements of the applications in this domain are given in Table 2.15. As there are only a few applications, it is difficult to make common conclusions from the summary table.

2.5.9. *Education and Training (ET)*

Another emerging application area of embedded systems is within education. It is possible to provide more attractive lab and classroom activities involving COs. Current activities aim at merging embedded systems into education methods.

Although the requirements of education and training applications are highly dependent on the context, they all must be cost-effective and affordable to users. Since these applications are meant to ease the training of users by taking the place of a human educator, they must have a high degree of automation. Other issues may be required more or less by the concept, context or type of the application. Table 2.16 summarizes the common and application-dependent characteristics of education and training applications.

	Smart Sight	Cyber Guide
Network topology	Single	Single
Indoor/outdoor	Both	Both
Scalability	Low	Low
Packaging	Low	Low
Fault tolerance	Low	Low
Localization	High	High
Time synch.	High	High
Security	Low	Low
Infrastructure	Ad hoc	Ad hoc
Prod-maint.cost	*High*	*High*
Mobility	Low	Low
Heterogenity	Low	Low
Data traffic flow	Medium	Medium
Automation	Low	Low
Power awareness	Medium	Medium
Real-time	High	High
Context-awareness	High	High
Reliability	Medium	Medium
Start date	Late 1990s	Late 1990s
Current stage	Available	Available

Table 2.15. *Tourism scenarios and their characteristics*

Common characteristics and requirements	Application-dependent characteristics and requirements
• Degree of automation • Production and maintenance cost	• Fault tolerance • Heterogenity • Localization • Mobility • Packaging • Power awareness • Real-time • Scalability • Security • Time synchronization

Table 2.16. *Characteristics of education and training applications*

Two projects were briefly introduced in section 2.4 for the education and training industries. The first project is *Smart Kindergarten* [39] which builds a sensor-based wireless network for early childhood education. It is envisaged that this interaction-based instruction method will soon take the place of traditional stimulus-responses based methods. *Probeware* [38] is the other education and training project example, and is an educational hardware and software tool used for real-time data acquisition, display and analysis with a computer or calculator. It is also known as Microcomputer-Based Labs (MBL). When it is used with a calculator, it is known as Calculator Based Labs (CBL). The characteristics and challenging requirements of the applications in this domain are given in Table 2.17.

	Probeware	Smart Kindergarten
Network topology	Single	Single
Indoor/outdoor	Indoor	Indoor
Scalability	Low	Low
Packaging	High	High
Fault tolerance	Low	Low
Localization	Medium	Medium
Time synch.	Medium	Medium
Security	Low	Low
Infrastructure	Ad hoc	Ad hoc
Prod./maint.cost	*High*	*High*
Mobility	Low	Low
Heterogenity	*Medium*	*Medium*
Data traffic flow	Medium	Medium
Automation	Low	Low
Power awareness	Medium	Medium
Real-time	Medium	Medium
Context-awareness	*High*	*High*
Reliability	Medium	Medium
Start date	Early 2000s	Early 2000s
Current stage	Testbed	Testbed

Table 2.17. *Education and training scenarios and their characteristics*

2.6. Scenario description structure

As stated, COs and WSNs are applicable to a wide range of environments and in various conditions. Also, each sensor network may have assorted roles depending on the task and the scenario. Therefore, each different task may be fulfilled by a set of different entities. Below, there exists a set of symbols to define the objects and to define the media used to describe scenarios.

However, human acting devices in a scenario, such as a fire truck with a GPS, will be an end-point. Characteristics of end-points based on human behavior will not be detailed further in the scenarios. These end-points are named as *Facilities* where humans meet the whole system.

Table 2.18 shows the symbol set for objects that are used in the scenario decompositions.

Basic Sensor Node collects information about the environment and transmits it to the collection point (maybe sink or base station, depending on the application). It may have different tasks and operational environments. Basically, there are two types of sensors. The first one is called *proactive* (identified by a "periodic send" symbol) which monitors the environment continuously. On the other hand, *reactive* sensors (identified by a "event-based send" symbol) need to transmit data to the collection point only when the variable being monitored increases beyond a pre-determined threshold. A sensor node is also expected to carry some sensors embedded on it. To indicate which types of sensors the node contains, one Latin character for each sensor will be placed around the figure. For example, a sensor node which senses temperature (T), acoustic (Q), photo (P) and also contains a barometer (B) on it, is given below.

Figure 2.1. *Node with sensors*

Letters to indicate sensor types are:

P: Photo
T: Temperature
A: Acoustics
X: Accelerometer
M: Magnetometer
H: Humidity
B: Barometer

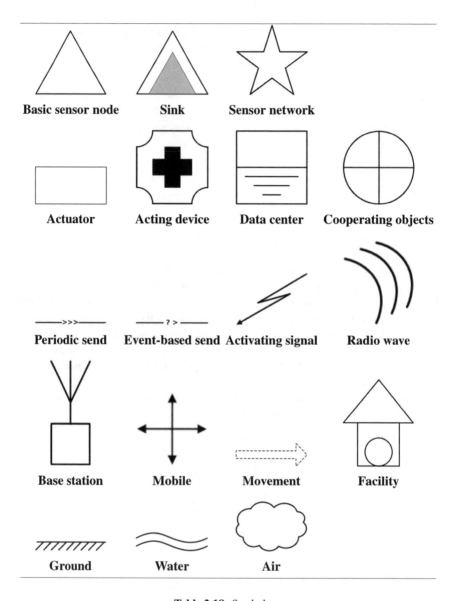

Table 2.18. *Symbol set*

Sink's task may vary from one application to another. It may be an actuator, a collection point or a base station. Various local sink nodes may collect data from a given area and create summary messages. It has more power and a longer lifetime than sensor nodes.

A sensor network may consist of hundreds or thousands of tiny nodes. Some of those nodes may be mobile. They do not have to be identical. Some of them may be superior to the rest in terms of lifetime, energy source, etc.

An actuator does not have to be included in all ubiquitous computing wireless sensor network applications. However, if we need to activate some devices according to the readings coming from sensor nodes, an actuator node is essential. It is worth briefly introducing a fire detection scenario as an example. Suppose that hundreds or thousands of sensor nodes are scattered in a forest and they measure the temperature of the environment continuously. If there were a sudden increase in the temperature of the environment, they send an emergency data to the actuator node as quickly as possible, and then it activates the *acting device*. To sum up, actuators extend control from cyber space into the physical world.

An acting device is the device which is activated by the actuator. If it is decided to act, the actuator sends activation signals to the acting device to interfere with the related event. An acting device can be a fire distinguisher or a traffic light.

A data center stores the data which is received from base stations for further analysis.

A cooperating object defines an object that may include all the components given above. In order to explain what "cooperating object means", it is worth taking a closer look at a Car Control scenario. In this example, each car is a cooperating object and cars cooperate and exchange information in order to provide more secure and less dense traffic conditions. Here, each car has a few sensors and an actuator. These sensors may have different tasks. The actuator makes the car behave according to information collected by sensors and sent by other cars. As a result, there is a cooperation between two or more entities and each entity is a cooperating object. The figure must be thought as slices of a pie. Figures of objects that are in cooperation will be inserted into the equally divided slices of the CO symbol.

Periodic (continuous) data sending is done by proactive sensor nodes that monitor the environment continuously and send periodic data to the parent nodes. This kind of communication is mostly seen in environmental monitoring applications. The direction of the arrow indicates the direction of the transmission.

Event-based data sending occurs when the variable being monitored increases beyond a threshold. Fire-detection application is an example where sudden temperature increase may happen and sensor nodes send an emergency signal to the actuator. It is a kind of event-based data transmission. The direction of the arrow indicates the direction of the transmission.

An activating signal is sent by the actuator to the entity that is to be activated.

A radio wave symbol describes an object having the ability of communicating via RF signals.

A base station can have more power and a longer lifetime than ordinary nodes. It may have a longer communication range in order to make it possible to communicate with remote sites. It receives sensor node readings and stores data.

Mobility can be a property of some objects in the scenario. For a mobile sensor node, this symbol will be added to the inbounds of the sensor symbol.

Movement and the direction of the movement can be predetermined or illustrated. The dotted thick arrow expresses that the object on the edges is moved in the direction of the arrow points. The moved object is also symbolized as a dotted form.

Facility is the end-point of the whole system. The user interacts with everything at this level.

For example, suppose a scenario, which contains an object that is an actuator node that can work as a sink, is mobile and has already moved in the right direction. Such an object will be described by a figure that is composed of all of the attributes, as shown in Figure 2.2.

Figure 2.2. *The usage of symbol set*

In a scenario, devices and related entities can all be in the same medium, or each one can be in a different medium. Whenever the medium needs to be referred, the *medium* symbols will be used to express that medium to be in the *air*, on the *ground*, or in the *water*.

2.7. Application scenarios

As noted, few readily available scenarios exist today, if academic testbeds do not count. Many projects still at testing stage determine the characteristics and require-ments. Therefore, real requirements are not available for many scenarios yet. In this section, we present six scenarios ranging from readily available to scenarios at design stages. The selected scenarios cover a wide spectrum of domains with different charac-teristics such as scalability, mobility, robustness, etc. In studying the readily available scenarios the following information is provided: user requirements, object decomposition, functional specification of each object in the scenario, architectural description, general characteristics and the current state. For the scenarios in early stages, basically, motivation behind the scenario and currently available object decomposition information are given. The scenarios given below are: forest fire detection, precision agriculture, safe transport, sow monitoring, intelligent surrounding and bridge monitoring.

2.7.1. *Forest fire detection scenario*

2.7.1.1. *Introduction*

Consider a scenario for forest surveillance and protection such as a natural park, with important ecological value, in fire-prone season. The park interacts with urban settlements. The full scenario has three different stages:

• Forest surveillance involving the detection of dangerous activities and the determination of the conditions that involves high forest fire risk.

• Fire detection and localization.

• Fire monitoring and aids to forest fire extinguishing.

In the following, only the second stage will be detailed.

Traditional forest fire detection is done by means of non-automatic procedures and particularly by human surveillance from watch-towers. The main drawbacks are errors due to the subjectivity of the operators, as well as the tiredness and de-motivation of the personnel, the considerable delay in detection (especially during the night) and errors in localization. Despite these drawbacks, these manned procedures are the most commonly used operational techniques. Several systems for automatic forest fire detection have been developed and a few of them are used in operational conditions.

The use of wireless sensor networks could contribute to overcoming some of the above mentioned drawbacks. Assume that a network of low-cost wireless sensors has been deployed in the natural park. Sensors could also be installed in vehicles or can even be carried out by personnel involved in surveillance and fire extinguishing activities. Then, mobile sensor issues are relevant in the scenario.

The application of wireless sensor networks to fire detection has been proposed by different authors. In fact, sensor nodes capable of detecting high temperature or heat exist. It could also be possible to use other sensors (i.e. carbon monoxide) to detect physical phenomena related to the presence of a forest fire. Then, dropping hundreds or thousands of these nodes in the fire-prone forest-park seems an automatic detection alternative. Fire detection has been considered in the FireBug project [4, 6, 8]. The objective of this project is the design and construction of a wildfire instrumentation system using networked sensors.

On the other hand, it should be noted that the scenario involves different kinds of COs of different sizes and characteristics. For example, Personnel Device Assistants (PDAs) can be used to update information from the environment and to guide the firefighters. Portable field computers and laptops on-board vehicles are useful for surveillance and fire fighting. Satellite positioning systems can also be used integrated with PDAs, portable computers and laptops to know in real-time the absolute position of people and vehicles.

Moreover, the application of autonomous objects is useful to provide information of areas poorly covered by the sensor network in the natural park and to calibrate the sensor nodes. Different autonomous systems can be used for surveillance, detection, localization and fire monitoring and measurement. These include computer vision systems for detection and monitoring [29] and autonomous vehicles and robots to acquire information or even to actuate in the extinguishing operations.

Autonomous ground vehicles can have significant sensing, communication and power infrastructure on-board. However, they have significant mobility constraints in the forest scenario. Then, it could be very difficult to access the desired locations to acquire information or to actuate. The application of unmanned aerial vehicles is also useful for the surveillance, detection and extinguishing activities since it may be difficult to deploy sensors in the whole area. The COMETS project of the European Commission (IST-2001-34304) on multiple heterogenous unmanned aerial vehicles considers the application of multiple unmanned aerial vehicles for forest fire detection and monitoring.

2.7.1.2. *Scenario characteristics*

Three main platforms can be used for automatic forest fire detection: ground platforms, platforms on aerial means and satellite-based systems.

Several automatic forest-fire ground detection systems have been developed in the last 10 years. In most cases they are based on the computer processing of images provided by infrared and/or visual cameras mounted on surveillance towers and other locations with good visibility conditions. Their main advantages when comparing to satellite based systems are the detection delay and the resolution.

Infrared cameras provide estimations of radiation intensity of the scene. For fire images in natural environments, fire is the source that produces the highest infrared radiation intensity. The main infrared band used for forest fire detection is the mid-infrared [3-5] m, but the far-infrared band ([8-12] m) or thermal infrared are also used for detection and false alarm discrimination. Detection systems using infrared cameras have demonstrated the ability to detect small fires (about 1 m) up to several kilometers (between 10 and 20). There are various commercial detection systems that have been in operation for several years in Spain and Italy using infrared technology (i.e. BOSQUE system and BSDS system). Early detection can be achieved when there is direct visibility of the fire. However, the application infrared systems have problems detecting forest fires in areas with dense vegetation and when the topography precludes direct vision of all possible sources of early fires.

Wildfire-detection systems using visual cameras have also been developed, including automatic detection systems such as ARTIST-FIRE which has been operational in France in recent years. These systems are usually based on smoke plume detection. The adequate selection of the optical system of the cameras is very important. Furthermore, the compensation of the variation of lighting conditions and the lack of contrast between the smoke and the background play an important role in automatic detection systems. Moreover, smoke plume detection by means of visual cameras cannot be applied at night.

Precise localization of the alarm is not easy if the number of observations is low. Thus, if the fire is only detected from one observation, the localization of the alarm relies on the accuracy of the available map and the modeling of the optical system. The accuracy also depends on the characteristics of the terrain and the relative position of the observatory. Moreover, in the case of smoke detection, when there is no direct vision of the fire, only the direction can be determined.

2.7.1.3. *Functional specification*

The use of sensor networks to cover large forests requires that nodes use multi-hop communication. If a node detects a fire, it should send an alarm message (along with its location) to a forest monitoring and control center.

The main practical problems are similar to those encountered in environment monitoring, as well as the following:

• Fire should be detected early enough otherwise the system will be useless.

• False alarms should be avoided. This depends on the sensor and also another parameter or information (i.e images) can be used to validate the event.

• Localization should be provided. The use of GPS and transmission of the position information along with the alarm is still a costly solution.

Currently, hybrid alternatives, combining wireless sensor networks (low cost and low energy consumption nodes) with cameras and other communication systems, seems to be an attractive solution. Some of the nodes should be mobile to overcome coverage limitations and should have power and communication resources to support cameras.

The main problems related to the application of autonomous systems are also related to coverage. One solution could be to use one long-endurance UAV for patrolling for fire detection in critical seasons. However, in the current state of technology this is an expensive solution. Another alternative would be to use several low cost aerial vehicles with appropriated coordination capabilities as in the COMETS project. The combination of autonomous systems and wireless sensor networks for environment monitoring and detection seems to be an attractive solution. The robotic nodes have much higher capabilities than the sensor nodes in terms of both hardware functionalities and networking throughput, overcoming the difficulties of the wireless sensor networks mentioned above. In particular, a limited number of mobile robotic nodes can cover a large scale sensor network.

2.7.1.4. *Object decomposition*

Wireless Sensor Network: the scenario requires the deployment of a wireless sensor network composed of a large number of cooperative sensor nodes.

There are different types of sensor nodes on the market. The Berkeley motes are the most widely known. These sensor nodes have the advantage of containing complete information about their software and hardware; furthermore, the software is developed under the open source license and there is enough information to develop sensor nodes or sensor expansion boards for particular applications. The main characteristics of these sensor nodes are:

• bidirectional communications in RF: at 400, 900 MHz and 2.4 GHz using technologies of spread spectrum and frequency hopping;

• TinyOS operating system: this is under an open source license and is implemented in NesC. This operating system has a small size and a modular structure, allowing it to implement light threads and implement communication protocols designed for sensor networks;

• based on microcontrollers of 8 bits with low calculation power;

• opportunity for over-the-air reprogramming;

• modular hardware system for its easy expansion. This expansion is made with sensor boards as the Mica Weather Board which provides sensors monitoring changing environmental conditions with the same functionality as a traditional weather station. The Mica Weather Board includes temperature, photo-resistor, barometric pressure, humidity and passive infrared (thermopile) sensors.

The thermopile may be used in conjunction with its thermistor and the photoresistor to detect cloud cover. The thermopile may also be used to detect occupancy, measure the temperature of a nearby object and sense changes in the object's temperature over time. If the initial altitude is known, the barometer module may be used as an altimeter. Strategically placed sensor boards with barometric pressure sensors can detect the wind speed and direction by modeling the wind as a fluid flowing over a series of apertures.

It should be noted that new COx and NOx sensors have recently been developed. These sensors could be used to monitor the smoke concentration, preventing harmful effects on fire fighters and the population in general. Both static nodes and mobile nodes carried by the fire fighting team could be used.

Ground vehicles: the use of ground vehicles is an alternative way of providing mobility to the nodes. In fact, these vehicles can have more communication and power resources. The interest in unmanned vehicles can be justified by the risk involved in fire fighting operations.

Autonomy of the vehicles in the natural terrain poses significant challenges. In fact, we may not be able to provide it with accurate models of the terrain. If maps are available, they may lack local detail, and in any case will not represent the changes that can occur in dynamic situations where transient obstacles (such as other mobile vehicles and humans) will be encountered, or where activities of the autonomous ground vehicle itself might alter its environment during accomplishment of tasks. On the basis of information obtained from its own sensors, the autonomous ground vehicle must be capable of building its own maps to represent the environment and flexibly support its own reasoning for navigational planning and, if necessary, re-planning. Moreover, as the world around the vehicle can change quickly, we want the software which implements control to be computationally efficient and supportive of real-time responses to events. In spite of the progress achieved in autonomous ground vehicles, it is clear that autonomy is still very difficult to achieve in operational conditions.

Unmanned aerial vehicles: unmanned air vehicles (UAVs) are self-propelled air vehicles that are either remotely controlled or are capable of conducting autonomous operations. Three different types of UAVs can be considered: airplanes, helicopters and airships.

Airships are interesting for surveillance missions in general. However, the use of airships is strongly constrained by the wind velocity. It should be noted that the deployment of UAVs for early detection could be recommended when the weather conditions favor the rapid propagation of the fire, which are strongly related to the wind velocity. Then, airships do not offer good characteristics to be used in this scenario.

Furthermore, unmanned airplanes are preferred to helicopters for surveillance and forest-fire detection tasks due to the greater flying range, coverage, ability of autonomous flight and easier teleoperation in case of failure of on-board automatic control systems. In the case of large areas, the use of several UAVs can be of interest. Multi-UAV surveillance strategies require appropriate planning and coordination algorithms.

2.7.1.5. *Step-by-step scenario description*

The scenario involves the application of the following objects:

- WSNs with nodes providing measures of temperature, humidity, NOx and COx;

- autonomous low cost aerial vehicles: airplanes and helicopters;
- conventional ground and aerial vehicles for patrolling and fire extinguishing;
- mobile computing objects: PDAs, laptops and portable computers;
- computer vision systems.

One sensor gives conditions near the limit of the alarm, but the threshold has not been passed. There are also coverage problems, so the managers have doubts about the existence of a fire.

One of the cameras sees a small smoke plume. However, due to the vegetation and terrain characteristics, the hot spot is not seen from the infrared camera and there are strong uncertainties about the location of the possible fire.

The patrolling ground vehicle cannot access good observation points. The managers decide to fly the autonomous airplane over the suspicious area.

The autonomous airplane has infrared and visual cameras, satellite positioning systems and software to determine the location of objects in the images by using the positioning and navigation sensors.

The autonomous airplane confirms the alarm and locates the fire. It is in an inaccessible area with a very high risk of propagation due to vegetation, slope and wind conditions.

The managers activate the fire extinguishing means.

The fire detection and localization scenario depicted by the standard symbol set is given in Figure 2.3.

2.7.1.6. *System requirements*

Some relevant system requirements in this application scenario are as follows.

Scalability: due to the fact that sensors have a small transmission range in the region of a few meters, the number of nodes in the network may be in the region of thousands to cover a large area. Therefore, scalability is another issue that needs to be solved in order to allow the algorithms to work properly without affecting the application performance.

Mobility: due to the fact that the system will have a number of wireless mobile robots, it must support mobility and ad hoc (infrastructureless) communication. Mobility leads to frequent changes in the number of nodes and the topology of the network, and also affects other requirements such as localization.

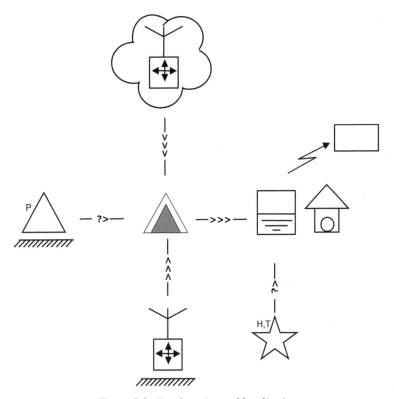

Figure 2.3. *Fire detection and localization*

Fault tolerance: the need for robustness is apparent due to the harsh operational conditions in an outdoor environmental monitoring application. Even when robustness is provided for, there will be some node failures, so fault tolerance should be provided for accurate monitoring.

Power awareness: efficient power management strategies are necessary for prolonging the network lifetime, especially in an application designed for environmental monitoring, since it is not usually easy to access that environment and replace the network with a new one. The network lifetime also depends on the communication range of a single node, the total number of nodes, and the total size of the area to be surveyed.

Real-time: it is desired that responsible authorities take necessary measurements against fire as quick as possible. Therefore, end-to-end delay must be low.

Localization: event localization, such as fire or smoke detection, is a significant requirement in this scenario. Some nodes may have GPS receivers, and geometrical and mathematical methods could be applied to estimate the locations of other nodes without GPS. Furthermore, the vehicles (aerial or ground) considered in this scenario are usually equipped with GPS and several algorithms have been recently proposed for the localization of the nodes of a WSN by using certain information from robots moving near their locations.

Those robots and other CO considered in this scenario are also provided with sensors (such as visual and/or infrared cameras) and algorithms that allow them to detect and locate the fire if direct visibility is available.

Synchronization: in this scenario, it is usually required to associate data coming from different COs with the same event, so data aggregation and data fusion techniques are relevant. Furthermore, a detection task over a broad area requires the cooperation among different COs such as nodes of a WSN and UAVs patrolling over the forest. Therefore, time synchronization is a relevant issue in this application scenario.

Heterogenity: it is practically impossible for all nodes to be identical in a wireless sensor network application. In fact, some nodes may be mobile whereas others can be static, and different sensors measuring different parameters are possible in this scenario. Furthermore, autonomous systems and particularly aerial vehicles with different architectures and levels of decisional autonomy are considered. It has also been mentioned that those robots can be ground or aerial robots. Therefore, the system should support a high level of CO heterogenity.

2.7.2. *GoodFood*

2.7.2.1. *Introduction*

[10, 25] are the main resources for the scenario. Recent developments in wireless sensor technology have made people aware of environmental changes and tackling them properly. The AmI paradigm explains the case where the user is surrounded by intelligent and intuitive interfaces able to recognize and respond to his/her needs. It is possible to integrate AmI and WSN technology in order to monitor different environments and control sudden changes.

Agriculture is a challenging area where newly emerged wireless sensors based on micro and nanotechnologies can be applied in order to increase the safety and quality of food products. The implementation of micro and nanotechnologies into agriculture is called precision agriculture, where a number of distributed elementary sensors communicating wirelessly are used to monitor the processes of food production and to detect the remaining chemical substances in products. WSNs can monitor every step of the food chain from farmer to consumer and introduce smart agro-food processing. Precision agriculture is a multidisciplinary application that includes the cooperation of the microelectronics, biochemistry, physics, computer science and telecommunications areas.

Precision agriculture applications aim at monitoring and controlling the parameters that affect plant growth. Various nutrients (zinc, poshphorus, nitrogen, etc.), soil moisture, temperature, pH, soil organic matter content, rooting depth, weed pressure, and pathogens have significant influence on food production. These soil parameters change within fields; therefore plant growth also varies within a field. The term "spatial variability" defines the variability of a parameter within a field and a precision agriculture application should obtain spatial variability maps of a field.

The integrated project GoodFood is a precision agriculture project that aims at assuring the quality and safety of food chain. Principally, it is based on the massive use of tiny detection systems capable of being close to the foodstuff and merged in AmI paradigm.

In GoodFood, high density sensors are deployed in the agricultural area desired to be monitored and wirelessly interconnected, capable of locally implementing computations based on predetermined mathematical models and forwarding information to a remote site.

The first GoodFood application scenario was implemented in spring 2005 in a pilot site in a vineyard at Montepaldi farm, a property of the University of Florence. Based on MICA motes and TinyOS, a WSN is disseminated in the area and it will primarily be used for detection of insurgence toxins in the vineyard.

2.7.2.2. *Scenario characteristics*

The main characteristic requirements of the GoodFood scenario are network lifetime and power efficiency, synchronization, security, localization, address and data centric communications, scalability, fault tolerance and node heterogenity. In truth, each characteristic is a challenging area for the scenario and requires research in order to provide proper operation. They will be investigated further in section 2.7.2.7.

2.7.2.3. *User requirements*

A precision agriculture application aims at monitoring and controlling the quality and safety of food products. It is crucial to measure not only the conditions of agricultural area but also the percentage of chemical substances within the food.

First of all, users of the system want to obtain some information about the soil such as nutrients (zinc, poshphorus, nitrogen, etc.), organic matter included in soil, moisture, temperature, and pH of soil. By using this data, it can be determined which food best fits this soil. Another user requirement is to determine remaining chemical substances in the products. With this information, hazardous foods are indicated and end-users know that these marked foods may cause severe health problems.

2.7.2.4. *Functional specification*

In general, there are four main components of GoodFood: WSN, sink node, base station and data center. The WSN is composed of hundreds or thousands of basic sensor nodes deployed close to the event. Each sensor node may have several nodes with different tasks. They measure the parameters related to the quality and safety of food and send this information to the sink node which is more powerful than elementary sensors. A sink node has computational capabilities and it runs distributed algorithms. The data sent by the sensor network are sorted and classified according to the measured parameters. It also has data aggregation capabilities. The base station acts as a relay between the sink node and the data center. In other words, it is like a gateway that delivers the information coming from the field to the remote side. The data center is a data pool where all information related to the food safety and quality is collected. The users accessing the data center may be food specialists, nutritionists, manufacturers, ordinary people, etc. The quality and safety of produced food is evaluated by users at this step.

2.7.2.5. *Object decomposition*

Generally, the scenario requires the cooperation of three different technologies given below.

The high density WSN consists of a large number of low power, low bandwidth components that collect the information about the parameters affecting the safety and quality of food. The network will consist of hundreds to thousands of nodes designed for unattended operation. It is possible to have a few mobile nodes for better coverage. The data rate is expected to be low (in the order of kbits). The WSN is divided into groups and each group is called a cluster where each cluster has a cluster head. The cluster head is also called a sink node. It is more powerful than basic elementary sensor nodes and has data aggregation capabilities. It also runs distributed algorithms.

The wireless base network acts as a relay between the sensor network and the fixed base network and consumes high bandwidth. It is not far from the agricultural area. A WSN is deployed in the area. The larger the distance that sensors have to send data over, the less efficient the power consumption strategy. Therefore, the distance between the sensor network and the base station should be as small as possible.

The fixed base network has a high performance computing platform to support truly scalable data processing. It is called the data center. The information coming from the source is stored and retrieved for the evaluation of the food safety and quality. Also, users may send various queries to the WSN.

2.7.2.6. *Step-by-step scenario description*

The GoodFood project scenario mainly consists of three components. The networked sensor nodes that include various sensors disseminated close to the area are the first part. A base station that relays information coming from sensor network to the remote data center is the second, and the data center that collects whole data sent by a WSN is the last component. The layered structure of the GoodFood scenario is given in Figure 2.4. The facility in the figure may be food analysts, nutrition specialists, etc., who can access the data center and evaluate the quality and safety of food.

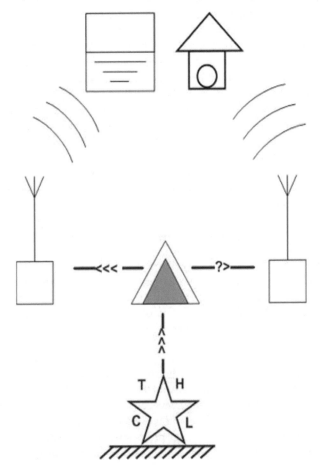

Figure 2.4. *The object decomposition of GoodFood*

WSN: this is based on MICA motes and run by TinyOS. Each sensor node may consist of several sensors varying from sensors measuring ambient conditions such as humidity and temperature of soil to sensors measuring chemical substances in food and pathogens that harm food products. Once sensors are deployed to the area, they construct a multi-hop WSN and start to monitor the environment. Sensor nodes collect temperature, pressure, moisture, and pesticies (bacteria, fungi, insects) data. That is, each sensor node includes several sensors on-board. They also attach time and location information to those data and send them to their sink node that is capable of analyzing data. They have computational capability and can interpret the information sent by sensor nodes. Each sink node has data aggregation and computation capabilities. They collect raw data from sensor nodes, process and aggregate it, then forward it to the data center via the base station. The WSN established in the field is called the monitoring infrastructure. Information collected by the monitoring infrastructure will be used to develop mathematical models for predicting the insurgence of pathogens such as botrys and pheronospera, or food toxins such as ochratoxin. Distributed mathematical models and algorithms in the WSN will be used to perform spatio-temporal correlations of data and detect the presence of mycotoxins and pathogens. This will make it possible to take quick and appropriate measurements. The clustering structure among sensors is used for more efficient power consumption and longer lifetime. According to clustering, the WSN is divided into small groups and each group has a leader called the cluster head that is more powerful than other nodes in the group. By dividing a large sensor network into small groups, the communication range of each node is decreased, thus power resources are more effectively used, because it is well known that the longer the communication range, the more power is needed to send data over that distance.

Base station: this acts as a radio relay that delivers sensor readings to the data center. It may also control the autonomous operation of the network. Data coming from basic sensors deployed in the agricultural area are sent to the data center via the base station. Also, users in the data centers may send various queries to the sensor network via the base station.

Data center: all data coming from the agricultural area comes to the data center. Users can access and control the network via the data center. Information sent by networked sensors is evaluated and the field conditions and remaining chemical substances in food products are determined. Thus, each step of the food production is controlled and monitored. Some foods may have higher levels of harmful chemical substances than pre-determined thresholds. In such a case, those products are marked and people are warned about them.

Generally, the GoodFood scenario has three phases: sensing, processing and decision making, and acting. Distributed networked sensor nodes monitor the environment and parameters related to the quality and safety of food. Data evaluation is fulfilled by sink nodes. In other words, sensor readings are collected and aggregated by sink nodes where distributed algorithms and mathematical models run on. This means that the information delivered to the data center is not raw data. Information sent by sink nodes via base station is stored in the data center where acting phase occurs. Data storage is used, as the name implies, for storing and retrieving information in a more organized way according to the sensor modeling language and the database model. Also, a data mining system for analyzing the information to find some demanding correlations is needed. What is more, users that have the access right to the system may want to make decisions according to the obtained knowledge about the quality and safety of food products. Therefore, a decision support system is needed. The file system is necessary for storing data structured in files as well. Users such as food quality specialists, nutritionists and consumers can develop strategies for the safety and quality of food according to this data. They may also send queries to the sensor network and demand specific information about the area such as what the risk level of a pathogen attack on the field is. For another example of data storage application, suppose that some kind of food has been monitored during the product process and any necessary information about the quality and safety of it has been obtained. The product may contain risky chemical substances. Now, any user that has access to the database may introduce the community if the food has any risk or not. The picturized scenario is given in Figure 2.5.

2.7.2.7. *System requirements*

The GoodFood project has several problems and requirements that need to be solved in order to provide proper operation. In order to present a complete definition of system requirements, some specific information and details about the GoodFood project which has been implemented in a pilot site on a vineyard are needed. For instance, the communication range of a single node, the number of nodes deployed in the vineyard, the total area of the pilot site, the planned system lifetime are among critical parameters that affect system requirements. Some important system requirements are as follows.

Network lifetime and power efficiency: the cost of radio communications in terms of energy is high. The larger the distance between two communicating entities, the higher the energy consumed. Therefore, in order to optimize energy consumption, a clustering structure should be used and information should be exchanged between neighboring nodes. In addition, energy saving measurements provides a longer lifetime. As a result, power efficient communication algorithms lead to a longer network lifetime. Due to the fact that a precision agriculture application scenario may be responsible for monitoring a food's production process for several months, an efficient power consumption algorithm is a must. The network lifetime of GoodFood is directly affected by the communication range of a single node, total number of nodes and total size of the area. Therefore, this information is needed for a precise definition of network lifetime.

Figure 2.5. *The GoodFood scenario*

Synchronization: time synchronization among entities is an important requirement in order to provide consistent coordination among nodes. The start and stop of the transmission between two devices should be defined correctly. In precision agriculture application, users can send queries to sensor nodes for some specific information. In this session, before starting communication, time synchronization is necessary. Synchronization between sensors and sink node should be provided.

Security: WSN applications face security requirements on different levels of the network. Whereas the loss of one or more nodes due to failure or damage can be tolerated in a specific case, in another situation data may need protection against deliberate attacks or accidental alteration. On the other hand, security measurements cannot make the system harder to access for users.

In this precision agriculture application scenario, it is important to have some security mechanism that controls the access to the system and information. Also, authentication and authorization are necessary among nodes to prevent fake nodes. It is possible to encrypt data with encryption algorithms that bring about low overhead. In order to provide system security, efficient user and key management schedules are needed.

Localization: if the data sent by sensors do not include the location information of the event, it is not valuable for the user. After the deployment phase, sensors should be able to estimate their locations in order to send reasonable data. It is possible to face a toxin attack on food in a specific region of the agricultural area. While sensor nodes notice this attack, they should give the position of the attack in order to allow responsible authorities to take necessary and efficient action. The deployment procedure of networked sensors determines the localization method of the GoodFood scenario. They may be placed manually, or there may be a few nodes superior to the rest which are deployed randomly. The superior nodes may have GPS receivers and therefore can define their positions via GPS. Other nodes may estimate their locations with geometrical and mathematical methods by communicating with those nodes that know their positions *a priori*. As a result, the minimum sensitivity of location information is mainly determined by the deployment method of the GoodFood project.

Address and data-centric communications: it is required that this precision agriculture scenario must support both address-centric and data-centric approaches. In the first model, each node has its unique logical address and communication is mostly point-to-point. This model will be used especially when queries for sensor readings are sent and configuration and diagnostic procedures are initiated. On the other hand, a data-centric model is used when exchanged data is more important than the participants of the communication. For instance, interests are propagated within the network and information is collected from nodes whose data match the requests.

Scalability: the number of sensors in the initial network topology may be in the order of tens or hundreds. However, the agricultural area to be monitored may be so large that it requires thousands of sensors. Mathematical and computational methods and the software of the scenario must be capable of working with this extremely high number of nodes. The design should allow the system to work properly even when the number of sensors is increased.

Fault tolerance: it is highly possible to lose some sensors of the network due to the environmental and operational conditions. However, the network should maintain its operation even if a few nodes fail to perform their task. That is, it should have a predetermined fault tolerance against possible errors.

Node heterogenity: it is practically impossible to have all nodes identical in a wireless sensor network application. Some of them should be more powerful than the others. Also, some nodes may be mobile while others are static. For this precision agriculture application scenario, we need sink nodes that are more powerful than elementary sensor nodes. It also has various sensors measuring different parameters of the soil and food products. The system should support node heterogenity.

2.7.3. CORTEX's Car Control

2.7.3.1. Introduction

[5, 37] are the main resources of the scenario. Car Control scenario aims at demonstrating the feasibility of the sentient object paradigm for real-time ad hoc environments. In this scenario, cars are able to operate independently and cooperate with each other to avoid collisions. It has been researching for the feasibility of future car systems that will be able to transport people without human intervention. The final destination and optionally the desired time of arrival are the only two parameters that need to be provided. Thus, the car control system will automatically select the optimal route according to those information for reaching the destination. Cars will cooperate with each other to move safely on the road, reduce traffic conditions and reach their destinations.

The principal target of this application scenario is to present the sentient object paradigm for real-time and ad hoc environments. It requires decentralized (distributed) algorithms.

Each car is a CO. However, they can be called sentient objects as well. They are autonomous and proactive. Sentient objects sense their environment via sensors and react to sensed information via actuators. Various sensors mounted on cars are responsible for monitoring the environment and sensing related information. They should receive both periodic events, such as position information broadcasted by nearby cars and sporadic events, such as emergency stop signals, which provide information about unpredictable situations.

2.7.3.2. *Scenario characteristics*

The Car Control scenario is an example of proactive applications relying on sensor-rich components. Each car is capable of acting based on the acquisition of information from the environment. Some of the main characteristic requirements of the scenario are sentience, autonomy, scalability, time and safety critical behavior and mobility. The scenario characteristics are introduced in the system requirements section.

2.7.3.3. *User requirements*

The main requirement of the scenario is the cooperation between the cars on the road. It is desired that each car contributes to the determination of traffic conditions. They should also obey the traffic lights without human intervention. If a car senses an obstacle ahead, it should broadcast an emergency signal, thus each car becomes aware of the situation. When there is a traffic jam in a particular region, other cars approaching the jammed area know the situation and have an opportunity of finding alternative ways.

2.7.3.4. *Functional specification*

In the Car Control scenario, each car is a CO and includes several sensors with different tasks and an actuator working as a control logic engine. The functionalities of each component of the scenario are given below.

Sensors: elementary sensor nodes are mounted on cars. They are not identical and each has different tasks. For instance, a sensor may collect information about traffic lights, while another one may measure the distance to another car or obstacle ahead. They do not interpret data, just sense their environment and send those data to the actuator.

Actuator: this is the brain of the car. The reaction to sensed information is organized by actuator. Principally, it is a control logic engine and decision maker.

2.7.3.5. *Object decomposition*

As stated above, the Car Control scenario is mainly composed of cars. In other words, each car is a CO. Also, traffic lights and the data center are members of the system. Each car has several sensors, each having different tasks and an actuator. The object decomposition of a car is given in Figure 2.6.

Sensors: these are mounted on the car. Each sensor has a different task. For instance, there may be a distance sensor, a speed sensor, etc. They collect the information for which they are responsible. They are not capable of interpreting the collected information.

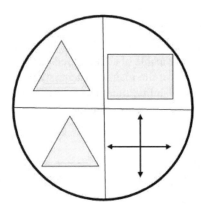

Figure 2.6. *The object decomposition of Car Control scenario*

The actuator: this provides the car with the ability of being independent of human control. The information collected by sensors are evaluated by the actuator. It determines the car's reaction to the sensed data. Also, it provides communication with other cars.

The data center: the Car Control scenario is based on decentralized algorithms run by cars on the road. However, there is a need for the data center even when the operation is distributed. The data center may build strategies and statistics about the traffic on the roads for particular time segments of the day and disseminate those information to the cars. It may also send weather and news information.

2.7.3.6. Step-by-step scenario description

The system mainly consists of cars where each car is a CO. They include a number of sensors that receive various periodic and event-based information from the environment. Sensors' output is only an approximation of the value sensed in the real world. Due to the physical measurement, uncertainty is inevitable in the data. A probabilistic sensor fusion scheme based on Bayesian networks can be used as a mechanism for measuring effectiveness of derivations of context from noisy data. Each car needs an actuator which is a control logic engine and interprets raw sensed data that needs to be transformed into meaningful information. Sensor readings also include context information such as the position of each car with respect to other cars and traffic lights. Each car needs to be aware of its safety zone and be able to detect any objects entering its safety zone. For instance, ultrasonic sensors fitted at each side of the car provide context awareness. The interaction between sentient objects differs from traditional interaction models. They communicate via an anonymous, generative communication abstraction. The messages are generated spontaneously rather than in request/response style. This leads to unplanned and spontaneous interaction.

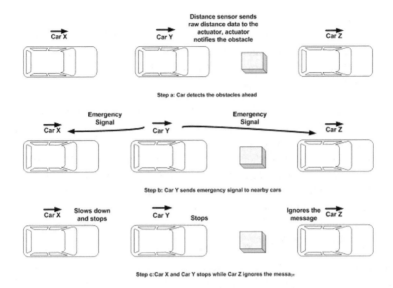

Figure 2.7. *The Car Control scenario*

Obstacle detection may be a scenario example of the CORTEX Car Control application. In order to avoid colliding with both obstacles on the road and other cars, it should be able to detect its position and the positions of other cars, and take action upon this information. If a car detects an obstacle ahead, it should rapidly apply an emergency stop and disseminate an emergency signal. Cars receiving an emergency stop signal from a car ahead should also apply an emergency stop themselves. The picturized scenario is given in Figure 2.7.

2.7.3.7. *System requirements*

Autonomy: as stated before, this scenario aims at searching for possible ways of providing traffic control without human intervention. Cars need to make autonomous decisions such as deciding when to brake or change the car speed. Decision-making takes place based on current and past context information. A control logic engine is necessary to fulfill such autonomous behavior.

Cooperation: in order to avoid of human-based traffic control, some degree of cooperation between cars is required. Each car notifies other cars about the actions it takes by dissemination of events. The events are propagated to all cars in its proximity. Event dissemination should be as real-time as possible. For instance, a car suddenly slows down and the car is followed by many other cars. The braking car should broadcast an event notifying brake action to cars which follow it, in a timely manner. The cars that have received the brake event on time take necessary measurements to avoid collision. Disseminated events should also include context information such as the location of the braking car for the correct perception of the event.

Scalability: the system should be designed to deal with scenarios where there are a large number of cars. The number of participant cars, or sentient objects, has a varying nature and depends on the congestion on the road. The number of cars cooperating with each other will be potentially large and each car will communicate in a one-to-many fashion without the help of centralized servers. The cars and traffic lights must use wireless networks that use an ad hoc and multi-hop model for communication. For an effective scalability strategy, location aware communication should be used to avoid message propagation beyond the geographical area of interest.

Time critical: the CORTEX Car Control scenario needs timely dissemination of events whereby higher priority events obtain higher priority channels and higher priority threads. It should provide real-time guarantees for assuring that critical messages such as an emergency stop signal have bounded delays.

Safety critical: in distributed systems based on wireless ad hoc environments, there may be an unpredictable number of application components that compete for a limited amount of resources such as CPU and communication resources, thus leading to unpredictable delays for executions. Due to the fact that a car control system has strict timeliness requirements, the system is only feasible if timing failure detection and QoS based adaptation are supported. A timing failure detection service can be used to detect timing failures, which makes it possible to take fail safe actions upon timing failures.

Mobility: the CORTEX Car Control scenario mainly includes both fixed and ad hoc structures. Traffic lights and GPS services provided by satellites are some elements of fixed structure, while the scenario mainly depends on cars that create a highly mobile environment.

2.7.4. *Hogthrob*

2.7.4.1. *Introduction*

Progressively, IT is finding its way into pig production, a leading industry in Denmark. Most farmers are now equipped with computer-based planning and control systems (including wireless networks and hand-held PCs). The use of IT for practical animal husbandry is still very limited; however, a recent project conducted by the National Committee for Pig Production in collaboration with DTU explored the usage of microphones together with a tailored voice recognition software to identify coughing pigs. Furthermore, a new law requires pregnant sows to move freely in a large pen. This is a challenge for farmers. They need to identify the pregnant sows that should be placed in the small pen. Sows are nowadays equipped with RFID tags. The farmer needs to use a tag reader (possibly physically applied on the RFID tag) to identify the sows. This solution is not practical in large pens.

2.7.4.2. *Scenario characteristics*

A sensor node with an integrated radio placed on each sow could transmit the sow's identification to the farmer's hand-held PC, thus alleviating the need for a tag reader. The use of sensor nodes on the animals could facilitate other monitoring activities: detecting the heat period (missing the day where a sow can become pregnant has a major impact on pig production), possibly detecting illness (such as a broken leg) or detecting the start of farrowing (turning on the heating system for newborns when farrowing starts).

2.7.4.3. *User requirements*

The farmer needs to be alerted when a sow is in heat. The farmer needs to be locate a sow given its identification.

The cost of a sensor node has to be in the order of 1 euro (due to the limited profit margin on a sow – note that a sensor node on a pig should not cost more than a couple of cents).

The lifetime of the sensor node should ideally be 2 years (the lifetime of a sow). A lifetime of 6 months and more is acceptable.

2.7.4.4. *Functional specification*

Hogthrob is a research project. Functional specifications are thus a bit of an overkill. Here is a description of the three main challenges tackled in the context of the project:

• The current generation of sensor nodes is assembled from off-the-shelf components. The next step is to integrate all the components of a sensor node on a single chip in order to minimize energy consumption as well as production cost. We propose building networked on-a-chip sensor nodes where processing, communication and sensing capabilities are integrated on a single chip. We opt for a system-on-a-chip design because the limited number of components guarantees a low production cost and because the small size of the device gives room for robust packaging.

• Early work on sensor networks has focused on demonstrating the potential of the technology. Only a couple of experiments have been led with actual applications. We propose to develop a sensor network infrastructure adapted to the needs of an actual application.

• The sow monitoring application will allow farmers to track sows in a large pen, to detect the start of the heat period and possibly to detect the start of farrowing as well as illness (such as broken legs). There is no such monitoring system available today.

2.7.4.5. *Object decomposition*

The object decomposition of Hogthrob is given in Figure 2.8

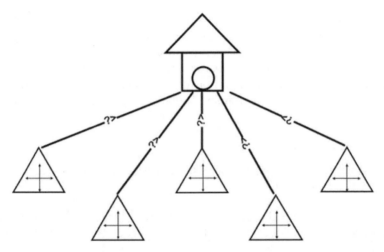

Figure 2.8. *Object decomposition of Hogthrob*

The behavior of sows has been studied over the years. It is a well known fact that the behavior of sows changes when they are in heat and when they are close to farrowing. Sows in heat become more active and go exploring more. If they are housed in a pen with contact to a boar pen, they more frequently approach the boar. Often sows will also mount each other when they are in heat. In the last hours before farrowing a sow exhibits characteristic nest-building behavior. The pattern is to a large extent prohibited by the confinement of the sow, but the restlessness behavioral pattern remains. The first challenge is to design a predictive model of the heat period, taking the movement of sows as input and providing as output the probability of the heat period. Validating such a model will require a body of statistical data about sow movements in relation with their heat period. To the best of our knowledge, such data do not exist. Field experiments will make it possible to acquire this. Similar models will also be built for predicting the start of farrowing and for illness when the activity level is lower. Because illness events are rare, our predictive model will be more speculative. It will have to be validated over a time period that is probably longer than the project. The second challenge is to capture the data needed to feed the model. What sensors should be installed on the sensor nodes? What kind of data should be collected from the sensor network? What kind of processing should be performed on the sensor nodes? What are the requirements in terms of response time? These questions are being tackled in close cooperation with the development of the sensor nodes and of the sensor network infrastructure.

2.7.4.6. *Step-by-step scenario description*

1. Sow localization: the farmer enters the pen with his PDA. He selects the ID of the sow he wants to locate. The sensor node on the selected sow emits a visual or acoustic signal that allows the farmer to locate the given sow.

2. Heat period: the farmer gets an alarm whenever a sow enters the heat period. Technology has been developed for monitoring the heat period of cows (in particular, the University of Twente did experiments in the context of the Eyes project). However, the technology developed for cows does not scale to sows in terms of cost, form factor and lifetime.

2.7.5. *Smart surroundings*

2.7.5.1. *Introduction*

[11, 23, 24] are the main resources of the study. The overall mission of the Smart Surroundings project is to investigate, define, develop and demonstrate the core architectures and frameworks for future ambient systems. In this context, ambient systems are referred to as networked embedded systems which support people in their everyday activities by integrating with the environment. They are aimed to create a Smart Surrounding for people to increase their living standards and their productivity at work. These systems differ from traditional computing systems by being based on an unbounded set of hardware and software, which will be embedded in everyday objects or appear as new devices. In this project, the aspects of ubiquitous computing, which is a rapidly developing area, are intended to be detailed and aimed to be implemented. Currently, most new technologies can be practically built into many objects to create smart networked environments. However, integrating these technologies in an open platform to be used in a wide range of applications is still an issue to be resolved.

Main objectives of this project can be explained as follows:

1. To open a platform for ubiquitous computing systems that integrates the required infrastructure components and provides an extensible set of universally installable tools, devices and services for the developers, operators and users of Smart Surroundings. The platform development will be driven with an engineering ethos of providing solutions that are practical and sustainable in the face of real world, and effective in reducing the cost for development and installation. The aim is to establish this platform as a standard for research and development of ubiquitous computing environments.

2. To lay the foundations for understanding interactions in ubiquitous computing with the conceptual frameworks, models and notations needed to describe the structure and behavior of system components from a variety of research perspectives. The work on foundations aims to overcome the current ad hoc nature of designs and evaluations. The expected result is a set of fundamental models and frameworks that will support evaluation and comparison of designs and systems.

3. To study ubiquitous computing in concrete and complex settings to ensure that development of platforms and foundations remains firmly grounded in reality. The concrete settings will investigate ambient system environments ranging from small and dense to large and sparse, and from digitally well provisioned to digitally impoverished. The scenarios explored in these settings will not be focused on selected applications as such but on the complex situations that arise from the interaction of various users with many different threads of activity. The project target is to design and implement real-life experiments that expose ubiquitous computing systems to the challenge of supporting a multitude of competing applications and user experiences.

2.7.5.2. *Scenario characteristics*

Settings

At present, the project is focusing on two main settings: *well-being* and *office*. The main application on the well-being setting is stress management, while in the setting the application is developing flexible office environments. Additionally, a short time ago a sub-project, *Home Care SenseNet*, was started to improve elderly care.

Stress management: stress is experienced by most people due to their work, sports, family responsibilities, etc. People believe they can use some assistance in dealing with stress. Medically, stress is defined as a "perturbation of the body's homeostasis". The common indices of stress include changes in biochemical parameters (such as epinephrine and adrenal steroids), physiological parameters such as increased muscle tension, heart rate and blood pressure, and behavioral effects such as anxiety, fear and tension. However, just as distress can cause disease, it seems plausible that there are good stresses that promote wellness. Stress is not always necessarily harmful. Increased stress at work, for instance, may result in an increase in motivation and awareness, providing the stimulation to cope with challenging situations. However, long-lasting periods of stress may result in negative stress. Excessive, prolonged and unrelieved stress can have a harmful effect on mental and physical conditions; for instance, it is associated with cardiovascular diseases, immune system diseases, asthma and diabetes. Common signs of negative stress are tiredness, concentration and memory problems, changes in sleep patterns, etc. People need to find an optimum between positive and negative stress. Since there is no single level that is optimum for all people, *personal stress-training systems* can assist in reducing elevated levels of stress. Controlling stress by means of this system contributes to the subject's well-being. Stress management should be individually adapted, able to learn from ongoing experiences, private, available anywhere and any-time, active, and continuously interacting with its surrounding to optimize feed-back. This addresses the need for a wearable or private product (clothes, jewelry, etc.) so the subjects can and will be trained in a variety of settings which are not physically bounded.

Stress will be determined using different methods: measuring physiological data (e.g. muscle tension, heart rate), monitoring the user's interaction with the envi-ronment (e.g. preasure excerted when using the mouse, playing with a pen or pressing a ball), letting the user provide feedback about his stress level with tangible interfaces, and taking into account the context of the user (e.g. room temperature, light level, activity performed, agenda, etc.).

The feedback will be provided modifying the ambient (e.g. lights, sounds, music) where the user is at present (e.g. office, home, car), and using wearable and private products such as PDA, watches, jewelry, clothes, etc.

Figure 2.9 provides a high-level picturized view of the stress management set-ting. The general hierarchical view of the Smart Surroundings' settings is shown in Figure 2.11.

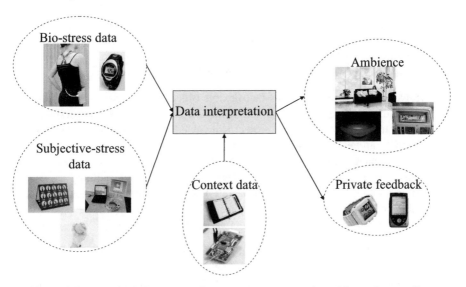

Figure 2.9. *Picturized abstraction of stress management setting of Smart Surroundings*

Flexible offices: nowadays, the trend of working in a more flexible way both while traveling to various locations and countries and within their own company build-ing becomes apparent. Governments and employers have already started seeing advantages of flexible office solutions. It saves building maintenance expenses and parking costs, reduces traffic jams and increases effectiveness of employees. This trend will continue in the future. Such a change will be motivated by economic reasons as well as a way to enhance the quality of work.

Smart Surroundings envisions the future of working in offices to be different from what we experience today. Both the environment and the working itself will differ from today's style of working. International cooperation and global activities of businesses, industry and research will involve more and more people working abroad. The mobility and temporary residence of working people promotes the idea of renting offices for short and mid-term periods instead of buying or continuously renting them. In this project, they picture the situation of a flexible office being rented for some weeks to months for people who work abroad. This setting outlines requirements and possible solutions for such flexible offices. Special attention will be on the personalization of the office space as well as practical flexibility and usefulness of furniture and electronic equipment.

People will start working in more flexible ways not only while traveling abroad but also within their own company building. A growing number of companies switch to open workspaces and flexible working style. An example of such a transformation is the Dutch insurance company, Interpolis. The biggest technical enabler for implementation of the flexible office concept was development and popularization of wide area networks, wireless communication and the Internet enabling remote access to the office. This change has impacted on technology as well as on the way teams work.

Smart Surroundings believes that flexible offices will be part of our future. It is clear that if not correctly implemented, it can result in a undesirable work environment. Therefore, the project is involved in providing a flexible and yet friendly and efficient way of working. They picture the situation of a flexible office, where no employee has his/her own, static office space. Every person has the means to work from any place within a company building as well as from another location (client, home, etc.). They outline requirements and possible solutions for such flexible offices. Special attention will be paid on the communication aspect as well as enablers for ad hoc meetings.

At present, they are working on an abstraction called "Mini-ME". Mini-ME is the user's doublet in the working and family life, providing and receiving information around the worlds from/to other Mine-MEs, and giving to employees desired data in order to make decisions, organize time and to get in contact with other people. Mini-ME will also allow users to control the environment and personalize it to fit their preferences as well as control the devices in the offices and outside them (e.g. coffee machines, copiers, overhead projectors). Envisaged scenarios are also supporting workers to hold informal meetings in coffee rooms or corridors, and supporting formal meetings with unknown people. Figure 2.10 gives a high-level view of the functionality of Mini-ME.

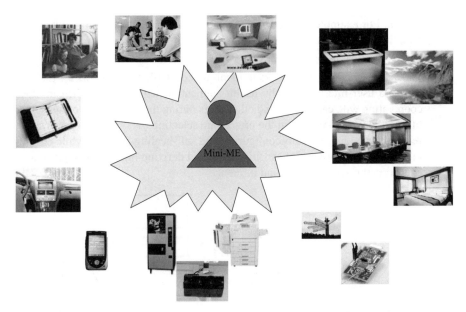

Figure 2.10. *Picturized abstraction of office setting of Smart Surroundings*

Home Care SenseNet: the Home Care SenseNet (HCSN) [3] aims to improve the quality of the life of patients in need of care and their care-givers. It achieves this by enabling longer independent living, and automating routine administrative task of care-givers.

The strength of the HCSN is in its simple deployment in existing private flats/ houses and care institutions because no expensive cabling infrastructure is necessary. Sensor nodes can be attached exactly where they are needed and can last for years without maintenance. Small ad hoc grid WLAN nodes, which require only a power outlet, provide second layer infrastructure connections that extend the lifetime and reach of the network in large buildings. The whole network is largely auto-configuring and requires no or little IT knowledge for set-up and maintenance.

The WSN will monitor a part of the physical environment, physiological state, current location of clients and, optionally, care-givers. The health-care application uses this context data to derive significant events, notify care-givers when appropriate, and log events for long-term analysis, administration, and accounting uses. Examples for significant events are: "change diaper of client"; "client's diaper was changed by care-giver"; or "client sleeping uneasily". In case of potential emergencies (e.g. "client has fallen"), the HCSN will use the location information to alert the closest care-givers. A significant use can be the recognition and monitoring of behavioral patterns and the detection of changes in that pattern. For example, the HCSN may monitor "normal" wake-up times, when a particular room is used in the house, at what times the fridge is opened or medicine is taken, etc.

The actual events or parameters to monitor acceptable form-factors for the sensors and user interfaces to the health application will be established in close cooperation with care givers and other users during the project. Important scientific challenges addressed in the project are selecting the optimal, least invasive sensors; providing an auto-configuring and self-healing network while enforcing security and privacy of sensitive data; and deriving high-level client-related context using sensor fusion and inference.

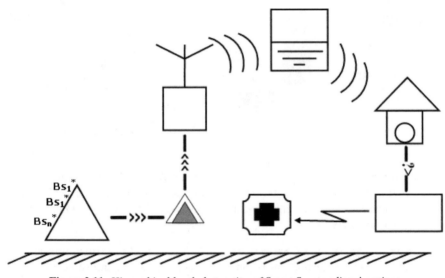

Figure 2.11. *Hierarchical level abstraction of Smart Surroundings' settings*

2.7.5.3. System requirements

Communication capability: the capacity to use the underlying infrastructure – wired and wireless – and create ad hoc networks when needed.

Sensing capability: sensing plays a very important role in Smart Surroundings. A wide variety of sensors will be used to determine phyisiological parameters, environmental data, activity levels, etc.

Wireless sensor networks: WSNs will be used both indoors and outdoors for sensing and location functions. Simple sensed data is combined in different ways to obtain more complex information, as explained in the stress management and the Home Care SenseNet settings.

Body area networks: to monitor physiological data and provide feedback.

Sense-making: how can systems make sense of user activity to provide ambient intelligence, and how can users make sense of environments that are characterized by dynamic and spontaneous composition of services?

Security and privacy: Smart Surroundings have to be designed to empower and support people in their activities, but in ways that would avoid the control or manipulation of non-authorized others.

Context awareness: context plays a very important role in the settings. The most important type of context information that have been identified are location, topology, neighboring, movement and activity, social networks, agenda, behavioral patterns and historical data, facilities and resources, mood and stress level, external sources of information, and special events.

Personalization: the environment and devices should adapt to the preferences of the users. Also, the way of controlling the devices and getting feedback should be easily personalized. The user should be able to manually set his preferences. Additionally, the system could try to learn from early behavior the preferences of the user.

Service discovery: both in the office setting and well-being setting, we need to control the environment and access services and resources available in the surroundings. Service discovery (and usage) is, thus, a must. The information provided by other Mini-ME and the environment systems (e.g. office system) are also viewed as services.

Adaptability: the system, services, interfaces and feedback have to adapt depending on the environment and available resources (e.g. energy, communication capabilities, storage and displays).

Resource management: Smart Surroundings envisages hundreds or thousands of users competing for the use of resources. Therefore, managing the resources efficiently is of crucial importance.

Energy-Efficiency: with devices embedded in the infrastructure and sensors everywhere in the environment, it is virtually impossible to be constantly changing batteries. Thus, energy-efficiency has to be addressed in the system architecture.

Innovative interfaces: a key motivation for this research is to find new ways for people to interact with computer-based services while giving primacy to the real world. A main thrust of this work will be aimed at innovative and embedded interfaces, with particular emphasis on supporting ad hoc composition and configuration of tangible interface components.

2.7.6. *Sustainable bridges*

2.7.6.1. *Introduction*

[9, 28] are the main resources of the scenario. This scenario examines the high-speed train railway bridges which are expected to meet the needs of the future. These needs are basically expected to be an increase on the capacities, heavier loads to be carried, and increments in the speeds of the trains concerning the increased railway traffic on the railways. All types of bridges are to be considered in this scenario. The network of the railways in Europe will be inspected primarily.

These improvements can basically be realized by assessing of the bridge structure, determining the true behavior of the structure, strengthening certain portions of the bridge or by monitoring critical properties.

The existing structures and designs of those bridges can detect the irregularities during their constructions or at any time while they are in service. However, it is very unclear whether they can carry much greater loads or handle much faster trains. That is why a new monitoring and alerting mechanism should be used. The overall goal is to enable the delivery of improved capacity without compromising the safety end economy of the working railway.

The most recent technique to determine irregularities in the bridges is the installation of wired monitoring systems to analyze the structural behavior. However, such monitoring systems use standard sensors and several other devices which are time-consuming to install and very expensive to spread around. Also, kilometers of wires are needed for their communication, which means a waste of money and a high risk of communication failure in the event of an irregularity in the medium.

An alternative way to monitor railways and bridges is equipping them with a wireless monitoring system and using micro-electromechanical systems. Hence, the set-up cost of a communication system will disappear (getting rid of wires), and using small and intelligent sensors carrying devices will dramatically reduce the production and maintenance costs.

2.7.6.2. *Application characteristics*

The basic principles of structural mechanics and materials science will be used and integrated with new techniques in monitoring, measurement and modeling in order to achieve the predetermined goal of this project. The idea is to analyze bridge types and their details for when they are critical for load carry capacity, safest maximum speed level and remaining lifetime.

Current traditional monitoring of civil structures has been performed either by visual inspection or by installation of a collection of sensors that communicate with each other using cable technology. In the first case, the interpretation and assessment of the structure is based on the experience of the expert, whereas the second method requires the expensive installation of kilometers of cables to cover the object to observe.

The main goal of this project is to develop a cost-effective solution for the detection of structural defects and to better predict the remaining lifetime of the bridges by providing the necessary infrastructure and algorithms. Due to the fact that this project aims to use WSN technology in order to reduce the costs of installation and maintenance of the traditional systems, a carefully evaluated infrastructure has to be formed.

The sensor network will be composed of several types of sensor nodes, temperature, humidity, vibration and material stress sensors, which will be located at specific points within and outside the bridge. A static network topology will be formed, where several types of nodes will cooperate with each other.

Each individual sensor is planned to be affixed to the structure, communicates with its neighbors and acts as a part of a cluster of nodes, which coordinates the decision process to evaluate whether a structural defect has occurred or an obvious change in the structural behavior has happened. The user (say an engineer) will have a global view of the network and will be able to specify which node will join a cluster or which will be the sink. In other words, the engineer will assign or modify the roles of each component in the system.

One requirement is the need of determining the roles of the components exactly. The roles of sensor nodes are clear: they can be vibration sensing, temperature sensing or relative humidity sensing sensors. It is also possible that one sensor node can have more than one role at the same time. Other than sensing, the sensor nodes can be assigned to other roles such as *cluster head* or *cluster member*, or *data aggregator*. However, these are only the predicted role types. There can be more or less than stated here. Figure 2.12 shows the hierarchical and the Figure 2.13 shows the picturized view of the system.

2.7.6.3. *System requirements*

Network configuration: the network configuration issue should be divided into two parts: assignment of the roles within the network, and determining the optimal routes. The first part is thought to be solved by human interference. The user will have enough technical knowledge to determine what kind of roles should be assigned to each node. The second issue requires the optimization of the network route in the most power-efficient way so that the monitoring system must be able to function for several years without changing the batteries.

Cluster management: the whole sensor network must be divided into clusters because of the necessity to decide if the sensed event is related to a structural

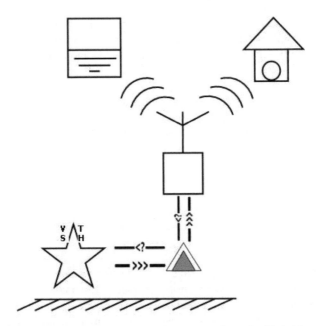

Figure 2.12. *Hierarchical level abstraction of sustainable bridges*

defect. In order to reduce the number of false positives, all members of a cluster that might have detected some changes to an event will decide that it should be reported to the engineer or not. Since the topology is static in this scenario, the user can best determine which node to act as the sink. This will be much more efficient and much easier.

Event localization: most sensor data is associated with the physical phenomenon that is sensed. The event localization is the determination of the position of the specific event. The use of signal strength techniques for event localization is not a problem here, because these techniques only determine the position of the sensor itself, not the origin of the event. The use of the measured transient acoustic event and its onset time with the combination of the knowledge about the specific positions and the sensing rangers of all nodes in the network will make the precise locating of events possible. This will also reduce the number of packets to be transmitted.

Time synchronization: nodes within a cluster will compare their readings about the events. Hence, each should be able to compare its readings in order to discard data that does not need to be forwarded. Thus, the overall data to be transmitted in the network will be reduced. In this scenario, a time synchronization technique at a precision level of 60 us is needed. Hence, new algorithms should be developed in order to achieve this level. This synchronization level is also

Figure 2.13. *Picturized abstraction of sustainable bridges*

important to localize the event more precisely. If this value can be achieved, the location of the event can be defined within a few meters of precision. Otherwise, the evaluated area of the event will be too big.

However, this level of precision is very difficult to achieve, and it seems to be nearly impossible with the current technology. This fact comes up with another issue: the need for developing better algorithms to handle a lower synchronization level while the precision level is high enough.

Data aggregation: to support energy conservation, the number of packets should be reduced as much as possible. So, data aggregation should be used to add more efficiency to the whole system. Although there are existing aggregation algorithms for such purposes, the scenario specific properties, like static topology information and role specification, can be used as an advantage to develop even more efficient algorithms.

2.7.6.4. *Functional specification*

A wireless monitoring system with microelectromechanical system sensors can reduce costs in comparison to traditional systems, as explained before. The referred system is described below.

Each mote, which itself is a complete small measurement system, has to be power and cost-optimized, so it can provide data only at small distances of 100 meters or less. For that reason, there is the demand to install a central processing unit on site in addition to the installation of the sensor motes. This central unit has to collect and store the data in a database, and to analyze the data from the sensor motes until this data is requested by the user or until a sudden event is detected which results in an alarm message. The central unit should also allow a calibration and a wireless reprogramming of the sensor nodes to keep the whole system flexible. A conventional computer equipped with a constant power supply and specific hardware and software is expected to fit all those purposes.

The sensor network will continuously collect sensed data from the bridges and the sinks will forward the whole data to a data center in a higher level in the architecture. The database will continuously be analyzed at the end-point, and if a critical change occurs, the immediate interference will be supplied. If the continuity of the information shows that maintenance became mandatory or just necessary, the required action will be planned for the future demands and the stability of the structure.

2.7.6.5. *Object decomposition*

Sensor motes: there are several tasks of the motes that have to be performed. These are collecting and processing data from various sensors, storing this data, acting as a pre-processor to analyze the data, sending and receiving selective and relevant data among other motes, and working for a long time without an externally wired power supply. Therefore, those motes should consist of a CPU or DSP, memory, a low power radio, an aligned ADC, a power supply and one or more diverse sensors:

a. *Acceleration sensors:* for the bridge-like structures, a bandwidth of 0.1 to 20 Hz is important for the related algorithms work best. Acceleration, velocity and sensitivity depend on these frequencies. Separate sensors to detect each of those changes are necessary to gain appropriate results.

b. *Humidity and temperature sensors:* the temperature and the moisture of the air and of the structure are general values, which are relevant for almost all monitoring tasks.

c. *Stress, strain and displacement sensors:* a high amount of strain gauges are in practical use to measure strain and also stress under static or dynamic loads.

Sink nodes: the sensor network is connected to one or more sinks, which may even be desktop computers located close enough to the observed structure, and will have the purpose of collecting data and interfacing the sensor network with the user.

Base station: eventually, the base station will forward the meaningful data to the data center and to the user, which do not have to be located near to the structure.

Data center: the forwarded data will be continuously collected in a database system to evaluate it in the long term. Hence, statistical information will lead us to develop better structures and even better monitoring systems in the future.

Facility: in the case of a sudden emergency event and hence the need for immediate interference, the user will independently be informed about the event which is continuously stored in the data center. The engineer then will decide what to do next.

2.8. Conclusions

State of the art projects clearly indicate that a wide spectrum of applications can benefit from the CO paradigm. The diversity of the characteristics and requirements make the field a challenging research domain. As a result, various enabling technologies such as energy-efficient schemes for networking, time synchronization, localization and tracking are being invented and studied. Table 2.19 summarizes the common and scenario-dependent requirements and challenging research issues for each CO application domain. Whereas $\sqrt{}$ stands for common characteristics of that category, X means that it is not relevant for that category. Application-dependent issues are represented by "AD".

	CA	HO	L	TA	EM	H	SS	T	ET
Network topology	AD	Single	AD	Multi	Multi	AD	Multi	AD	AD
Indoor/outdoor	AD	In	Both	Out	Out	Both	Out	Both	In
Scalability	AD	X	√	√	√	X	√	AD	AD
Packaging	AD	X	AD	AD	√	X	√	AD	AD
Fault tolerance	√	AD	√	AD	√	AD	√	AD	AD
Localization	AD	√	√	√	√	√	√	AD	AD
Time synch.	√	AD	AD	√	AD	AD	√	AD	AD
Security	AD	√	AD	AD	AD	AD	AD	X	AD
Infrastructure	AD	AD	AD	Ad hoc	AD	AD	Ad hoc	AD	AD
Prod-maint.cost	√	√	√	X	√	X	AD	√	√
Mobility	AD	AD	√	√	AD	AD	AD	√	AD
Heterogenity	√	AD	AD	√	AD	√	AD	X	AD
Automation	√	AD	X	AD	AD	AD	AD	AD	√
Power awareness	AD	AD	√	AD	√	AD	√	AD	AD
Real-time	√	AD	AD	√	AD	√	√	AD	AD
Context-awareness	AD	√	X	X	X	√	X	√	X
Reliability	X	X	√	X	X	√	X	X	X

Table 2.19. *Characteristics of CO application domains*

Moreover, as illustrated in Figure 2.14, almost all CO projects have emerged in the last five years and only few real sensor applications are readily available today, if academic testbeds do not count. Most projects are at testbed stages; mainly, verifying node prototypes or identifying real requirements of the applications. Today, current deployed applications share some common characteristics: raw sensor data transmission over wireless connection, mostly data processing at collection points, simple routing schemes and best-effort data transport delivery. However, these characteristics do not reflect the real requirements of most application domains. Therefore, common characteristics given in section 2.3 are still challenging research issues especially in multi-hop, scalable and real-time environments.

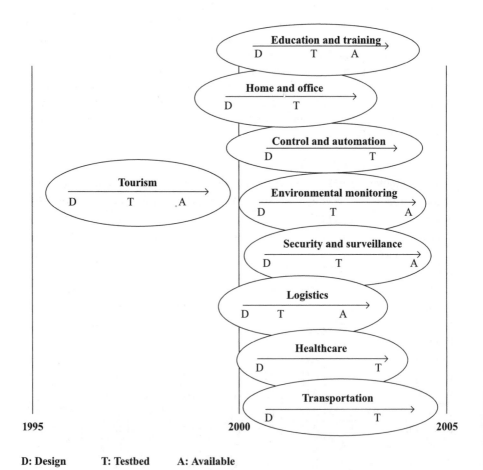

D: Design   **T: Testbed**   **A: Available**

Figure 2.14. *Timeline for CO application categories*

The field is analogical to the situation of the Internet 30 years ago. The major difference is that CO research is oriented towards application-specific solutions. General-purpose systems, standards for open interconnections, services and interfaces have yet to come into wider use. Another fact to point out is that the requirements and constraints of various applications are not fully understood yet. As a result, most of the applications are not ready for the real world.

In conclusion, this new CO paradigm has a potential of improving the quality of our lives considerably. For measurable outcomes, research in the field of common features needs to be matured. With the progress on sensor fabrication techniques and multi-disciplinary research cooperation, we can expect that real-world CO applications containing wireless sensor networks will come into being in the near future.

2.9. List of abbreviations

A	Acoustics Sensor
ADC	Analog-to-Digital Converter
AICIA	The Association of Research and Industrial Co-operation of Andalucia
AmI	Ambient Intelligence
AOA	Angle-of-arrival
APIT	Approximate Point-In-Test
B	Barometer Sensor
CBL	Calculator Based Labs
CD	Compact Disc
CO	Cooperating Objects
CoBIs	Collaborative Business Items
CORTEX	Co-operating Real-Time Sentient Objects
CPU	Central Processing Unit
CVD	Cardio-Vascular Diseases
DSP	Digital Signal Processing
DV	Distance Vector
DVD	Digital Versatile Disk
EPA	Environmental Protection Agency
EU	European Union
GPS	Global Positioning Systems
H	Humidity Sensor
I/O	Input/Output
IEEE	Institute of Electrical and Electronic Engineers
IST	Information Society Technologies
KU	Kopenhag University
M	Magnetometer Sensor
MBL	Microcomputer-Based Labs
MP3	MPEG Audio Layer 3

P	Photo Sensor
PCR	Polymerase Chain Reaction
PDA	Personnel Device Assistants
PIT	Point-In-Test
RFID	Radio Frequency Identification
RiscOFF	Rapid Intelligent Sensing and Control of Forest Fires
RSSI	Received Signal Strength Indication
SCOWR	Scalable Coordination of Wireless Robots
SNS	Sensor Network Server
T	Temperature Sensor
TDOA	Time Difference of Arrival
TOA	Time-of-Arrival
UAV	Unmanned Aerial Vehicles
UT	University of Twente
UWB	Ultra Wide Band
VCR	Video Casette Recorder
WLAN	Wireless Local Area Networks
WSN	Wireless Sensor Network
X	Accelerometer Sensor
YTU	Yeditepe University

2.10. Bibliography

[1] *Project Proposal: Rapid Intelligent Sensing and Control of Forest Fires – RISCOFF*, July 2003.

[2] Jakob Bardram, Hospitals of the Future-Ubiquitous Computing Support for Medical Work in Hospitals, in *Proceedings of UbiHealth 2003, The 2nd International Workshop on Ubiquitous Computing for Pervasive Healthcare Applications*, 2003.

[3] Hartmut Benz and Paul Havinga, *Home Care SenseNet – Proposal for a Sub-project within Smart Surroundings*, 2005.

[4] M.M. Chen, C. Majidi, D.M. Doolin, S. Glaser and N. Sitar, *Design and Construction of a Wild Reinstrumentation System Using Networked Sensors*, 2003.

[5] CORTEX:, *Analysis and Design of Application Scenarios*, Deliverable D8, Lancaster University, May 2003.

[6] D.M. Doolin, S.D. Glaser and N. Sitar, *Software Architecture for GPS-Enabled Wildfire Sensorboard*, TinyOS Technology Exchange, University of California, Berkeley CA, February 2004.

[7] D.M. Doolin and N. Sitar, Wireless Sensors for wildfire Monitoring, in *Proceedings of the SPIE-Smart Structures and Integrated Systems*, vol. 5765, May 2005, pp. 477–484.

[8] S.D. Glaser, Some Real-World Applications of Wireless Sensor Nodes, in *Proceedings of SPIE Symposium on Smart Structures and Materials*, March 2004.

[9] C. Grosse, F. Finck, J. Kurz and H.W. Reinhardt, Monitoring Techniques Based on Wireless AE Sensors for Large Structures in Civil Engineering, in *EWGAE Symposium*, 2004, 843–856.

[10] D. Guerri, M. Lettere and R. Fontanelli, Ambient Intelligence Overview: Vision, Evolution and Perspectives, in *1st GoodFood AmI Workshop*, Florence, July 2004.

[11] P. Havinga, P. Jansen, M. Lijding and H. Scholten, Smart Surroundings, in *Proceedings of the 5th Progress Symposium on Embedded Systems*, 2004.

[12] http://corporate.traffic.com/ntdc/.

[13] http://grvc.us.es/comets/.

[14] http://grvc.us.es/cromat/.

[15] http://meteora.ucsd.edu.

[16] http://nesl.ee.ucla.edu/projects/smartkg/.

[17] http://robotics.usc.edu.

[18] http://www robotics.usc.edu.

[19] http://www.ascet.com.

[20] http://www.aubade group.com.

[21] http://www.ubisense.net/.

[22] http://www.uk.research.att.com/bat/.

[23] Rianne Huis in't Veld, Miguel Bruns Alonso, Aga Matysiak, Hermie Hermens, Frans Grobbe, Ana Ivanovic, Caroline Hummels, David Keyson, Albert Krohn, Maria Lijding, Joost Meijer, Klaas Sikkel and Felix Smits, *Deliverable 5.1 – Settings: The Future is Not What It has Been...*, Tech. report, Smart Surroundings consortium, 2005.

[24] Rianne Huis in't Veld, Miguel Bruns Alonso, Aga Matysiak, Hermie Hermens, Frans Grobbe, Ana Ivanovic, Caroline Hummels, David Keyson, Albert Krohn, Maria Lijding, Joost Meijer, Klaas Sikkel and Felix Smits, *Deliverable 5.2 – Requirements and Early Implementation of Prototypes*, Tech. report, Smart Surroundings consortium, 2005.

[25] R. Husler, Goodfood-Security, in *1st GoodFood AmI Workshop*, Florence, July 2004.

[26] R.H. in't Veld, M. Vollenbroek and H. Hermens, *Personal Stress Training System: From Scenarios to Requirements*, Tech. report, Roessingh Research and Development.

[27] M. Maroti, G. Simon, A. Ledeczi and J. Sztipanovits, Shooter Localization in Urban Terrain, *IEEE Computer Magazine*, vol. 37, no. 8, pp. 60–61, 2004.

[28] P.J. Marron, O. Saukh, M. Kruger and C. Grose, Sensor Network Issues in the Sustainable Bridges Project, in *European Project Session of the 2nd European Workshop on Wireless Sensor Networks*, 2005.

[29] A. Ollero, J.R. Martínez de Dios, B.C. Arrue, L. Merino and F. Gómez-Rodriguez, A perception system for forest fire monitoring and measurement, in *Proceedings of the Int. Conf. on Field and Service Robotics (FSR)*, July 2001.

[30] A. Ollero, S. Lacroix, L. Merino, J. Gancet, J. Wiklund, V. Remuss, I. Veiga Perez, L.G. Gutierrez, D.X. Viegas, M.A. Gonzalez Benitez, A. Mallet, R. Alami, R. Chatila, G. Hommel, F.J. Colmenero Lechuga, B.C. Arrue, J. Ferruz, J. Ramiro Martinez-de Dios and F. Caballero, Multiple eyes in the skies – architecture and perception issues in the comets unmanned air vehicles project, *IEEE Robotics and Automation Magazine*, vol. 12, no. 2, pp. 46–57, 2005.

[31] robotics.eecs.berkeley.edu/pister/SmartDust/.

[32] K. Römer and F. Mattern, The design space of wireless sensor networks, *IEEE Wireless Communications*, vol. 11, no. 6, pp. 54–61, 2004.

[33] Lee Seon-Woo, Location Sensing Technology for Pervasive Computing, in *Proceedings of the 17th Annual Human-Computer Interaction Conference, HCI2003*, 2003.

[34] Steven Shafer, The New Easyliving Project at Microsoft Research, in *Proceedings of the 1998 DARPA/NIST Smart Spaces Workshop*, 1998, pp. 127–130.

[35] C. Sharp, S. Schaffert, A. Woo, N. Sastry, C. Karlof, S. Sastry and D. Culler, Design and Implementation of a Sensor Network System for Vehicle Tracking and Autonomous Interception, in *Second European Workshop on Wireless Sensor Networks*, January-February 2005.

[36] Dana A. Shea, *The Biowatch Program: Detection of Bioterrorism*, Congressional Research Service Report RL 32152, Science and Technology Policy Resources, Science and Industry Division, November 2003.

[37] T. Sihavaran, G. Blair, A. Friday, M. Wu, H.D. Limon, P. Okanda and C.F. Sorensen, Cooperating Sentient Vehicles for Next Generation Automobiles, in *Proceedings of the MobiSys 2004, 1st ACM Workshop on Applications of Mobile Embedded Systems (WAMES 2004)*, June 2004.

[38] Michele W. Spitulnik, *Design Principles for Ubiquitous Computing in Education*, Tech. report, Center for Innovative Learning Technologies, University of Berkeley.

[39] M.B. Srivastava, R.R. Muntz and M. Potkonjak, Smart Kindergarten: Sensor-Based Wireless Networks for Smart Developmental Problem-Solving Environments, in *Mobile Computing and Networks*, 2001, pp. 132–138.

[40] Arne Svenson, *Executive Summary – The Safe Traffic Project*, February 2005.

[41] G. Werner-Allen, J.B. Johnson, M. Ruiz, J. Lees and M. Welsh, Monitoring volcanic eruptions with a wireless sensor network, in *Proceedings of Second European Workshop on Wireless Sensor Networks*, January-February 2005.

[42] www.cartalk2000.net.

[43] www.cc.gatech.edu/fce/cyberguide.

[44] www.concord.org.

[45] www.ctit.utwente.nl/research/projects/international/streps/cobis.doc.

[46] www.envisense.org.

[47] www.greatduckisland.net.

[48] www.hitech projects.com/euprojects/myheart/.

[49] www.rec.org/REC/programs/telematics/enwap/gallery/waternet.html.

[50] www.robocup.org.

[51] www.zebra.com.

[52] J. Yang, W. Yang, M. Denecke and A. Waibel, Smart Sight: A Tourist Assistant System, in *ISWC '99*, 1999.

Chapter 3

Paradigms for Algorithms and Interactions

3.1. Summary

This chapter is intended to provide a survey of the algorithms and paradigms for systems based on cooperating objects (COs). The aim of the chapter is to identify the areas that need further research and the most promising design approaches in the context of CO-based systems. The chapter is structured in three main parts. The first part, which encompasses section 3.2 and section 3.3, introduces the subject considered in this chapter and describes the thematic areas that have been considered in the analysis of the literature. The second part, consisting of sections 3.4–3.7, provides an overview of the most common and interesting solutions adopted in the thematic areas previously introduced. Finally, the third and last part of this chapter, which extends over sections 3.8–3.10, provides a survey of the literature and identifies the critical issues and the most promising design approaches in the context of COs.

3.2. Introduction

3.2.1. *Aim of the chapter*

The CO paradigm embraces different fields, such as traditional embedded systems, wireless sensors networks (WSNs), ubiquitous and pervasive computing systems, and so on. In general, such systems have very different functional characteristics, though some common features do exist, such as the hardware and software modules they are built upon, the use of wireless communication, the dynamic and unpredictable system topology, and so on. As a consequence of such heterogenity, the research activity

Chapter written by Andrea ZANELLA, Michele ZORZI, Elena FASOLO, Anibal OLLERO, Ivan MAZA, Antidio VIGURIA, Marcelo PIAS, George COULOURIS and Chiara PETRIOLI.

on CO systems has been generally performed with regard to specific aspects of specific systems, mostly ignoring the aforementioned common features. In this way, the research effort risks being partially wasted, since studies are duplicated in different, though similar, contexts. Even worse, solutions that might be of general interest for systems adhering to the CO paradigm could remain isolated within the borders of the specific system they were designed for.

Therefore, to boost the diffusion of CO technologies in the near future, it is necessary to change the perspective by considering the CO systems as a single, though varied, subject. This first requires a deep understanding of the current state of the art on the subject that will help to identify and classify the various facets of the topic.

This chapter aims at providing a rather comprehensive overview of the algorithmic solutions proposed for CO systems, thus making it possible to clearly identify common features and differences. Furthermore, a classification of the algorithms according to the requirements provided by Chapter 2 is proposed. This classification will make it possible to assess the matching between the features of an algorithm and the requirements of an application, thus providing a tool to evaluate the suitability of a solution for a specific scenario. It will also be emphasized which functions require vertical integration, i.e. messages passing among entities of different protocol layers. Such functions are thoroughly described and characterized in Chapter 4. Finally, Chapter 5 will provide an overview of the mechanisms and strategies that can be adopted to realize in practice the algorithms that are described by this book.

3.2.2. *Organization of the chapter*

For the sake of clarity, the subject has been divided in the following four *thematic areas*, which are representative for a number of possible application scenarios and differ for characteristics and requirements:

• **Wireless Sensor Networks for Environmental Monitoring** (WSNEMs): characterized by a large number of stationary sensor nodes, disseminated in a wide area and few sink nodes, designated to collect information from the sensors and act accordingly. Sensor nodes are often inaccessible, battery powered and prone to failure due to energy depletion or crashes. Furthermore, network topology can vary over time due to the on/off power cycles that nodes go through to save energy.

• **Wireless Sensor Networks with Mobile Nodes** (WSNMNs): characterized by the use of mobile nodes which offer many advantages, for example, fewer nodes needed to cover a given area, dynamic adaptation to environmental changes and dynamic topology variation to optimize communication.

• **Autonomous robotics teams for surveillance and monitoring** (ART): multiple-robot or in general multiple-CO systems can accomplish tasks that no single CO can accomplish, since ultimately a single CO is spatially limited. Autonomous robotic teams are also different from other distributed systems because of their implicit "real-world" environment, which is presumably more difficult to model and reason about than traditional COs of distributed system environments (i.e. computers, databases, networks).

• **Inter-Vehicular Networks** (IVNs): networks composed by an ad hoc organization of wireless-powered vehicles, each able to transmit or receive data on a highway or in a city street.

Table 3.1 shows how the reference scenarios can be associated with one or more thematic areas. As can be observed, the mapping is not one-to-one, since the scope of the thematic areas is wider than that of the reference scenarios. In fact, the thematic areas have been selected with the purpose of covering as many applications scenarios as as possible, thus giving a fairly complete overview of the huge variety of applications that involve COs.

	WSNEM	WSNMN	ART	IVN
Control and automation	✓	✓	✓	✕
Home and office applications	✓	✓	✕	✕
Logistics	✓	✓	✕	✕
Transportation	✓	✓	✕	✓
Environmental monitoring	✓	✕	✕	✕
Healthcare	✓	✓	✕	✕
Security and surveillance	✓	✓	✓	✓
Tourism	✓	✕	✕	✓
Education/training	✓	✕	✓	✕

Table 3.1. *Relation between reference scenarios and thematic areas
(tick = related; cross = unrelated)*

The rest of the chapter is organized as follows: in section 3.3, we will first define the concepts considered in the chapter, in order to gain a common understanding of the covered topics. Sections 3.4–3.7 are dedicated to WSNEM, WSNMN, ART and IVN, respectively. Each of these sections opens with a detailed description of the characteristics and requirements of the considered thematic area. Then, a rather comprehensive overview of the most interesting algorithmic solutions proposed in other works for that thematic area is provided. Section 3.8 provides a taxonomy of the works, according to the applications requirements identified by Chapter 2. From this classification, we will extract the general trends that characterize the design of algorithms for COs systems and we identify the issues that require further investigation in the future, as reported in section 3.9. Finally, section 3.10 wraps up the study and provides some final remarks on the research trends that are expected to play a significant role in the future development of CO systems.

3.3. Definition of concepts

This section is aimed at providing a common understanding of the different concepts involved in the chapter. To this aim, in the following we propose a short definition for each one of the concepts considered.

Paradigm In this context, the term *paradigm* refers to the methodologies and strategies that can be followed to approach a problem and define the solution.

Algorithm An *algorithm* is the description of a step-by-step procedure for solving a problem or accomplishing some goals. Several different types of algorithms can be defined according to the purposes they are designed for. In particular, this chapter deals with the following types of algorithms:

• *Medium Access Control:* MAC algorithms define the mechanisms used by the objects to share a common transmission medium.

• *Routing:* generally speaking, a routing algorithm provides a mechanism to route the information units (usually data packets) from the source object(s) to the destination object(s).

• *Localization:* localization algorithms are mechanisms that allow an object to determine its geographical position, either with respect to an absolute reference system or relatively to other objects in the area.

• *Data processing:* data processing includes both data aggregation and data fusion techniques. Data aggregation algorithms are methods to combine data coming from different (and possibly heterogenous) sources into an accounting record that can be then forwarded, reducing the number of transmissions, overhead and energy consumption of the system. A possible example is the aggregation of temperature and pressure data produced by two different sensors located in the same area into a single compounded packet that will hence be delivered to an environmental-monitoring station.

Data fusion algorithms are used to merge together information produced by different sources, in order to reduce redundancy or to provide a more syntectic description of the information. For example, the temperatures measured by several sensors located in a given area can be fused into a single average value for that area.

• *Synchronization:* synchronization protocols allow nodes (or a subset of nodes that perform a common task) to synchronize their clocks, so that they all have the same time, or are aware of offsets of other nodes. Time synchronization is essential for numerous applications where events must be time-stamped. Moreover, protocol design is eased when nodes share a common clock (time division techniques for channel access, design of sleep/awake schedules, etc.).

Time can be absolute (i.e. referred to an external, well-known measure of time), or nodes can agree a common time reference. This last case is useful in those cases when time is needed to compare the occurrence of events.

- *Navigation:* robots using sensor networks open a new research area which includes the navigation of autonomous robots using distributed information as a relevant issue. The navigation of the robot is possible even without carrying any sensor and just using the communications with the WSN. Furthermore, it should be mentioned that these algorithms can also be used for the guidance of people with a suitable interface.

Interactions This refers to the exchange of information among objects that enables the realization of the coordination and cooperation of objects:

- *Coordination* is a process that arises within a system when given (either internal or external) resources are simultaneously required by several components of this system. In the case of autonomous robotic teams, there are two classic coordination issues to deal with: spatial and temporal coordination.

- *Cooperation* can be defined as a joint collaborative behavior that is directed toward some goal in which there is a common interest or reward. Furthermore, a definition for *cooperative behavior* could be: given some task specified by a designer, a CO system displays *cooperative behavior* if, due to some underlying mechanism (i.e. the "mechanism of cooperation"), there is an increase in the total utility of the system.

Taxonomy Taxonomy consists of the classification of concepts according to specific requirements and principles.

3.4. Wireless sensor networks for environmental monitoring

Technological advances as well as the advent of 4G communications and of pervasive and ubiquitous computing have fostered a renewed interest in multi-hop (ad hoc) communications. In particular, the interest is in self-organizing wireless multi-hop networks composed of a possibly very large number of nodes. These nodes can be either static or mobile, and are usually constrained in terms of power and computational capabilities.

A typical example of these kinds of networks are the WSNs [1, 2]. In this case, the well-known paradigm of ad hoc networking specializes in considering a higher number of nodes (in the thousands and more) that are heavily resource-constrained. Rather than being on mobility, the emphasis is now on data transport from the sensors to other sensors, or to specific data collection nodes (sinks). Sensor nodes are usually irreplaceable and become unusable after failure or energy depletion. It is thus crucial to devise protocols for topology organization and control, MAC, routing and so on, that are energy conserving, scalable and able to prolong the overall network longevity, especially in networks with a large number of devices.

3.4.1. *Application scenarios*

The thematic area named *Wireless Sensor Networks for Environmental Monitoring* (*WSNEM*) deals with the deployment of WSNs in static environments. For instance, WSNEM may be used in greenhouses to monitor the environmental conditions, such as air humidity and temperature, light intensity, fertilizer concentration in the soil and so on. The information is then delivered to central nodes that trace the dynamic of the environmental conditions and determine the actions to perform. The WSNEM may also be used to perform some actions, by activating specific actuators such as watering springs, sliding shatters (to darken the glasshouse), air humidifiers and so on.

Another possible use of the WSNEM is for monitoring the integrity of buildings, bridges or, more generally, structures. In this case, sensors are displaced on strategic points of the structure and detect any significant variation in the structure form, by showing variations of pressure, positions of landmark nodes, or relative position of the surrounding sensors. The sensors may periodically read the data and send it to a controller node, either spontaneously on in response to a direct request from the controller. Thus, the controller nodes may generate an alarm when a significant variation in the structure is detected.

WSNEM also involves the realization of the so-called *Ambient Intelligence* or *Smart Environments* (home/market/building/park), where the environment is equipped with sensor nodes that allow interaction with people. For instance, an art exhibition could be equipped with radio nodes and sensors that can detect the proximity of a visitor and provide (upon request or spontaneously) a set of information regarding a particular painting or, maybe, the location of the closest toilet with facilities for the disabled. Similarly, commercial malls equipped with sensors can inform the nearby clients about on-going commercial promotions and the direction to specific stores from the current position of the user. Home networking provides yet another example of a smart space. Different appliances can be interconnected, often wirelessly, and depending on the personal profile of the users in the home, act on the environment to provide personalized features such as the ideal room temperature for a specific user, or the redirection of incoming calls to the phone that is closest to the user's current position.

Therefore, the range of scenarios that can be mapped onto this thematic area is rather large. In particular, referring to the sectoral areas proposed in Chapter 2, the application areas that may be ascribed to WSNEM are the following:

- Environmental monitoring
  - GoodFood
  - Habitat monitoring on Great Duck Island
  - Smart mesh weather forecasting
  - Waternet

- Home and office
  - Smart Surroundings
  - Oxygen
- Logistics
  - CoBIs
  - Smart-dust inventory control
- Security and surveillance
  - Sustainable bridges
  - Under water acoustic sensor networks
  - FloodNet
  - Monitoring volcanic eruption with WSNs
  - Cooperative Artefact and handling of storage of chemicals

3.4.2. *Peculiarities of WSNs*

To some extent, a WSN can be considered as a special case of an ad hoc network, for it inherits the characteristics of (quasi-) random topology, multi-hop wireless communication, absence of backbone or core structure, and distributed control. However, WSNs and ad hoc networks differ in some important features. Some of the specific characteristics are the following:

• The network is composed of a large number of stationary sensor nodes, disseminated in a wide area and few sink nodes and designated to collect information from the sensors and act accordingly.

• After placement, the sensor nodes are often inaccessible, so they need to be autonomously powered and controllable by a remote connection.

• In order to save energy, the sensors nodes alternates between periods of activity and sleeping, so that the network topology is time variant.

• The connectivity of the network should be guaranteed when some nodes are inactive due to a sleeping phase, breaking or battery depletion.

• Traffic flows are mostly unidirectional, from sensors to one or more sinks. Sink to sensor traffic can be generated in some cases, such as whenever the sensors are explicitly solicited for reading and communicating their data.

• The traffic generated by sensor nodes usually has a very low bit rate and it can be continuous and constant (e.g. temperature monitoring) or bursty (e.g., event driven).

• Nodes close to the sinks are required to relay the traffic generated by peripheral sensors, so that the traffic density increases in the proximity of sink nodes.

• There may be a strong correlation among data generated by sensors that lie in a common region. For instance, more than one sensor node may detect the same event, such as the starting of a fire. In this case, it is not necessary that all nodes transmit the same information. Priorities can be defined in order to eliminate redundancy and to reduce traffic burst, preferring, for example, that only nodes which are nearest to the source send their information. Therefore, data aggregation can be considered to reduce the energy expense of the network.

Moreover, it is worth pointing out that WSNs are usually data-centric in nature, in that data is requested based on certain preferred attributes, which are characteristic to a data query. From this point of view, addressing functions in data-centric sensor networks may be performed by an attribute-value pair, so that if a station is interested in detecting movements that happen at a speed greater than 10 km/h, then it will issue a query that resembles {movement-speed $\geq$ 10 km/h}, and only the sensors that detect such a movement speed need to report their readings. This is also a way to simplify addressing and save energy.

Therefore, typical performance indexes considered in ad hoc networks, such as QoS, throughput and protocol fairness, might not apply to WSNs. On the other hand, other metrics such as the consumed energy per event, the event delivery ratio and so on, acquire a new important role in the performance evaluation of WSNs. For this reason, although many protocols designed for ad hoc networks can be adapted to the WSN, a different approach is advisable in order to obtain higher performance.

The existing literature on protocol design for sensor networks highlights many design issues and relevant problems that are to be solved or handled before efficient communication and management is achieved in WSNs. We recall that a sensor network works under the general concept that the network lifetime needs to be extended as much as possible while obtaining efficient information forwarding and preventing link disconnections due to node failures. As the primary source of node failure is identified as battery depletion, network connectivity degradation may be prevented by the use of effective energy conservation techniques. On the basis of previous observations, all the network layers should be carefully designed, taking into account, in particular, the following issues:

- MAC and packet scheduling;
- routing;
- sensor data aggregation;
- node discovering and localization;
- self-hierarchical organization and/or clustering.

In other works, these problems are widely considered, though not all solutions are suitable for the specific scenario of WSNEM. Generally, two different approaches are taken into account: *layered* and *cross-layer*.

The layered approach considers a classical network architecture organized according to the OSI model, where each layer is separately developed and optimized.

In the cross-layer approach, all possible interactions and integrations among layers are considered in order to optimize the entire system. Such an approach appears particularly valuable in the context of WSNs, for the high specialization of this type of solutions. Indeed, in this context, it appears reasonable to sacrifice the flexibility of the layered approach in order to better exploit the peculiarities of the systems at every layer and design more efficient solutions.

In the following, we consider the main issues concerning WSNs, also taking into account cross-layer solutions, and we underline the protocols, algorithms and proposals that are particularly suitable for environmental monitoring.

3.4.3. *Medium Access Control*

The Medium Access Control (MAC) mechanism defines the strategy used by the wireless node to access the common transmission resource, namely the radio channel. Ideally, the MAC mechanism should have the following properties:

- high energy-efficiency;
- low access delay;
- support to different access priorities.

In a wireless network there are four fundamental causes of power wasting:

- *Collisions*: a receiver is in the reception range of two or more transmitting nodes and is unable to cleanly receive signal from either node.

- *Overhearing*: nodes also sense transmissions addressed to other nodes.

- *Overhead*: nodes have to transmit/receive control traffic.

- *Idle listening*: sensing channel is also performed when the channel is idle (many measurements have shown that idle listening consumes 50–100% of the energy required for receiving).

In order to minimize the power consumption, an efficient random access mechanism should reduce all these factors.

Unfortunately, in many cases there is a trade-off between energy-efficiency and access delay, which is determined by many factors, such as power-saving techniques based on active/sleep cycles, medium contention, time synchronization and so on. In general, however, the access delay is considered a secondary issue in WSN, with respect to energy-saving.

Finally, in several application scenarios for WSNs, access fairness among contending nodes is not a primary issue, due to the high data redundancy.

Many of these aspects are conflicting, so that different MAC algorithms have been defined according to the specific purposes of the network. A first, rough classification of the MAC mechanisms is in two categories:

- *random access* and
- *deterministic access*.

The random access algorithms are based on a medium contention policy and on a carrier sensing mechanism. Such mechanisms achieve good performance in the case of wide and dynamic networks, since they do not require synchronization among the nodes and they make use of local topology information only. Therefore, random access mechanisms gain in flexibility, to the detriment of energy-efficiency.

The deterministic access algorithms are based on a time division mechanism (TDMA) that enables each node to transmit on a different time slot. This approach can potentially achieve very high energy-efficiency, since it makes it possible to schedule the sleeping and transmission phases of the nodes in an appropriate way and to avoid collisions. On the other hand, deterministic methods are inefficient for a high number of nodes, since the time frame required to allocate all the nodes becomes excessively long. Furthermore, heavy signaling is required to maintain the synchronization among the nodes.

In most cases, MAC protocols for WSNs are a mixture of deterministic and random access.

In other works there are many examples of MAC protocols for WSNs. In the following, some of the most important algorithms are briefly presented, taking into account their relation with environmental monitoring applications.

3.4.3.1. *Random Access Protocols*

CSMA: most of the random medium access protocols are based on the Carrier Sense Multiple Access (CSMA) strategy. The CSMA requires every station to be able to sense the wireless medium before transmitting. If the station reveals energy above a given threshold, then it defers transmission to a later time (which depends on the specific CSMA scheme adopted). The CSMA mechanism, however, is not sufficient to solve the problem of collisions, as the carrier sensing is performed in the vicinity of the transmitter while collisions occur at the receiver. Indeed, the CSMA mechanism increases the *hidden node problem* and the *exposed node problem*. The first happens when a node (B) lies between the transmission range of two other nodes (A and C), which are mutually hidden, i.e. which cannot sense each other's transmissions [3]. In this case, collisions may occur at the intermediate node B, since node C will keep sensing an idle channel even during the transmission of node A. Hence, node C may start transmitting when B is still receiving a valid packet from A, causing severe interference on node B. The exposed node problem, instead, occurs when a station (A) is prevented from transmitting by the transmission of a nearby station (B) that occupies the channel, even though no interference would be generated to the receiver (C) [3, 4, 5, 6, 7]. A graphical representation is given in Figure 3.1.

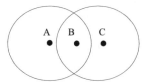

Figure 3.1. *Example of hidden and exposed node scenarios. C is a hidden terminal when it starts transmitting while B is receiving a message from A. A is an exposed terminal when it is kept from starting a transmission cause B is transmitting to C*

To alleviate these problems, a Collision Avoidance (CA) mechanism is often adopted. CA makes use of two special control packets, namely request-to-send (RTS) and clear-to-send (CTS), which are much shorter than usual data packets. Before attempting a data packet transmission, nodes can try to reserve the channel by exchanging a pair of RTS/CTS packets. Clearly, collisions may occur, but they will be limited to the short RTS packet, thus saving time and energy. Nodes that successfully receive either the RTS or the CTS packet are required to refrain from transmitting for the time period declared in the packet header. If the RTS/CTS handshake is successful, all nodes within the coverage range of both transmitter and receiver units will remain silent for the entire duration of the data transmission. Hence, the CSMA/CA mechanism helps to alleviate the effect of strong interference on data communications, and at the expense of an overhead that is heavier as the data transmission rate increases. Little can be done against *weak* interference, i.e. nodes that lie in the border of the sensing region [6, 8]. Such nodes will have a weak probability of receiving any RTS/CTS or packet header transmission, being outside of the reception region. Thus, they may transmit without any limitation, causing interference on the target nodes.

HIGHLIGHTS

The CSMA mechanism achieves energy-saving by reducing collision probability. On the other hand, the mechanism requires idle listening to the channel and does not alleviate the problem of overhearing. The mechanism is completely distributed, robust against topology variations and provides low delay in case of light traffic.

MACA: *MACA* (Medium Access with Collision Avoidance) [9] is one of the first MAC protocols designed for wireless LANs. The MACA protocol is largely based on the CSMA/CA mechanism, from which it inherits the CA strategy (based on the RTS/CTS handshake), while dropping the carrier sense feature (hence the acronym MACA). The sender first transmits an RTS packet to the intended receiver to elicit the transmission of a CTS reply. Nodes overhearing the RTS packet do not occupy the channel during the immediate successive time period, in order to not interfere with the reception of the CTS packet returned by the intended receiver. A specific field in the CTS packet carries the expected duration of the pending transmission. Therefore, nodes overhearing the CTS packet refrain from transmitting for such a time period, thus avoiding interference with the receiver. Since nodes do not perform carrier sensing, collisions are likely to occur over RTS packets. If the sender does not receive a valid CTS packet within a fixed time after the RTS transmission, it will schedule a new transmission attempt after a random time delay, picked up in the contention window $[1, CW]$. After every failed retransmission, the contention window size, CW, is exponentially increased, whereas after every success it is reset to the initial value. The backoff strategy is operated over a slotted time domain, where each slot corresponds to the transmission time of an RTS or CTS packet (of approximately 30 bytes). Notice that MACA does not encompass any acknowledgment (ACK) at the MAC layer, so that erroneous data packets have to be retransmitted by the upper layers.

HIGHLIGHTS

The MACA algorithm largely inherits the pros and cons of CSMA. However, the RTS/CTS handshake helps to alleviate the energy consumption due to collisions and overhearing.

MACAW: *MACAW* [10] is largely inspired by the MACA protocol that is modified in order to alleviate some inefficiencies. In particular, the authors of [10] observe that the binary exponential backoff (BEB) strategy adopted in MACA in case of collisions may lead to an unfairness probability of medium access in the presence of heavily loaded nodes. Indeed, nodes that undergo a series of collisions, thus exponentially increasing their contention windows, have a lower probability of winning the contention with other, less-backed-off terminals. To alleviate this problem, in MACAW all nodes in the neighborhood have the same contention window size. The contention window size is embedded in the header of each packet transmitted over the channel. Whenever a station hears a packet, it copies that value into its own backoff counter. Furthermore, the backoff window is adjusted in a gentler way: upon a collision, the backoff interval is increased by a multiplicative factor (1.5), while upon success it is linearly decreased.

In order to further improve the fairness of the channel access, MACAW also suggests the use of the multi-stream model. Basically, authors propose to keep, in each station, separate queues for each stream and then to run the access mechanism independently for each queue. Collisions that could potentially occur among the traffic flows of the same terminal will be internally resolved by the terminal by choosing a winner and backing off the other colliding streams. Notice that the multi-stream model is also considered in the standard *IEEE 802.11e*, which aims at providing different access classes.

Finally, MACAW encompasses the use of a link layer ACK packet. This makes it possible to run an Automatic Retransmission Query (ARQ) algorithm at the MAC layer, thus increasing the reliability of the link layer transport service and reducing the inefficiency due to higher layer retransmissions. Therefore, MACAW introduces the RTS-CTS-Data-ACK access paradigm.

HIGHLIGHTS

MACAW has been originally proposed for ad hoc networks. It is based on CSMA/CA and it aims at providing access fairness among users, which might be self-defeating in WSN, as previously discussed.

PAMAS: *Power Aware Multi-Access* (PAMAS) [11] is a protocol for ad hoc networks based on MACAW, with the addition of a separate signaling channel with the aim of reducing power consumption.

The RTS/CTS message exchange takes place over the signaling channel that is completely separate from the data channel used for data-packet transmissions. If the RTS/CTS handshake is successful, the data transmission begins in the data channel. Simultaneously, the receiving station transmits a busy tone over the signaling channel to inhibit any other transmission in its neighborhood. The length of the busy tone is greater than twice the length of a CTS, therefore neighbors that transmit an RTS while reception is still occurring will not receive back the expected CTS packets.

The mechanism is also used to reduce the power consumption of the terminals that are not actively involved in communication or are blocked by nearby transmitting stations. To this end, if at least one neighbor of a node is transmitting (data channel is busy) or receiving (busy tone is activated), the node can turn off its radio because it cannot receive or transmit packets. Generally, after overhearing a CTS or RTS transmission, a node enters the sleep mode for the time period signalled by the control packets or the busy tone. However, a node that wakes up after a long sleeping period may hear an ongoing transmission without knowing its duration. In this case, a binary search to determine the end of the current transmission is performed by sending short question messages over the control channel that are replayed by the transmitting nodes with longer residual transmission times.

HIGHLIGHTS

PAMAS is still a CSMA-based mechanism. It aims at enhancing the energy-efficiency by reducing the overhearing of nodes. However, this is accomplished by means of a second radio transceiver, which might increase the complexity and cost of the devices.

SMAC: *Sensor MAC* (SMAC) [12] is a MAC protocol designed for saving energy in sensor networks. It is based on the assumption that sensor networks are often deployed in an ad hoc fashion with nodes alternating long periods of inactivity and short, aperiodic and (sometimes) unpredictable periods of rush activity when an event is detected.

SMAC inherits from PAMAS the powering-off strategy during transmissions of other nodes but, unlike PAMAS, it does not require a separate signaling channel. In order to reduce energy consumption and message delivery latency, SMAC resorts to three techniques:

1. periodic listen and sleep periods;
2. collision and overhearing avoidance;
3. message passing.

The basic scheme for *periodic listen and sleep* is simple. Each node sleeps for some time and wakes up periodically to listen for any transmission request by neighboring nodes. The time duration of the sleeping and listening periods can be selected according to the constraints of the specific application scenario considered. All nodes are free to choose their own listen/sleep schedule. However, control overhead is reduced by forcing neighboring nodes to synchronize on a same schedule. The schedule is thus broadcasted to the immediate neighborhood so that each node can maintain a table with the information about the waking-up periods of all its known neighbors. Such a synchronization is required to allow the communication between in-range nodes. If a node receives a schedule that differs from its own, the node merges the two schedules (this event should be fairly unlikely). Therefore, neighboring nodes form a sort of *virtual communication cluster*.

Collision and overhearing avoidance is obtained by adopting a RTS/CTS-like mechanism, including both virtual and physical carrier sense. In particular, SMAC protocol tries to reduce the energy cost caused by the reception of packets directed to other nodes, by powering off the interfering nodes, i.e. the nodes that overhear an RTS or CTS packet. Interfering nodes are all immediate neighbors of either the sender or the receiver node, so that they are not allowed to occupy the radio channel until the ongoing transmission is over. Therefore, interfering nodes can set the Network Allocation Vector (NAV) with the duration of the transmission and sleep for the entire NAV period.

Message passing is aimed at ensuring an application-level fairness instead of a per-node fairness. It is largely due to the IEEE 802.11 fragmentation mode that enables the transmitting of a burst of packets after a single, successful RTS/CTS handshake. In IEEE 802.11, however, the RTS and CTS only reserves the medium for the first data fragment and the corresponding ACK. The first data fragment and ACK then reserve the medium for the next fragment and so on. When a node receives a fragment or an ACK, it knows that there is one more fragment to be sent, but it cannot know the total number of pending fragments. Furthermore, if an ACK packet is not received by the sender, it releases the channel and re-enters the contention phase. This mechanism is intended to provide fairness among the stations.

On the other hand, in an SMAC protocol, RTS and CTS reserve the medium for all data fragments, so that an interfering unit can sleep for the entire message transmission. Moreover, if the owner of the channel fails to receive an ACK, it extends the reserved transmission time for one more fragment, by adjusting accordingly the duration field in the header of the following fragments and ACKs. Hence, the node that wins the contention for the channel access can use the medium until it has transmitted all its fragments. Notice that neighboring nodes that wake up or join the network while the burst transmission is still ongoing can set their NAV by overhearing the duration field in the header of the first fragment or ACK packet that they see.

In this way, the per-node fairness is not guaranteed anymore. However, in WSNs it is more important to provide application-level fairness.

HIGHLIGHTS

The major SMAC protocol advantage is the reduction of the energy waste obtained by the scheduling of the sleep periods. On the other hand, this time scheduling may result in high latency and it is not efficient under variable traffic load. Furthermore, SMAC is a first attempt of cross-layer optimization, in that it considers both aspects of physical and application layers.

Sift: as previously noted, nodes in sensor networks often encounter spatially-correlated contention, where multiple nodes in the same neighborhood sense the same event. In other words, multiple sensors sharing the wireless medium may all have messages to send at almost the same time in response to a common external event. In the case of environmental monitoring, it would be sufficient that merely some of such nodes report the event to the controller (sink) node.

Therefore, in event-driven sensor networks, the main objective is to reduce the latency of the first event reports, rather than maximizing the network throughput.

Sift [13] is a medium access control based on a randomized CSMA which takes into account previous observations. Unlike other CSMA protocols such as IEEE 802.11, it does use a fixed size contention window from which a node randomly chooses a transmission slot. To reduce collisions, nodes try to estimate the number of contending neighbors and adapt the transmission probability accordingly in each slot within the contention window.

Suppose a sensor network is run by Sift. Every node with something to transmit competes for any slot $r \in [1, CW]$ (with $CW = 32$) based on a shared *belief* of the current population size N. The estimation of N is adjusted after each slot in which no transmission has occurred. For instance, if no node transmits in the first slot, then the estimated number of competing node is reduced, while the transmission probability for the second slot is increased, and so on for every silent slot.

In [13], results show that Sift outperforms IEEE 802.11 in terms of latency. One of the disadvantages of Sift is that it does not take into account any energy-saving techniques except for collision reduction.

HIGHLIGHTS

Sift is based on a CSMA mechanism and aims at reducing the collision probability while limiting the access delay. To this aim, the nodes determine the contention backoff on the basis of the estimation of the node density. This makes the algorithm topology-dependent, though adaptable to topology variations. No explicit energy-saving mechanisms are considered.

STEM: STEM [14] is an acronym for *Sparse Topology and Energy Management*.

At the beginning of the network operations, STEM selects a sensor node that is called *target*. This node is of paramount importance from a forwarding perspective, since all data and control messages that a node needs to send must be directed to its own target node.

In the first version of this protocol, namely STEM-B (where B stands for *Beacon*), nodes wishing to initiate data transmission, i.e. to inject a new packet into the network, explicitly send beacon or *wakeup* messages containing the *target* address until the target wakes up and is able to answer the message. When this happens, a link between the initiator node and its target node is established, so that the data packet may be forwarded from the initiator to the target. If the packet needs to be relayed further, the target node will start sending beacon messages to its own target node and so on until the final destination is reached.

Note that aggressive wake-up message transmission is needed to avoid excess latency in link establishment, but, on the other hand, a persistent wake-up message transmission may disrupt data transmission efforts, since it would generate collisions at receiving nodes with high probability. In order to cope with both problems, the authors in [14] suggest making use of two radios operating on two separate frequencies, one for data transmission and the other one for control signaling.

Another version of the STEM protocol, namely STEM-T (where T stands for *Tone*), is devised in [14]. In this case, nodes use tone-based signaling to wake up their target node. Since the wake-up process is initiated by the simple overhearing of the tone, all sleeping nodes in the neighborhood of the sender leave their sleeping status even if they are not important for the packet relaying phase. This type of signaling has the advantage of waking up the target sensor in an online manner, thus reducing the mean latency affecting data packet transmissions, but it also raises the overall network energy consumption, as it forces more nodes to wake up than are really useful.

HIGHLIGHTS

To some extent, STEM is a cluster-based algorithm in that each node selects a target node that is in charge of data delivering. STEM incurs a high access delay since nodes have to wait for the target node waking up. To alleviate this problem, a two-radio solution has been proposed that, however, worsens the energy-efficiency of the network since a transmission wakes up all the over-hearing nodes.

DB-MAC: the primary objective of *Delay Bounded Medium Access Control* (DB-MAC) [15] is to minimize the latency for delay bounded applications, while also considering the power consumption reduction by means of a path aggregation mechanism.

DB-MAC adopts a CSMA/CA contention scheme based on a four way RTS/CTS/DATA/ACK handshaking. It is defined for a scenario in which different sources contemporarily sense an event and they have to send their information to the sink node. Generated data flows can be dynamically aggregated in the path towards the sink, giving rise to an aggregation tree. Intermediate nodes in the path may aggregate several flows into a single flow to reduce transmissions and amount of data to be sent. Also, in the case in which the aggregation does not reduce the effective amount of data to be transmitted, it can reduce the overall transmission overhead (for example, the contention overhead is reduced).

The MAC protocol scheme is very similar to the IEEE 802.11 RTS/CTS access with some modifications: RTS/CTS messages are exploited to perform data aggregation and the backoff intervals are calculated by taking into account the priority assigned to different transmissions. In particular, each node takes advantage of transmissions from other nodes by overhearing CTSs in order to facilitate data aggregation and gains access to the medium with higher probability if it is close to the source. According to this policy, the relay nodes are selected from among those nodes that already have some packets to transmit.

HIGHLIGHTS

DB-MAC, based on the CSMA/CA principle, is a cross-layer solution that also encompasses data aggregation functionalities to reduce traffic and energy consumption. The algorithm, however, requires rather complicated data processing.

3.4.3.2. *Deterministic access protocols*

Below are two examples of MAC protocols based on a TDMA approach rather than on random access.

Energy-aware TDMA-Based MAC: [16] the main objective of *Energy-aware TDMA-Based MAC* protocol is to extend the lifetime of the sensor through topology adjustment, energy aware routing and MAC. Message traffic between sensors is arbitrated in time to avoid collisions and to allow the unneeded sensors to be turned off. The network is organized in such a way that some gateway nodes assume responsibility in their cluster for sensor organization and routing/MAC management.

This protocol is based on a time division multiple access (TDMA) whose slot assignment is managed by the gateway. It informs each node about the slots that the node can use for its transmission. This type of mechanism requires clock synchronization among all nodes within the same cluster, but collisions are completely avoided if each node correctly receives its slot assignment.

The protocol consists of four main phases: data transfer, refresh, event-triggered rerouting and refresh-based rerouting. In the refresh phase, every node uses its preassigned slots to inform the gateway about its state (energy level, state, position). Taking into account this information, the gateway decides how to allocate cluster resources, which nodes should be used as relay and so on, and on the basis of a cost function assigns the time slots for the transmissions. In particular, rerouting is performed when the sensor energy drops below a certain threshold, after receiving a status update from the sensors and when there is a change in sensor organization.

HIGHLIGHTS

Energy-aware TDMA-Based MAC aims at reducing the energy cost of nodes by superimposing a cluster structure on the network, where cluster heads are in charge for managing the channel access of the other nodes in the cluster. The algorithm is a cross-layer solution, in that it provides both MAC and routing functionalities. It suffers of typical TDMA drawbacks, such as need for nodes synchronization and tight cooperation, topology dependency and so on.

TRAMA: the *TRraffic-Adaptive Medium Access* protocol (TRAMA) [17], a TDMA-based algorithm, is introduced for energy-efficient collision free channel access in WSNs. TRAMA reduces energy consumption by ensuring that unicast, multicast, and broadcast transmissions have no collisions, and by allowing nodes to switch to a low power, idle state whenever they are not transmitting or receiving. It is similar to NAMA (Node Activation Multiple Access) where for each time slot, a distributed election algorithm is used to select one transmitter within a two-hop neighborhood. This kind of election eliminates the hidden terminal problem and hence ensures all nodes in the one-hop neighborhood of the transmitter will receive data without collision. However, NAMA is not energy-efficient and causes overheating.

TRAMA assumes that time is slotted and uses a distributed election scheme based on information about the traffic at each node to determine which node can transmit at a particular time slot. TRAMA avoids the assignment of time slots to nodes with no traffic to send, and also allows nodes to determine when they can become idle and not listen to the channel using traffic information. TRAMA is shown to be fair and correct, in that no idle node is an intended receiver and no receiver suffers collisions. The performance of TRAMA is evaluated through extensive simulations [17]. Delays are found to be higher compared to contention-based protocols due to higher percentage of sleep times. On the other hand, the advantages of TRAMA are the higher percentage of sleep time compared to contention-based protocols and a lower collision probability achieved, especially in broadcast communications.

HIGHLIGHTS

TRAMA is a completely distributed mechanism, where the slot assignment is decided by nodes on the basis of the traffic at each node. TRAMA outperforms contention-based and scheduling-based protocols with significant energy-savings.

TSMA: *TSMA* (Time Spreading Multiple Access) [18, 19] is a robust scheduling protocol which is unique in providing a topology transparent solution to scheduled access in multi-hop mobile radio networks using a TDMA fashion. The traditional requirement that the schedule guarantees a collision free transmission in every slot is not a necessary condition. Instead, the following requirement is considered in TSMA: for each node and for each one of its neighbors, there is at least one time slot in the frame, in which the packet can be received successfully by the neighbor. In order to satisfy this requirement, the TSMA mechanism proceeds in the following way. It assigns each node one or more slots in a frame, so that while collisions at some of the neighboring nodes may occur in every slot, requirement is satisfied by the end of the frame. To obtain such a solution, slot assignment is performed by using of the mathematical properties of finite (Galois) fields and it is based on an estimation of the mean neighbor number of a node. TSMA adds the main advantages of random access protocols to scheduled access. Similarly to random access, it is robust in the presence of mobile nodes. Unlike random access, however, it does not suffer from inherent instability, and performance deterioration due to packet collisions. Unlike current scheduled access protocols, the transmission schedules of the TSMS solution are independent of topology changes, and channel access is inherently fair and traffic adaptive.

HIGHLIGHTS

TSMA is a TDMA-based algorithm which is topology-independent and traffic adaptive. It make use of the mathematical properties of finite fields to assign each node a set of slots.

3.4.4. *Routing and forwarding algorithms*

Sensor nodes are known to possess very low data processing and storage capabilities. From such a perspective, the design of algorithms and protocols that are supposed to let nodes forward packets towards a final destination should obey the following rules:

- not require complex calculations;
- rely on easy to calculate and easy to store metrics;
- be easy to program;
- be designed with attention to energy conservation techniques like employing on/off duty cycles.

The characteristics listed above are essential to efficiently manage the scarce memory capabilities and to avoid overwhelming the sensor node's microcontroller with too many operations, because this would lead to a fast exhaustion of the available energy. Moreover, it may turn out to be difficult to align with the ISO-OSI model when designing protocols for WSNs: it is easier (and sometimes more affordable) to think of the MAC and routing layers as a monolithic entity so that there is no need for explicit communication between the two layers and information can be easily shared.

Following [20], we summarize here some general WSN features which are important for routing protocol design:

1. Nodes are typically deployed in very large networks. This makes the use of IP-like global addressing schemes unfeasible, as maintaining a global identification may result in high energy expenditure. Furthermore, getting the data transmitted by the sensors is sometimes more important than getting the exact ID of the node that sent the data. This is particularly of interest as sensors may not always be supervised by a centralized attending unity and need to organize themselves in an ad hoc manner, forming connections and coping with the subsequent nodes distribution.

2. Almost all applications of sensor networks involve data communications from multiple sources to a particular base station called the *sink*, even if other kinds of traffic are indeed supportable.

3. Sensor nodes are constrained in terms of energy, processing capabilities and storage capacity, as described before, and therefore need wasteless resource management.

4. Sensors in most applications are supposed to be stationary, which enormously simplifies the prediction of the behavior of routing protocols. However, in some cases, mobility could be allowed, even for a small amount of nodes inside the network. Routing protocols need to be aware of node mobility to properly engage forwarding decisions. We stress that if an alternation between a low power and an operating state is independently implemented at each node, then sender nodes could find a different network topology each time they want to transmit a packet, so that issues related to duty cycle and mobility management are expected to converge.

5. Sensor networks are considered to be application-specific, as different applications require different configurations to meet specialized constraints. For instance, low latency requirements of tactical surveillance are different from long network life and connectivity requirements of a periodic temperature sensing system.

6. Data collection is performed on a location basis in many applications. Position awareness is thus a major issue in sensor networks, as it is difficult to obtain with high precision. Typical approaches include the use of GPS hardware and the estimation of a node's location by measuring incoming signal strength.

7. Many data collection tasks, in particular during environmental monitoring, are originated by common phenomena, so that the amount of sensed data may be redundant. This may generate a large amount of redundant packets which could be processed by data aggregation or data fusion methods to decrease the total amount of data sent, thereby enhancing energy-efficiency.

The main design problems incurred in creating routing protocols for use in WSNs may be summarized as follows.

Random topology. First, nodes are deployed to form a network following a usually random displacement. This is not always true, but is indeed the most frequent situation (manual deployment is more affordable for some applications: this simplifies the routing problems as fixed data forwarding paths may be preemptively set up at the beginning of network operations). With random deployment, nodes are required to form an ad hoc infrastructure in an unattended fashion. When the node position distribution is non-uniform, clustering may turn out to be a valid choice to reduce the overall energy consumption in data communication. Moreover, as networks are large in size, multi-hop communications (i.e. communications formed by many independent transmission-reception phases that packets undergo on their way to the final destination) are to be employed and will require an efficient way to manage them.

Wireless communication. A second problem is tightly bound to the various radio communication methods. In a WSN, nodes are linked by a wireless medium that is used at typically low speeds, in the order of 1–100 kbps. The traditional problems associated with a radio channel (e.g. fading, high error rate, hidden terminals, collisions) may affect the general operations of a WSN. To this extent, routing protocols for WSNs should be appropriately designed to cope with radio impairments.

Node mobility. Third, a WSN is often assumed to be formed of fixed nodes. However, in many applications both the nodes or the sinks may move [21]. Movement brings into account topology stability issues. In addition, the phenomenon under tracking may be mobile in nature (e.g. a moving objective during tactical surveillance), therefore requiring proactive or periodic reporting from nodes to the base station, whereas fixed events may require a reactive (i.e. generating traffic when reporting) behavior. Dynamic nodes or phenomena also generate area coverage issues. In fact, the sensing capabilities of a sensor node are generally circumscribed to a limited area, so that the trajectory of the sensors in the monitored area has to be carefully planned in order to avoid some spots remaining out of the monitored perimeter. Thus, coverage conservation issues must be treated in protocol design.

Scalability. Fourth, as the number of sensors in a network may vary from tens to hundreds to thousands of units, a routing protocol must be able to scale from a small to a large number of nodes. It is worth highlighting that scalability may also be referred to as the efficiency or quality in event reporting: when nothing particular is occurring to be sensed, most sensors can spare energy by spending the majority of the time in a power saving mode, while other active sensors provide a coarse detection quality to be improved when needed, i.e. when an event occurs.

Application-awareness. A fifth issue stems from the data reporting methods used. These methods are strictly application-dependent and are also to be selected depending on data time criticality. We can identify three major methods of data reporting:

• *Time driven*: to be used in applications that require periodic monitoring, like temperature and weather sensing. With this method, sensors will periodically wake up, sense the environment, generate related data and send a packet to a base station.

• *Event driven*: to be used when events are sensed automatically, but not in a continuous way. In this configuration, sensors are expected to be in sleep state most of the time, and to switch on only when a significant event occurs. The rationale behind this is that rare events such as earthquakes and bridge breaks do not require continuous sensing, but are to be continuously followed when they happen.

• *Request driven*: to be used when events are sensed, but data are predominantly generated following upcoming requests from a base station. For instance, we may be tracing the level of soil humidity and other relevant environmental parameters in an agricultural context, and may be interested in sending queries regarding the parameters to sense rather than having them sensed automatically.

Quality of service (QoS). Finally, quality of service may be a fundamental issue for some applications. If data are to be delivered within a certain amount of time from the moment they are sensed, then low latency routing protocols are to be implemented. If, on the other hand, network lifetime is a major concern, energy-aware routing protocols should be implemented to prolong network operativity as much as possible.

Note that hybrid configurations implementing more than one of the previous methods are also possible. The routing protocol is greatly influenced by the selected method in terms of route calculation and energy consumption. For the sake of clarity, we classify the approaches taken in the design of routing protocols for WSNs into three macro-categories:

- *location-based* routing, that exploits geographical information to route data inside the network;

- *data-centric* routing, where nodes are assigned essentially the same functionalities, and

- *hierarchy-based* routing, where nodes play different roles in the network.

Obviously, this is not the only possible classification. Routing protocols with peculiar features may be defined better by terms like *multipath-based, query-based, negotiation-based* or *QoS-based*. Moreover, a further classification is possible into *reactive* protocols which calculate the routing paths only when they are requested and *proactive* protocols which preserve all routing paths; into *adaptive* and *non-adaptive* protocols, which can or cannot tune certain parameters to adapt to current network or residual energy conditions, and into *cooperative* and *non-cooperative*, which are or are not able to forward data to central nodes (or concentrators) which are endowed with the task to process, aggregate and possibly further process incoming packets, so that the overall amount of data packets in the network is reduced.

3.4.4.1. *Location-based routing*

We start by discussing location-based routing, as algorithms of this kind are sometimes easier in concept.

Geographic Adaptive Fidelity (GAF): this protocol represents an effort to use system-level and application-level information to steer unnecessary nodes into a sleeping status, while keeping active the nodes that are important from a forwarding point of view.

The protocol is based on the concept of *equivalence* of forwarding nodes. Each node assumes that all sensors are distributed over a virtual planar grid subdivided into square sections: nodes inside a grid square section are all equivalent from a routing perspective, in the sense that it is not important whether the data packet is forwarded to one of the nodes or another inside the same square. This feature lowers the energy expenditure since all nodes that lie in the same region of the selected relay may be turned off to save energy, but implies an inherent reduction factor of $\sqrt{5}$ in the maximum allowable transmission range, which is needed to allow for communication between the two farthest possible nodes in two adjacent grid squares [22]. Thereby, the mean number of hops needed to reach a destination increases.

The aim of GAF is to have a single active node inside each grid square, while all other nodes are sleeping. This is obtained through periodic discovery operations, the broadcasting of neighborhood information and through the tuning of the sleeping time of each node.

As an additional feature, GAF explicitly takes into account node mobility through the knowledge of a system-level node speed parameter from which it measures the probability that an active node moves out of the square it is in charge of and, in turn, the mean allowable sleeping time a node can afford without losing contact with a previously known active node.

SPAN: *SPAN* [23] is a protocol oriented to power saving via topology information maintenance. It is based on the selection of some nodes to the role of *coordinators*, so that forwarding is only allowed from the initial source to a coordinator and then only between coordinators up to the sink.

If the network is sufficiently dense, it is possible to discover a set of nodes which ensures total network coverage and connectivity: this set of nodes is to be periodically substituted with a totally disjoint set, so as to redistribute energy consumption.

Every node may become a coordinator: this choice is made as the sensor node wakes up, and is based on the residual node energy and the perceived advantage of the node's neighborhood if the node decides to become a coordinator.

HIGHLIGHTS

GAF and SPAN are topology-dependent location-based algorithms. The reduction of energy consumption is obtained by selecting a connected subset of nodes as relays. These nodes are determined by means of negotiation procedures that take into consideration the neighbors' positions and other physical layer parameters, following a cross-layer paradigm.

MFR, DIR and GEDIR: in [24], basic geographic routing algorithms are described. The main concern here is to advance towards the final data destination through the selection of a next hop that has some desirable properties which are derived from the geographical locations of the sink, the sender and the intermediate relay. Most Forward within Radius (MFR) tries to select the relay which offers the maximum advancement towards the destination by (the minimization of) the dot product $\overline{DR} \cdot \overline{DS}$, where D is the sink, R is the relay, S is the source and, for instance, $\overrightarrow{DR}$ is the Euclidean distance between D and R. DIR is a "compass" method that maximizes the inner product $\overrightarrow{SR} \cdot \overrightarrow{SD}$, i.e. tries to forward the packet in a direction that is closest to the line that ideally joins S and D. Geographic DIrection Routing (GEDIR) is a greedy algorithm that, in its basic version, forwards the packet to the closest neighbor to the destination among all available neighbors. There may be situations which force a node to forward a received packet towards the same node that sent it: in this case, the algorithm fails.

We note that none of the algorithms described above is able to guarantee the delivery of the packet.

Geographic Random Forwarding (GeRaF): with this protocol, the sensors manage MAC and routing operations in a distributed manner, based on a contention to select the next hop for packet forwarding. The routing operations are conducted by means of geographical information, i.e. the position of the active relay node, of the final destination (the sink) and of the selected next hop. It is assumed that this information is already available the moment a node has to make MAC/routing operations, hence a preliminary phase of network set-up with flooding of localization messages has to be taken into account.

Since sensor nodes have limited storage capability, it is not feasible for each sensor to keep a trace of every other node, since this would lead to a waste of memory resources. In order to make nodes aware of the geographic position of all the other sensors implied in a contention or forwarding phase, it is sufficient that every needed positional information (namely, the transmitting relay, the sink and the chosen next hop) is exchanged on time by means of control messages. Since GeRaF sets up a collision avoidance mechanism to increase the global network throughput, the use of control messages to send and piggy-back positional information comes at no cost.

From an operational point of view, GeRaF tries to select the relay node that guarantees the maximum advancement towards the sink. This is done by subdividing the advancement zone into regions: nodes belonging to regions nearer to the transmitting node offer a lesser advancement towards the destination.

When a node has a packet to send, either to relay another node's packet or to transmit a packet generated on its own, it interrogates advancement regions using an RTS message, containing the sender's geographical coordinates: the nodes in the first advancement region (all of them) reply to the message by a CTS packet, including their own coordinates. Should a single node reply to the RTS, it is automatically selected as a relay. If multiple nodes reply and a collision occurs, the sender node issues a COLLISION message to start up a collision-resolution scheme at the eligible receiving nodes: in particular, from then on, nodes will reply to subsequent solicitations using a probabilistic bisection rule, that is, they send back control messages with a fixed probability of 0.5.

On the other hand, if no nodes reply to the first RTS message, the sender assumes there are no relays available in the first advancement region and transmits a CONTINUE message to solicit the nodes in the next advancement region. If there is still no reply, the sender keeps issuing CONTINUEs until the last region is reached, or until one or more nodes reply: in this case, the sender and the relays behave in the same way as described before, when CTS transmission was cited.

Note that all the previously described operations are transparent to on/off duty cycles.

This protocol needs geographical information to be available, for instance via control messages: it is worth noting that this information has to be processed to make routing decisions, namely calculating one or more Euclidean distances. Since these calculations imply root-square extraction, they may turn out to not be possible for nodes with low processing capabilities, unless simplifying assumptions of some kind are made.

GeRaF was initially designed for sensor nodes with two separate radios to be used for a data and a control channel [25, 26], but there also exists a version for single radio sensors [27].

HIGHLIGHTS

GeRaF is a position-based algorithm, designed according to a cross-layer approach (MAC/routing). The energy-saving mechanism consists of an on/off cycle. The algorithm is quasi-topology independent, since nodes are required to know their own spatial coordinates and those of the target node only. The next relay is chosen to maximize the progressive advancement of the message towards the destination. The contention mechanism used to select the relay node is robust to topology variations due to on/off cycles. GeRaF is developed to work either with one radio only or two radios.

Geographic and Energy Aware Routing (GEAR): This algorithm [28] addresses the problem of query dissemination to appropriate data regions, since often queries contain geographic information. GEAR is similar to a protocol discussed below, directed diffusion, in that it propagates queries on an interest basis, but only to a geographically restricted portion of the network, hence conserving more energy than directed diffusion. In GEAR, each node keeps an *estimated cost* and a *learned cost* which are related to the energy to be spent and the distance to be covered in sending the data to the destination through neighboring nodes. The learned cost also includes an estimation of the additional cost due to routing around network connectivity holes. Learned costs are propagated one hop back every time a packet is forwarded, so that the next route to the same destination could be adjusted.

The first phase of the algorithm covers the forwarding operation towards the region data are claimed from. Each node in the path to the queried region selects the neighbor closest to the destination as a next hop and performs cost learning if connectivity holes occur.

The second phase is then introduced once the packet reaches the geographic region it was directed to: at this point, further query dissemination may be performed with any sort of flooding. For example, flooding as in directed diffusion could be employed or, more efficiently, nodes could engage recursive geographic forwarding (the region to cover would be subdivided into four smaller regions, sending a copy of the packet to each one and so forth, until regions with only one node are left).

In [28], GEAR is also compared to a non-energy-aware routing protocol, Greedy Perimeter Stateless Routing (GPSR), which can also solve network hole problems. GPSR is simpler in that it reduces the number of states a node must keep. However, it was designed for general mobile ad hoc networks, and is in fact outperformed by GEAR, which offers higher packet delivery rates and lower energy consumptions under both even and uneven traffic distributions.

HIGHLIGHTS

GEAR is an application-aware algorithm that aims at gathering information from a target region rather than a specific node. The query is forwarded to the nodes in the target region by means of a distributed algorithm, which takes into consideration both energy cost and connectivity problems in order to reach the destination. The algorithm requires nodes to maintain a fairly large state vector.

Adaptive Self-Configuring sEnsor Networks Topologies (ASCENT): ASCENT is another topologically-driven protocol. It is based on the distinction between *active* and *passive* nodes. An active node may forward packets while passive nodes eventually process the collected information but do not participate in any forwarding procedure.

If a node finds itself caught in the impossibility of forwarding a message, it sends a HELP message to neighboring passive nodes, which may in turn choose whether to become active. If so, they inform their neighborhood of the state transition and begin forwarding packets.

HIGHLIGHTS

In general, location-based algorithms are completely distributed and require no or only local topology information. Energy-saving is usually obtained by allowing each node to schedule on/off cycles. Most of them are designed according to a cross-layered approach using physical parameters or implementing data aggregation techniques. All of them introduce some control traffic overhead. They do not work properly in the case of sparse networks.

3.4.4.2. *Data-centric routing*

The data-centric routing category encompasses algorithms that are aimed at getting information about a given event or from a specific region, rather than creating a path to a specific node.

Ideas behind these algorithms come from the fact that as nodes tend to be deployed in large networks, it is inefficient to assign each node a global identifier, whereas it is better to query regions or nodes which have access to data of interest, then wait for the nodes to answer back.

Directed Diffusion: Directed Diffusion (DD) [29] is a popular data routing and aggregation paradigm for WSNs which enables efficient multiple-to-single node transmission.

In DD, when a certain station is interested in harvesting data from nodes in the network, it sends out an *interest*, which describes the task to be done by the network. Each node receiving the interest forwards it to its neighbors in a broadcast fashion. As broadcasts are received at a node, it sets up *interest gradients*, i.e. vectors that describe the next hop to send the query results back to the requesting station. Generally, if a node S sends an interest which reaches node A and B and both forward the interest to node C, then node C will set up two vectors indicating that query results matching the interest should be sent back to A and/or B. The gradient modulus (or *strength*) is different towards different neighbors, which may result in different amounts of information being redirected to each neighbor. Each gradient is related to the interest it has been set up for. As the gradient set-up phase has been finished for a certain interest, future requests interested in the same attribute are used to reinforce the best (i.e. strongest) gradients throughout the network, so as to convey information flows through the best paths, hence avoiding future flooding. As nodes may receive related information flows from different nodes in the network, they perform a sort of data aggregation or fusion to reduce the total amount of forwarded information. This process may be done efficiently as all nodes in DD are application-aware and may then be able to use the best aggregation method. Finally, the base station will be eventually reached by the information flows. We note that the base station must periodically re-send interests to the network in order to avoid interest loss due to unreliable wireless transmission.

Another DD application is to spontaneously send important detected events to some portion of the network.

Because of the hefty gradient set-up operations, DD is only suited for applications that require nearly continuous query answers, whereas it is too energy-expensive to implement if the gradients are to be set and used only rarely.

Sensor Protocols for Information via Negotiation (SPIN): the SPIN protocol family [30, 31] runs on the concept that information must be conveyed through the network assuming that all nodes are potential base stations. SPIN assigns each data to be routed a metadata, i.e. a high level name that enables negotiations before sending data, and that is totally configurable depending on application needs. SPIN is also energy-aware and makes routing decisions based on residual node energy levels.

The handshake prescribed by SPIN obeys the following sequence. First, if a node has data to be shared, it sends an ADV message to announce to all neighbors of the upcoming communication. If a neighbor is interested in the data, it sends back a REQ message and then waits for DATA transmission.

A second SPIN version, namely SPIN-2, performs as described before if the available energy is abundant, but when it approaches a low threshold level, then the protocol first verifies if the residual energy is sufficient to terminate a handshake without depleting the energy. Other protocol versions are SPIN-BC (for broadcast channels), SPIN-PP (for point-to-point connections), SPIN-EC (similar to SPIN-PP with additional energy heuristic) and SPIN-RL (used over lossy for channels). All of these protocols are suited to applications where the sensors are mobile, since all forwarding decisions are based on local neighborhood information.

The advantages of SPIN include the metadata negotiation, which substantially reduces the amount of redundant data, and the need for only local neighborhood information for routing. On the other hand, a strong disadvantage is that SPIN cannot guarantee data delivery, since uninterested nodes on the path to the destination will not participate in the communications.

SPIN is different from DD, since DD issues data queries and builds up interest gradients for query answers, whereas in SPIN, data sources are the first to initiate communication. Moreover, all communication in DD is towards neighbors with data aggregation features, whereas SPIN needs to maintain global topology information to route towards the sink. Indeed, SPIN is more suited than DD to applications that require continuous data reporting to a base station, since every communication is driven by nodes that have access to the requested data. Moreover, matching queries in DD to the available data may require extra overhead and thus extra energy consumption with respect to SPIN.

Rumor Routing: Rumor Routing (RR) [32] exploits the fact that for many applications, geographic routing is infeasible and it is not necessary to find the shortest path to the destination; an arbitrary path is sufficient. As explained before, DD floods queries over the whole network to reach nodes that can provide data matching the query. With RR, however, nodes detecting an event record it into a local event table and then send a long life packet called an *agent*. When a node is interested in a certain event, it forwards a query that eventually reaches a sensor that has the event recorded in its local table. If this happens, the node aware of the requested event can respond to the query, since a hit in the event table may happen only if the node had previously been reached by an agent that informed it of the event. This way, there is no need for global network flooding and energy may be saved.

A drawback of RR is that it is only suitable for applications where few events are generated, otherwise the network would experience saturation due to excess agent generation. Moreover, the protocol is very dependent on the heuristics used to propagate agents hop-by-hop, and is vulnerable to node mobility and sparse regions.

Minimum Cost Forwarding Algorithm (MCFA): MCFA [33] exploits the fact that the direction of routing is known since it is towards a certain sink. Thus, a node need not maintain a routing table, but instead can maintain a minimum cost value containing the estimate of the least forwarding cost to the base station. Such a minimum path-cost is enclosed in the header of each transmitted packet. Furthermore, the packet also carries the total cost that has been consumed along the path, from the source to the current intermediate node. However, no node identification fields are required to be enclosed in the forwarded message.

When a node has a message to transmit, it broadcasts the message to its neighbors. Hence, each node checks whether the sum of its own minimum cost towards the base station and the cost consumed so far by the packet (enclosed in the packet header) equals the minimum path cost declared for the packet (also enclosed in the packet header). If so, the node belongs to the transmitter's least cost path to the base station and, hence, it rebroadcasts the message. The process continues this way until the message is delivered to the base station.

In order to obtain the least cost estimate, the base station floods the network with a message with an initial cost zero, while each node sets its cost estimate to ∞. When a node receives a cost estimate, it checks if the sum of the estimate with the cost of the link from which the packet was received, is less than the actual known estimate. If this is true, it updates and broadcasts the new estimate, otherwise it does nothing. A backoff algorithm is implemented to avoid collisions at nodes distant from the sink that are more likely to receive a greater number of estimates.

Information-Driven Sensor Querying and Constrained Anisotropic Diffusion Routing: these two protocols (IDSQ and CADR) [34] aim to be a general form of DD, which queries sensors in such a way that information gain is maximized, while energy consumption and delay are minimized. CADR, in particular, only activates the sensors that are close to a required event and selects the route towards the sink in order to balance the path forwarding cost and the amount of information that can be collected along the way. Moreover, IDSQ sets up a communication between the sender node and the neighbor that offers the highest information gain, but also best balances the energy cost. As IDSQ does not provide instructions on how to forward data requests from the base station to the nodes, it may be seen more as a complementary optimization procedure.

The CADR-IDSQ approach is more efficient than DD, since it avoids the excess energy expenditure due to isotropic query forwarding.

COUGAR: this protocol [35] abstracts from the format used to compile a query (which is left to be application-dependent) and utilizes data aggregation to reduce the amount of transmission in the network. It differs from DD in that it makes nodes select a *leader* to perform data aggregation. The leader itself is in charge of data forwarding to the sink.

The base station provides a query that is compliant with the application-specific format and which incorporates a method to select leaders among the nodes. The main advantage of this procedure is that the network itself operates in a way that is independent of the application, but the architecture also has some drawbacks: first, the nodes need query processing functions which may in turn generate extra overheads and energy consumption, and second, the leaders need to be dynamically changed in order not to become bottlenecks or to deplete available energy too fast.

Routing protocols with random walks: routing based on random walks [36] tries to achieve load balancing through multipath routing in WSNs.

It assumes that the network is made of a large number of nodes with limited or no mobility which may go to sleep or wake up at random times. A regular grid is assumed so that nodes fall exactly on a crossing point, even if the topology is irregular (i.e. nodes may not cover all of the grid crossing points). To find a route towards the sink, nodes apply a distributed asynchronous version of the Bellman-Ford algorithm and select the next hop to be the one which is closer to the base station with a certain probability. By adjusting this probability of node choice, some load balancing may be obtained. The next hop is changed any time a new packet has to be forwarded.

This protocol has its main drawback in its assumption about the topology of the nodes.

HIGHLIGHTS

These algorithms use flooding mechanisms to propagate the requests over the networks and to find the best routing path to deliver the queried information to the sink. They are distributed and strictly topology-dependent. They are often designed according to a cross-layer approach, since data aggregation techniques are combined with path discovery to reduce data redundancy and energy consumption. Notice that this additional data processing does not require trivial storage and computational capabilities.

3.4.4.3. *Hierarchical-based routing*

Hierarchical protocols exploit the advantages of clustering techniques, especially from a scalability and a communication efficiency point of view. Moreover, the packet forwarding may also be more energy-saving, since high-energy nodes may be selected to process and send information in place of low-energy nodes which may limit sensing operations.

Low Energy Adaptive Clustering Hierarchy (LEACH): LEACH [37] is a protocol based on clustering which selects and subsequently rotates cluster heads (CH) on a periodic basis to evenly distribute the energy consumption throughout the network. CHs set up communication with a TDMA method, making use of CDMA to reduce inter-cluster interference. Energy-savings come from data compression, data fusion and the random rotation of CHs.

The operation of LEACH is separated into two phases: one where the cluster heads are selected and a second phase where data transfers are performed. During the first phase, all nodes independently decide whether to become a CH: the selection depends on the desired percentage of CHs, on the actual rounds and on the set of nodes that has not become a CH in the previous rounds. Selected CHs broadcast an advertisement to neighboring nodes. During the second phase, sensing is performed and data received by CHs is aggregated and sent to the sink. Periodically, CHs are re-selected.

This protocol indeed suffers from a number of problems, i.e. it leads to a fast exhaustion of the energy of nodes that lead from a "hot spot" to the sink, where a hot spot is a place where multiple events take place in rapid succession. Moreover, it is not suitable for time-critical applications. Also it assumes that nodes always have data to transmit and that, if needed, each node is sufficiently close to the sink to be able to communicate directly with it.

LEACH could also be combined with metadata negotiation [37].

Power-Efficient Gathering in Sensor Information Systems (PEGASIS): PEGASIS [38] is derived from LEACH [37] and shares its advantages and problems. In particular, it assumes that all nodes possess the location information of all other network nodes and are all able to transmit directly to the sink. Furthermore, only static nodes are considered. These assumptions lead PEGASIS to outperform LEACH in the sense that it does not need any more dynamic cluster formation, and only needs to equilibrate the overhead due to the communication between the leader and the sink by means of token passing.

Threshold-Sensitive Energy-Efficient sensor Network protocol (TEEN): TEEN [39] is a forwarding protocol that is most suitable for time-critical applications. It forces nodes to avoid transmission if the value output from their sensing apparatus, or the variation from the last sensed value, is below the threshold. Specifically, clustered network structure is maintained as in LEACH, but the CH also broadcasts to the nodes a pair of threshold values, namely a *soft* and a *hard* threshold. Thus, as a new value is sensed, the node compares it with the hard threshold. If it passes the test (i.e. if it is in the range specified by the threshold), then the variation from the last sensed value is evaluated and compared with the soft threshold before transmitting. If this test is also passed, then transmission is allowed. At cluster changes, new parameters are broadcast.

APTEEN (Adaptive Periodic TEEN) [40] is a hybrid protocol that modifies TEEN in order to achieve more adaptability to application needs, as it allows to scale periodicity and threshold values. It engages a TDMA schedule between nodes to manage multiple transmissions in the same cluster and to let parallel transmissions by hybrid networks take place without excess interference. Its main drawback is the additional complexity required for maintaining the TDMA schedule.

Small Minimum Energy Communication Networks (SMECN): the work in [41] devises a protocol (MECN) that computes an energy-efficient subnetwork that is made by neighboring nodes towards which transmission is more efficient than direct transmission to the destination of a packet. These regions may reconfigure automatically to adapt to node failures or to the deployment of new sensors.

SMECN [42] is an extension to this protocol that constructs smaller subnetworks and thus operates on smaller graphs representing the network helping in constructing minimum-energy paths to send messages. The algorithm is local in that it does not calculate the global minimum energy path, but builds a subnetwork in which it is guaranteed to exist.

Self-Organizing Protocol (SOP): SOP [43] provides a hierarchical architecture where each node willing to send a packet to a base station first forwards it to router nodes. Thus, only routers need to be addressed uniquely and each sensor in their coverage range is identified by the address of the router they report to. Routing is performed by a local Markov loop algorithm (i.e. a random loop on spanning trees) that lets the network be robust towards node faults.

As nodes are directly reachable through their router address, this protocol is particularly suited to applications that interrogate single nodes. However, some overhead is required for the organization of clusters.

Virtual Grid Architecture Routing (VGA): the work in [44] regards an energy-efficient algorithm that is best applied to static networks. A localization method that does not make use of GPS [45] is implemented to build clusters that are equivalent in shape and non-overlapping. A square shape is chosen in [44]. Inside each zone, a node is optimally elected to cover the role of CH. Two-level aggregation is used as a form of data redundancy removal, both in a local and a global context. The optimal choice of local aggregators (LA) is NP-hard, hence some heuristics have been proposed in [44] to solve this problem. Moreover, another work [46] presented a way to reduce the energy consumption by optimally selecting global aggregators (GA) among the LAs that minimize the overall energy consumption.

These algorithms have proven to be fast and scalable for large networks, and to produce LA and/or GA selections that are not far from the optimal solution.

Hierarchical Power-Aware Routing (HPAR): the HPAR [47] protocol divides the network into smaller groups of sensors that are clustered in a geographic zone. Each zone is left to manage packet routing in such a way that the energy consumption in the cluster is minimized. Routes are chosen so that the formed path has the maximum over all minimum remaining powers (the *max-min path*). In [47], this algorithm was approximated by a *max-min* zP_{min} approach. As a matter of fact, the algorithm first calculates the shortest path with the Dijkstra method using power consumption as a link metric and then finds a path that maximizes the minimum residual energy in the network. This is obtained by relaxing the constraint on the P_{min} power consumption of the shortest path through the multiplication of P_{min} by a factor $z \geq 1$. Thus, the algorithm consumes at most zP_{min} while maximizing the residual power in the network.

Zone-to-zone routing is then implemented to find a path from zone to zone. Each node inside a zone participates in the routing process by estimating the zone power level.

Two-Tier Data Dissemination (TTDD): TTDD [21] realizes effective data delivery to multiple mobile base stations. Sensors are assumed to be stationary and location-aware. Each data source builds a grid structure that is used to disseminate data. Its formation is as follows. A future data source sends an announcement to its four nearest grid crossing points. When the message reaches the node closest to a crossing point, the node stops it and further propagates another message to all its nearest grid crossing points except the one the announcement had come from. Stopping nodes will then act as dissemination points.

With this structure, a base station may flood a query which will then propagate through dissemination points up to the queried node. The message is then sent back in the same fashion using the reverse path to the sink. The base stations forward trajectory information to let nodes predict their position for routing purposes.

This protocol does not calculate the shortest path to the destination, but is indeed very scalable, and the authors in [21] think this advantage is worth the lack of path optimality. Furthermore, another concern about this protocol is the maintenance of grid information and the necessity for an accurate position information.

HIGHLIGHTS

The main drawback of cluster-based algorithms is the resulting high cost, in terms of control traffic and energy consumption, to maintain the hierarchical architecture. On the other hand, cluster-based algorithms are suited to perform data aggregation and manage residual network energy in an efficient way. Some of them also require localization capabilities.

3.4.5. *Sensor data aggregation*

Data aggregation and in-network processing techniques have been recently investigated as efficient approaches to achieve significant energy-savings in WSNs by combining data arriving from different sensor nodes at some aggregation points enroute, eliminating redundancy and minimizing the number of transmissions before forwarding data to the sink. This paradigm shifts the focus from traditional address-centric approaches for networking (finding short routes between pairs of addressable endnodes) to a more data-centric approach (finding routes from multiple sources to a single destination that allows in-network consolidation of redundant data).

This approach is particular useful in many WSNEMs.

In [48] there is a study of the energy savings and the delay trade-offs involved in data aggregation and how they are affected by factors such as source-sink placements and the density of the network.

There are two different types of data aggregation:

- data fusion or aggregation with size reduction;
- aggregation without size reduction.

The first form of aggregation can be used, for example, in sensor networks intended for temperature monitoring when we are only interested in the averaged regional value of the temperature. In this case, an intermediate node that receives two values of temperature can calculate the average and forward it to the sink. The resulting data field has the same length as the incoming packets.

Aggregation without size reduction, on the other hand, occurs when two packets received from different sources are merged in a single packet with a longer data field.

In both cases, the MAC layer leverages data aggregation since the overall transmission overhead can be reduced. Furthermore, since the MAC layer is invoked any time a packet has to be transmitted over the radio channel, the contention overhead is reduced when a node contends only once to transmit a longer packet with respect to multiple contentions for shorter packets.

In general, it is possible to examine three schemes of data aggregation:

- *Center at nearest sources*: all sources send their data directly to the source which is nearest the sink. This source then sends the aggregated information to the sink.
- *Shortest path tree*: each source sends its information to the sink along the shortest path. If there are overlapping paths, they are combined to form aggregation tree.
- *Greedy incremental tree*: this is a sequential scheme. At the first step, the aggregation tree consists of only the shortest path between the sink and the nearest source.

At each step after that, the next source closest to the current tree is connected to the tree.

One of the first approaches to the data fusion problem is LEACH [37]. The network is divided into clusters, each having a CH that aggregates the data gathered by the cluster members. This scheme makes it possible to limit the amount of data transmitted over the air, thus reducing the network energy consumption and increasing the efficiency in the channel access. In [49], an analytical study of the data aggregation problem in a multi-hop geographical routing scenario is considered. The work presents an aggregation scheme where spatially correlated data is aggregated at the node called the "cluster" which, subsequently, forwards the information to the sink node through multi-hop routing. The authors show the trade-off between energy consumption and network lifetime when the number of nodes in the cluster is varied. The optimal cluster size as a function of the spatial data correlation is, finally, derived.

The most common approach for data aggregation is based on aggregation trees. Some examples of this approach are illustrated in [35, 50, 51]. In each of these studies, the goal is to find the tree that produces the best performance in terms of energy consumption and time delay. The tree-based approach, though, is sensible to variations of the network topology and link failure.

Such weaknesses are partially compensated by multipath-based schemes. However, [52] demonstrated that multipath routing often results in message duplication, which would cause a higher overhead (energy consumption). To alleviate this problem, authors have introduced an operator capable of annihilating duplicated measurements.

An evolution of [52] is presented in [53]. In this case, multipath routing [52] is used in some regions, while tree-based schemes are used in others. The dimension of the regions is modified depending on the data aggregation requirements and on the link and node failure probability. The disadvantage of this algorithm is due to the message overhead.

In [15], a cross-layer solution for routing and data aggregation is proposed.

HIGHLIGHTS

In summary, data aggregation and data processing techniques can be valid instruments to reduce the amount of data sent over the network and, consequently, to reduce the energy consumption. On the other hand, they might require higher storage and computational capabilities, resulting in a higher power consumption at nodes that perform data processing. Cluster-based network architectures are particularly suitable to support data aggregation, since CHs are ideal candidates to perform data fusion/aggregation. According to this perspective, it is advisable to use more powerful nodes as CHs in order to alleviate the problems that have been pointed out.

3.4.6. *Clustering and backbone formation*

3.4.6.1. *Clustering for ad hoc networks*

The notion of cluster organization has been investigated for ad hoc networks since their appearance. The first solution is aimed at partitioning the nodes into clusters, each with a CH and some *ordinary nodes*, so that the CHs form an independent set, i.e. a set whose nodes are never neighbors among themselves. In [54, 55], a fully distributed Linked Cluster Architecture (LCA) is introduced mainly for hierarchical routing and to demonstrate the adaptability of the network to connectivity changes. The basic concept of LCA is adopted and extended to define multilevel hierarchies for scalable ad hoc routing in [56]. With the advent of multimedia communications, the use of the cluster architecture for ad hoc network has been revisited by Gerla *et al.* [57, 58, 59]. In these latter works, the emphasis is on the allocation of resources, namely bandwidth and channel, to support multimedia traffic in the ad hoc environment. These algorithms differ on the criterion for the selection of the CHs. For example, in [54, 55, 59] the choice of the CHs is based on the unique identifier (ID) associated with each node: the node with the lowest ID is selected as CH, then the cluster is formed by that node and all its neighbors. The same procedure is repeated among the remaining nodes, until each node is assigned to a cluster. When the choice is based on the maximum *degree* (i.e. the maximum number of neighbors) of the nodes, the algorithm described in [58] is obtained.

The DCA algorithm generalizes these clustering protocols in that the choice of the CH is performed based on a generic "weight" associated with a node. This attribute basically expresses how fit that node is to become a CH. All these protocols produce a set of CHs that are independent, and the criteria for joining them to form a connected backbone must be defined. A possible rule adopted by DCA is that defined by Theorem 1 in [60]: in order to obtain a connected backbone it is necessary (and sufficient) to join all CHs that are at most three hops apart via intermediate nodes (called *gateways*). CHs and gateways form the backbone. Different choices and definitions for the weights and the effects of the particular choice on the DCA and similar protocols have been investigated in [61, 62, 63]. The effects of mobility on the basic clustering produced by the DCA, as well as methods for reducing role changes (protocol overhead), have been considered in [64, 65, 66, 67]. CH selection and backbone formation, although different from the methods used in the mentioned solutions, are also the two fundamental steps of the WAF algorithm [68]. More specifically, WAF starts by selecting a node that will serve as the root of a tree. The tree construction leads to the selection of some nodes as CHs, which are then interconnected to form quite a slim backbone.

Other protocols based on constructing an independent set of CHs and then joining them to form a backbone are defined in [69], where the choice of the CHs is performed based on the normalized link failure frequency and node mobility. Two rules are defined for joining the selected CHs. The first rule utilizes periodic global broadcast messages generated by every node but forwarded only by the CHs. Non-receiving re-broadcasts from all neighboring nodes via its CH make an ordinary node aware of a disconnection. The ordinary node becomes a backbone node providing connectivity. The second rule is similar to that used by the DCA algorithm to build up a backbone. In [70, 71], DCA is used as basic clustering and rules are then defined to limit the size of the clusters.

Once the nodes are partitioned into clusters, techniques are described on how to maintain the cluster organization in the presence of mobility (clustering maintenance). Mobility has been the driving design parameter for some clustering algorithms, such as those presented in [72]. Wang and Olariu discuss the problem of clustering maintenance at length in [73], where they also present a tree-based clustering protocol based on the properties of diameter-2 graphs.

Algorithms directly concerned about building a backbone which is a connected dominating set have been presented in [74, 75, 76, 77, 78]. The idea in this case is to seek a dominating set and then shape it into a connected dominating set. The emphasis here is to build a routing structure, a connected *spine* that is adaptive to the mobility of the network nodes. In contrast to the solutions mentioned above, which distribute and localize the greedy heuristic for finding a maximum (weight) independent set, these solutions are a distributed implementation of the Chvátal heuristic for finding a minimal set cover of the set of nodes. A similar approach is followed in [79] where a minimum set cover is built in a distributed and localized way: nodes in the set cover are databases that contain routing information. A somewhat different approach is adopted in the WuLi protocol [80]. Instead of constructing a dominating set and then joining its nodes to make it connected, a richer connected structure is built, and then redundant nodes are pruned away to obtain a smaller CDS. Several different pruning rules are investigated in [81]. One of the most effective methods in removing redundant nodes has been introduced in [82]. The number of nodes in the CDS can be further reduced by using nodal degrees and the node GPS coordinates, as proposed in [83].

Most of the clustering protocols mentioned so far generate clusters of diameter ≤ 2: the cluster head always dominates its cluster members. There have been advocates for larger, possibly overlapping clusters. For instance, [84] describes routing for dynamic networks (such as ad hoc networks) which is based on overlapping k-clusters. A k-cluster is made up of a group of nodes mutually reachable by a path of length $k \geq 1$ (1-clusters are cliques). Clustering construction and maintenance in the face of node mobility are presented, as well as the corresponding ad hoc routing. A cluster head election protocol, with corresponding cluster formation is described in [85]. The focus in this paper is to efficiently build disjoint clusters in which each node is at most $d \geq 1$ hops away from its cluster head. The network is clustered in a number of rounds which is proportional to d, which favorably compares to most of the previous solutions when d is small. Finally, the mentioned clustering protocol presented in [72] produces clustering of a variable diameter. In this case, the diameter depends on the degree of mobility of the nodes: the more the nodes move, the smaller the clusters (easier to maintain), and vice versa.

More recent work for clustering and backbone set up is described for networks quite different from the general ad hoc model considered in [72] and in the other works already mentioned. In [86], some nodes are assumed to have "backbone capabilities" such as the enhanced radio interface that makes it possible to communicate with other backbone nodes at much longer distances than those covered by generic sensor nodes. Thus, in this case, the network becomes heterogenous, since it comprehends different types of nodes. The solution proposed in [87] constructs a CDS relying on all nodes having a common clock (time is slotted and nodes are synchronized to the slot).

3.4.6.2. *Clustering for WSNs*

We now review clustering and backbone formation protocols that have been proposed explicitly for WSNs. The main problem here is that of devising energy efficient techniques to transport data from the sensors to the sink. The overall goal is to increase the network lifetime. Hierarchical solutions like those provided by clustering and backbones appear to be viable for accomplishing this task, as demonstrated in [88, 89, 90]. Rather than belonging to one of the general classes of protocols for clustering and backbone formation described above, papers on clustering for WSNs are often specific for a given scenario of application, and, as already mentioned, are designed to achieve given desirable goals, such as prolonged network lifetime, improved tracking, etc.

Among the protocols that use clustering for increasing network longevity, one of the first is the LEACH protocol presented in [91]. LEACH uses randomized rotation of the cluster heads to evenly distribute the energy load among the sensors in the network. In addition, when possible, data are compressed at the cluster head to reduce the number of transmissions. A limitation of this scheme is that it requires all current cluster heads to be able to transmit directly to the sink (single-hop topologies). Improvements to the basic LEACH algorithms have been proposed by [92, 93] where multi-layer LEACH-based clustering is proposed and the optimal number of cluster heads is analytically derived to minimize the energy consumption throughout the network. An alternative method for selecting CHs for a LEACH-like model is presented in [94]. Particle swarm optimization (PSO) is used for dividing the network into clusters. A CH is then selected in each cluster based on its mean distance from all the other nodes in the cluster.

Localized clustering for WSNs has been proposed by Chan *et al.* in [95]. The Algorithm for Clustering Establishment (ACE) has the sensor nodes iteratively talking to each other until a clustering with cluster heads and followers (the ordinary nodes) is formed where the clusters are quite uniform in size and mostly non-overlapping. This minimizes the number of CHs. No backbone formation among the cluster heads is described. A tree-based clustering algorithm for sensor networks is presented in [96]. A selected node starts the BFS-based process of building a spanning tree of the network topology. Clusters are then formed by those nodes whose sub-tree size exceeds a certain threshold. The protocol is quite message-intensive, and there is no explicit description of backbone formation.

Clustering protocols have been proposed for networks in which some nodes are capable of long-haul communications (heterogenous networks). This is the case of clustering proposed in [97] where special, more powerful, nodes act as CHs for simpler sensor nodes, and transmit the sensed data directly to the sink. The authors have further explored their idea in [98, 99, 100, 101, 102, 103]. Gerla and Xu [104] proposed to send in swarms to collect data from the sensors, and to relay the data to the sink via intermediate swarms (multi-hop transport). Clustering for sensor networks of heterogenous nodes is also explored in [105]. Clusters are formed dynamically, in response to the detection of specific events (e.g. acoustic sensing of a roaming target). Only a certain type of node can be a CH, and methods are presented for selecting the more appropriate CH for target identification and reporting. A comparison among sensor networks with homogenous nodes and with heterogenous nodes is presented in [106], both for "single-hop" (à la LEACH) and for multi-hop topologies. The same authors make the case for heterogenous sensor networks in [107]. The authors also propose M-LEACH, a variation of LEACH where intra-cluster communications are multi-hop, instead of having ordinary nodes directly accessing their cluster head.

Finally, the construction of a backbone of sensor nodes is considered in [108] where the nodes need to know their position (such as their GPS coordinates). This work is more along the lines of sensor network topology control (GAF [109]) rather than on the hierarchical organization of WSNs as considered in this chapter. Further references on CDS construction in sensor and ad hoc networks can be found in [110].

HIGHLIGHTS

Overall, clustering has been proven effective in solving many of the problems encountered in the management and deployment of ad hoc and WSNs. While paying little in terms of overhead for clustering construction and maintenance, and for the natural increase in route length, recent works have demonstrated that the benefits of imposing a hierarchy in the network are multifold. These benefits include reduced route maintenance cost, reduced information maintained at the node, and the corresponding overhead to update it, a natural way to identify the best nodes where to aggregate the data, or the nodes to be switched off for conserving energy, and overall, increased network lifetime. The application of clustering and backbone formation methods to large networks of resource-constrained nodes (e.g. sensor nodes) requires simple, fully distributed solutions. Recent investigations show that fully localized approaches to clustering (e.g. the WuLi protocol described above) are very advantageous, being able to significantly reduce the overhead, and result in low energy consumption while building small backbones.

3.4.7. *Localization in ad hoc and WSNs*

One of the major challenges for sensor network deployment is the *localization* of a sensor node. Each sensor should be able to infer its position with respect to some global (or relative) system of coordinates. This information is essential to all those applications that require spatial mapping of the sensed data for further processing, and is also useful for the overall system performance as the availability of positioning information enables the adoption of low overhead protocols such as geographic-based routing schemes. GPS provides an immediate solution to the problem of localizing a node. However, GPS is often not a viable solution for providing localization to sensors in many scenarios due to its cost, the associated energy consumption and its inapplicability in indoor or heavy foliage scenarios. Therefore, GPS-free solutions for localization are of great interest.

Existing localization protocols may be classified by the information used to estimate a position. In *range-based* protocols, a node uses estimates of the range or angle from its neighbors to compute its own position. When nodes have no capability for computing such estimates, the corresponding solutions are called *range-free*. In the latter case, the use of beacons is required. Both categories present centralized and distributed solutions.

3.4.7.1. *Range-free localization*

Range-free localization enables the use of very simple network node hardware. In short, there is no need in most nodes for on-board GPS receivers, compasses, antenna arrays or signal-strength indicators. Instead, nodes localize with the transmitted coordinates from other nodes, including special nodes called *landmarks* or *beacons*, which are usually scattered randomly over the network area. These landmarks have extra features with respect to common network nodes. Among these, they know their position – via GPS or manual set-up – and they have a greater coverage area than other nodes. Landmarks provide information to the nodes by *beaconing* their position over the network. A non-landmark collects this information and infers its position from this information, using a range-free algorithm.

In [111], the position is estimated using a simple *coordinate centroid method*. Once a particular node, labeled 0 here, has received coordinates from the in-range landmarks, it estimates its position as:

$$\hat{\mathbf{c}}_0 = \frac{1}{k} \sum_{i=1}^{k} \mathbf{c}_i \qquad (3.1)$$

where $\hat{\mathbf{c}}_0$ is the estimated position of node 0, and $\mathbf{c}_i$ denotes the received coordinates from the ith beacon. The accuracy of this method depends on the distribution of landmarks around the node of interest, as well as the number of landmarks a node can hear. Increasing the number of *symmetrically-placed* landmarks will increase the accuracy of the estimated position. In particular, the accuracy of the coordinate centroid depends on a uniform placement of landmarks with respect to each receiver node. In practice, the solution proposed by Bulusu *et al.* produces a coarse positioning of the nodes that makes the algorithm suitable only for application in which accurate localization is not required.

In [112], Niculescu and Nath propose a solution called *DV-Hop* which uses a method similar to distance routing. A certain number of landmarks flood the network with their location information. The message containing this information also contains the landmark identification and a counter that is incremented at each hop. Thus, range is estimated by the hop count. Once landmark i receives location information from landmark j, $j = 1, \ldots, N$, it computes the average one-hop (physical) distance as follows:

$$\text{AvgHopDistance} = \frac{\sum_{j=1}^{N} \|\mathbf{c}_i - \mathbf{c}_j\|_2}{\sum_{j=1}^{N} h_{i_j}} \qquad (3.2)$$

where h_{i_j} is the distance, in hops, from landmark i and landmark j, and $\mathbf{c}_i$ and $\mathbf{c}_j$ are the coordinates of landmark i and j, $i \neq j$. This information is then flooded through the network. During the first phase of the algorithm, nodes store the location and distance information from each landmark. When a node receives the average one-hop distance, it uses the stored information to estimate its own coordinates by applying *multilateration*. The main drawback of this method is network flooding, which results in a considerable waste of energy at the nodes. To reduce this energy cost, each node may decide whether or not to forward received information. While this technique performs well in uniform networks, it is unacceptable in non-uniform networks where the real distances have a high variance around the average one-hop distance value.

APIT, the localization protocol proposed by He *et al.* [113], is based on the intersection of areas. Unlike [111], APIT uses triangular regions to derive the area in which it is highly likely to reside. An *APIT network* is a heterogenous network in which the landmarks are *super nodes* that are different from sensors in that they are capable of high-power transmissions and equipped with GPS receivers. Each node considers triples of landmarks and decides whether or not it is in the triangular area whose vertices are the three landmarks via a test called *Point-In-Triangulation (PIT)*. Repeating this operation for each triple, a node is able to determine a series of areas whose intersection gives the area in which the node itself lies with high probability. The main weakness of this approach is due to the reliability of the PIT test. This test requires a node to move in order to determine if it had been previously inside or outside the triangle. This is possible only if the node has some mobility and it is able to determine in which direction it is moving. For these reason, the authors provide an approximate version of the PIT test which overcomes the need for mobility and direction-sensing capabilities but has an intrinsic approximation error which depends on the number of neighbors of the node.

In a recent work by Hu and Evans [114], range-free localization is performed by applying the Monte Carlo localization method [115]. This method is based on a first step in which a node guesses a set of possible locations based on the position distribution at time $t - 1$ and its movements in $[t - 1, t]$. A second step follows in which the node uses information received from the neighbors to eliminate from the set, the first step inconsistencies with the neighbors' observations.

Previous work on range-free approaches also includes [116, 117, 118]. All of these works employ area-based techniques for the estimation of a position. In [116], each in-range constraint to neighbors is mapped to a linear matrix inequality, and a feasible region was found by semidefinite programming. The estimation technique in [116] is centralized, and the algorithm does not enable uncertainty in the coordinates of the reference nodes. [117] and later [118] consider decentralized, area-based estimation of a position. However, their approach employs a discrete grid, the L_∞-norm for distance, and again assumes that reference nodes have perfect knowledge of their position.

3.4.7.2. *Range-based localization*

The main difference between range-based and range-free methods is the availability of a range estimate between the node of interest and a transmitting node. This range estimate is provided by some physical-layer measurement and ancillary radio hardware.

Priyantha *et al.* [119] proposed *Cricket*, a solution based on a heterogenous network in which particular devices provided with ultrasonic transceivers help the nodes in the network to decide their positions based on a time-difference of arrival between a radio frequency (RF) signal and an ultrasonic signal. In [120], an improvement of cricket is proposed through the use of compasses while in the more recent [121], the presence of noise on range measurements is investigated and an algorithm working on this noisy environment and without use of beacon node is proposed and tested.

In [122], the authors propose an RF-based solution for indoor applications in which distances are measured using RSSI (Receiver Signal Strength Indicator) techniques. During an initial calibration phase, the algorithm creates a signal strength map of the indoor environment based on the RSSI received from landmarks placed in the building. Based on this map a centralized system is used to determine the position of a node based on its received signal.

In [123] Savvides *et al.* propose a method called *AHLoS (Ad Hoc Localization System)* in which sensor nodes (called Medusa nodes) are equipped with arrays of ultrasound microphones, which enable measurements of acoustic time of flight. In the first phase of this algorithm, node 0 estimates the distance between itself and all transmitting, in-range landmarks i, $i = 1, \ldots, N$. During the second phase the node 0 estimates its position by minimizing a mean-squared-error metric. Each node applies multilateration when it receives information from at least three landmarks. Once a node estimates its coordinates, it becomes a landmark in order to make the algorithm converge faster. While location is estimated accurately in the nominal case, it is prone to error propagation if landmark positions are inaccurate. Secondly, this method assumes an acoustic line-of-sight path, which may not be possible in some sensor network deployments. A similar method is proposed in [124] where *N-Hop Multilateration* is introduced. This method estimates the coordinates of a node in a distributed way so that a node can use information provided by landmarks not directly in the range of the node.

In [125], Niculescu and Nath propose two range-based protocol alternatives to DV-Hop. The first one is called *DV-Distance* and is similar to DV-Hop but uses RSSI to measure range. Provided there are accurate range estimates, DV-Distance yields more accurate location estimates compared to DV-Hop. However, received power is not only a function of range, but also of path-loss exponent, shadowing and multipath fading, and so range estimation is not necessarily accurate. The second range-based alternative to DV-Hop is based on the Euclidean distance between nodes. A node 0 computes its coordinates when it has the distance of at least two neighbors that know their distance from a landmark and between them. With this information, node 0 infers its distance from the landmark, and hence its coordinates. Simulations in [125] show that the two range-based solutions perform worse than the range-free method because of the error due to the measurements of the distance and the propagation of this error.

An alternative to the RSSI-based range estimate used in DV-distance algorithm is the *AoA* (Angle-of-Arrival) [126], and an improvement of this last algorithm is proposed in [127] in which an error propagation control technique based on a simplified Kalman filter is applied.

A solution similar to DV-Hop, called *Hop-TERRAIN*, is proposed by Savarese *et al.* in [128]. In this localization algorithm, landmarks first broadcast their information, and each node keeps a table with the distances in hops from each landmark. Once a landmark receives a packet from another landmark, it calculates the average one-hop distance and floods this calculation through the network. When the distance information reaches node 0, it infers a coarse position estimate via multilateration and then refines its position estimate iteratively through 1-hop sharing of location information, for a fixed number of iterations.

This solution has two main drawbacks. First, the algorithm does not provide a method to contain location estimate errors which propagate through the network. Second, the energy cost for information sharing is high due to flooding and repeated local broadcasts.

Fretzagias and Papadopouli [129] presented a voting algorithm for determining the position of nodes in a grid-based representation of the terrain. Starting from the measured distances of in-range landmarks, a node deduces the area in which it could be located. These areas are circular crowns, obtained by taking into account the error on the distance. By superposition of the crowns corresponding to different landmarks, certain cells gain a relatively higher probability to host the node, and the node itself determines its position by maximizing this *a posteriori* probability. After node 0 estimates its coordinates, it broadcasts the information in order to allow other nodes to continue the process. The accuracy of this method depends on the grid resolution: a smaller grid size leads to increased accuracy. The main problem in increasing the grid resolution depends on the available hardware: small sensor nodes cannot support the required hardware for highly defined grids.

In [130], Chintalapudi *et al.* investigate the performance of the ad hoc localization schemes based on ranging and bearing information presented in [124, 125]. They observe that information about the angle of arrival or about the bearing sector lead to good performance and accuracy of localization algorithms even in sparse networks. When only the range information is used, the same accuracy is achieved only when the nodes are densely deployed.

A range-based approach in which no beacon node is used is presented in [131] for localizing sensor nodes. The protocol, called *Anchor-Free Localization (AFL)*, is a two-step algorithm. AFL provides an anchor-free solution to the problem of localization by using a *concurrent* approach. Instead of proceeding *incrementally*, every node guesses its initial coordinates based on local information. As is typical of concurrent methods, this may lead to *false minima*, i.e. each node believes it is in the optimal (correct) position but the global configuration is incorrect. This motivates a second step, in which, as is typical in these cases, the process converges to a "correct" localization by applying a force-based relaxation procedure [132, 133]. Simulation results show that AFL achieves acceptable accuracy even in sparse networks, as opposed to the majority of previous approaches that achieve the same accuracy at higher network densities.

Two interesting ideas appeared in [134] and later in [135]. While these works address target tracking in sensor networks, the estimation of target position is closely related to localization. Sequential Bayesian filtering is used to update, at each time, the posterior joint density of the target's coordinates. Confidence volumes may be easily determined by density contours, as was done in [135], or more generally by integrating the posterior density. The sequential Bayesian filtering approach in [135] requires the one-step state transition probability density function for the target's position, as well as the conditional joint density of the observations given the target's state. The first density requires an accurate model for target motion, and also a stochastic characterization model noise. The second density requires a stochastic model relating the target's position to received measurements (received power, etc.), and this would require stochastic characterization of the physical medium which corrupts the measurement.

HIGHLIGHTS

Among the several proposals for localization in WSNs, it is increasingly clear that range-based approaches are able to provide a reasonable localization of a node in terms of error and of energy consumption. The challenges now are mainly to provide an increasingly accurate determination of range, and possibly angle of arrival measurements, and new methods for limiting the propagation of the errors.

3.5. Wireless sensor networks with mobile nodes

3.5.1. *Introduction*

From a communication perspective, WSNs are expected to be deployed as support (transport) networks in that they allow the movement of data to/from the sensors from/to the sinks. For instance, data collected by a node may need to be conveyed to a number of infrastructures such as monitoring collection points, public networks, the Internet, etc., located at the *periphery* of the networked area for reliable storage and more sophisticated elaboration. Conversely, data and code can be sent from designated (query and tasking) centers to relevant sensor nodes.

The main objective of a sensor network is to support (a) efficient and robust data transport, and (b) uninterrupted coverage of the geographic area in which the sensors have been placed arbitrarily. Given that the sensors are irreplaceable, protocols that implement (a) and (b) should be designed so that important WSN performance parameters are optimized. These parameters include: extended network lifetime, energy consumption, data throughput, the amount of data to be collected, data fidelity and security and data transfer delay.

Most of the research on data delivery in WSNs concerns networks whose nodes do not move and are irreplaceable. The sensed data is delivered to static sinks. In this scenario, it has been observed that the nodes closer to the sinks have their energy drained from data transmissions more than all the others. These nodes relay data for all the other nodes in the network as well as possible packets from the sinks to the sensors. As a result, sinks are soon disconnected from the rest of the network, which determines the end of the network functionalities (network lifetime).

A trend of the research on data dissemination in WSNs has recently started trying to exploit the mobility of some of the network components in order to facilitate the delivery of the sensed data to the sinks and enhance the system's performance. For instance, some mobile agents (such as robots or mobile sinks) can move over the area covered by the static sensor nodes for many purposes, such as collecting data from the peripheral sensors, replacing exhausted nodes, synchronizing the nodes, or, in general, increasing the energy-efficiency network lifetime.

Also, when sensor nodes are, themselves, not expected to be mobile, mobility can still arise from the fact that the sensor nodes are located either in an environment that makes them move over time (e.g. underwater sensor networks), or into moving users or objects (e.g. buses, pedestrian users and so on).

The applications for WSNs with mobile autonomous nodes are multifold. Here are just a few examples showing why mobility has to be accounted for in WSNs:

• Wireless networks of mobile sensors that can be dispersed in a quasi-random fashion, e.g. from airplanes, intended to extend the ability of data collection, monitoring, and control of the physical environment from remote locations; for instance, submerging a network of sensors in an ocean bed to detect debris from plane crashes for recovery and identification purposes.

• Networks in support of mobile sensing/measuring devices, such as small-scale robot squads. While performing their tasks, the robots are enabled to exchange information among each other and/or to transmit the collected measures to final collection/information centers at the periphery of the inspected area.

• To collect information about the location of a user/piece of equipment, etc. in networks of small, mobile radio transmitters that enable wireless connection among heterogenous devices (cellular telephones and laptops, printers and PDAs, etc.) which can be placed arbitrarily by the user at home, office, etc.

This section deals with WSNs with mobile nodes. The main advantages of using mobile nodes are as follows:

• smaller number of nodes to cover the same area;

• dynamic adaptation with environment triggers or changes;

• dynamic change of the topology to optimize communications in the network.

Mobile nodes require specific paradigms and algorithms. Thus, localization of the mobile nodes is an important issue. Furthermore, in order to exploit the use of these nodes, algorithms for planning the motion of the nodes could be needed.

In the following sections, we will better characterize the WSNs with mobile nodes. Furthermore, we will provide an overview of the algorithms regarding the most interesting aspects of WSNs with mobile nodes.

3.5.2. *Types of mobile nodes and networks*

The above section introduces mobile nodes without discussing the nature of these nodes. In practice, these mobile nodes may consist of static sensor nodes installed in a suitable mobile object. These objects can be people, animals, vehicles or robots.

In fact, one person can carry a mobile node and possibly additional hardware for communications and data storage.

As far as the motion is concerned, the mobile nodes can be classified as:

• nodes with uncontrolled and non-predictable motion;

• nodes with controlled and predictable motion.

The first case corresponds to nodes that are moving in a scenario without explicitly considering their role as mobile nodes of the network (e.g. carried out by animals or persons). The motion can be considered random, and the time required for approaching a given static node can be considered as a random variable too. Obviously, this time depends on the density of mobile and static nodes and the motion characteristics.

In the second case, the motion of the nodes can be controlled.

On the other hand, there are two different types of WSNs with mobile nodes:

• **Static sensor networks with mobile nodes**: this type of sensor network is composed of a large number of static nodes and few mobile nodes. It has the characteristics of a static sensor network, plus some other advantages that come from the use of mobile nodes: sensor calibration, reprogramming nodes capability, coverage extension, dynamic density, etc.

• **Mobile sensor network**: this sensor network consists of a number of mobile nodes. A special interest case arises when the nodes are autonomous objects, e.g. autonomous ground, aerial or underwater vehicles. These types of networks are also referred to as *WSNs with autonomous mobile nodes*.

In the following sections, the characteristics of static sensor networks with mobile nodes and WSNs with autonomous nodes will be considered.

3.5.3. *Static sensor networks with mobile nodes*

The use of mobile nodes in sensor networks increases the capabilities of the network and allows dynamic adaptation with changes in the environment. The different applications of mobile nodes are the following:

• **Collecting and storing sensor data in sensor networks**: mobile nodes are used to deploy networks with conventional nodes having short range communications. The mobile nodes can approach distant static nodes to collect sensor data and store them. In this case, the mobile nodes should have enough storing capabilities to sequentially collect the information from the static nodes while moving to other nodes or to the base station (which could be provided by wired connections). Notice that storage capabilities are also required in the static nodes to wait for the visit of the mobile node. Alternatively, mobile nodes may be used to alleviate the power consumption due to multi-hop data forwarding. The disadvantage is the delay required to collect data in distant locations when the mobile node has to move sequentially to approach the static nodes. Then, in general, the solution could be applied only in case of delay tolerant scenarios.

• **Sensor calibration**: static nodes could be calibrated or re-calibrated for a particular application using mobile nodes with different and possibly more accurate sensors. An example could be the calibration of temperature sensors using a mobile platform with an infrared camera, or vice versa.

• **Reprogramming nodes**: the same can be done reprogramming "by air" static nodes for a particular application. The program or parameters can be downloaded to these nodes from a mobile node at a convenient distance. The functionality of the static nodes could be changed depending on the environment or any other factor. The sensor network will have more capability of adaptation, increasing its utility.

• **Network repairing**: when the static nodes fail to sense and/or to communicate, the mobile node could move to the static node location to replace it when required. Additionally, a robot could deploy some static nodes to fix the network and achieve a better communication between the static nodes. This could also be used for a particular application, as will be described later. In addition, a mobile node could be used as a gateway to communicate different parts of the sensor network that are too far away to be communicated directly by the static nodes.

3.5.3.1. *Nodes with uncontrolled and non-predictable motion*

Considering mobility as a blessing rather than a curse for network performance has been widely discussed for ad hoc and sensor networks in different contexts [136, 137, 138, 139, 140, 141, 142]. The primary objective of these works is to deliver messages in disconnected ad hoc networks. Examples include WSNs for environment monitoring or for traffic monitoring.

The work by Chatzigiannakis *et al.* [137] explores the possibility of using the coordinated motion of a small number of users in the network to achieve efficient communication between any pair of other mobile nodes. A fraction of the network nodes act as forwarding agents carrying packets for other nodes: the packet is exchanged when the source node and the agent are neighbors (i.e. in the radio vicinity of each other), and it is then delivered to the intended destination when the agent passes by it.

This basic idea has been introduced to WSNs by Shah *et al.* in their works on *data mules* [143]. Mobile nodes in the sensor field, called mules, are used as forwarding agents. The idea here is to save energy by having single-hop communication (from a sensor to the mule that is passing by) instead of the more expensive multi-hop routing (from the sensor to the sink): it is the mule that will eventually take the sensed data to the sink. The data mule architecture is effective for energy conservation in delay tolerant networks. Energy is traded-off for delay, i.e. the energy needed to communicate a packet to the sink is decreased at the cost of waiting for a mule to pass nearby (and at the cost of waiting for the mule to move to the vicinity of a sink).

This approach has been further investigated by Kim *et al.* [144] who propose a dissemination protocol in which a tree-like communication structure is built and maintained, and mobile sinks access the tree from specified sensor nodes in the tree (access nodes). The protocol, termed SEAD (Scalable Energy-Efficient Asynchronous Dissemination), demonstrates via simulation the effectiveness of deploying mobile sinks for energy-saving as opposed to keeping the sinks static. SEAD is shown to be more effective for conserving energy than other solutions for data dissemination in WSNs such as DD [145], TTDD [146] and ADMR [147].

Common to all these works is that the mobility of the sink is unpredictable and *uncontrollable*. For example, in [144] sinks move according to the random waypoint model.

3.5.3.2. *Nodes with controlled or predictable motion*

On the other hand, mobile nodes can use the static nodes to locate themselves or even to follow a path. The use of mobile sinks with *predictable*, or even deterministic mobility has been more recently proposed in [148, 149, 150, 151]. In these works, the sinks (airplanes) fly over the sensor field and gather the sensed data periodically. While the movement of the sink is fully controllable, it is external to the network infrastructure, i.e. the trajectories are not determined by the network components and activity. The main contribution of these papers concerns the energy-efficient transmission to the passing sink [148, 149], but no implementations with real vehicles have been provided. In [150], the authors consider heterogenous sensor networks made up of two types of nodes, and determine the densities of each type and the battery energy needed to achieve a given network lifetime.

Inherent patterns of the sink movement are exploited in [152] for the design of robust and energy-efficient routing. This paper assumes that there is a certain degree of predictability in the sink movement, such as the routine route of a ranger patrolling a forest. Based on statistics and distributed reinforcement learning techniques, the sensor nodes learn about the sink's whereabouts at given times and use this information to find routes to the mobile sink.

A model for sink movement is proposed in [151], where *observers* (i.e. the sinks) move along the same path repeatedly. The sensed data are collected while the observer traverses the network. When passing by sensor nodes, the observer wakes them up and receives their data (if any). The authors describe a prototype system developed at Rice University where the observers are carried by campus shuttles, and the sensors are spread out throughout the university property. In particular, the authors determine the transmission range needed to collect data from a predefined percentage of the sensor nodes, given the observer speed, the time required to transmit a piece of information, and the traffic pattern. The correlation among the various system parameters is investigated analytically.

An in-depth discussion on the advantages of incorporating *controllable* mobile components into the network infrastructure is presented in [153]. The authors present an implementation of a sensor network with an autonomous mobile router (a robot) that visits the (static) sensors, collects their data and reports them to the sink. The idea of collecting data in a single-hop fashion (i.e. when the robot approaches a sensor) is similar to that of data mules. The key difference here is that the motion of the robot is *controlled*: the movement of the robot adapts to data collection performance parameters, and it is determined by the network application priorities. The robot is part of the system. The testbed-based experimental results in this paper concern the evaluation of methods for controlling the speed of the robot for optimizing data collection. The robot traverses networks with different densities following a straight trail and collects the data that are then brought to the sink.

The controlled mobility of the sensor nodes (rather than the mobility of an autonomous router or of the sink) is explored in [154]. The idea here is to have the sensors move into positions that minimize the energy cost of reporting streams of data to the sink.

The problem of reducing energy consumption and of maximizing the lifetime of a sensor network by exploiting controllable sink mobility within the network has been tackled in [155] and, more recently, in [156]. In these two works, it is the sink that moves directly among the (static) sensor nodes and, while sojourning at given locations, collects data that are sent to it via multi-hop (ad hoc) routing.

The first paper is mostly concerned with energy minimization. The authors present an ILP (Integer Linear Programming) model to determine the locations of multiple sinks and the routes from the sensors to the sinks. Time is divided into rounds. At the beginning of each round, information on the nodes' residual energy is centrally gathered and the ILP problem is solved to determine new, feasible locations the sinks should travel in order to minimize the maximum energy consumption spent at the nodes during that round. Minimizing the energy consumption results in increased network longevity. No constraints are forced on the sink movements and successive location, and there is no relationship between the number of the sinks and their position in subsequent rounds.

The problem of network lifetime maximization through controlled sink mobility is explicitly addressed in [156] for networks with a single sink. Via a new LP formulation, both sink locations and sink sojourn times at those locations are determined to maximize the network lifetime. The experiments performed in the paper refer to scenarios where $n = L^2$ nodes are arranged in an $L \times L$ grid. The sink can visit each node at its location. Improvements on network longevity are obtained that are almost five-fold when the sink sojourns at the nodes located at the four corners and in the central area of the grid.

3.5.4. *WSNs with autonomous mobile nodes*

In this section, WSNs with mobile nodes will be treated in the particular case when the mobile nodes are robotic. However, the same algorithms can be useful for people as mobile nodes if a Human Machine Interface (HMI) is used. An example of an HMI between a person and a sensor network is the interactive device *Flashlight* [157], used to guide a person with the information acquired from a sensor network. The device uses a pager vibrator and an LED to guide the person, i.e. when the device is in the right direction the vibrator and the LED turn on. Finally, other types of HMI could be used: for example, a graphical interface, where an arrow on a map displayed in a PDA shows somebody the direction to follow.

The robotic nodes of the sensor network should have some autonomous navigation capabilities such as following a path or finding a sensor source from the sensors signals.

The most popular robotic mobile nodes are low cost and small sized nodes based in *Xbow* products, like the sensor node MICA2. These mobile nodes are usually applied in indoor environments.

Several types of robotic mobile nodes can be found in the literature:
- MICAbot [158];
- CostBots [159];
- Robomote [160];
- Millibots [161].

In some applications, it is possible to use autonomous vehicles (usually developed for outdoor environments) with more powerful locomotion capabilities, which use a sensor node on-board to become part of the sensor network. For example, in [162] an autonomous helicopter is integrated with a sensor network and applied for the deployment and repair of the network [162]. Figure 3.2 shows the sensor deployment in a recent experiment of the AWARE project funded by the European Commission (http://www.aware-project.net).

Figure 3.2. *Node deployment with the Marvin autonomous helicopter in the AWARE project*

It is also possible to combine autonomous nodes with other mobile nodes, like persons, animals or vehicles, in the same network.

3.5.5. *Algorithms*

3.5.5.1. *Localization algorithms*

As was previously described, the localization of the nodes in a sensor network can be solved using two classes of algorithms: range-free and range-based localization. Most of the algorithms that will be presented are range-free and they are based on the use of a global positioning system such as GPS by some mobile nodes. It is not possible to use GPS with every node of the sensor network due to the high cost and energy consumption of GPS.

However, the use of GPS in few mobile nodes, i.e. carried by people, animals or vehicles, makes the solution feasible in cost and energy. If the mobile nodes were autonomous nodes they could always recharge their batteries somewhere.

The following algorithms [163] compute an estimate $\hat{c}_i$ of the position of a static node from the positions $p_{ij} = (x_j, y_j)$ transmitted by the mobile nodes with a strength s_j. These algorithms could be based on the following principles:

• The static nodes take the strongest received message so far, as the best estimate of node position:

$$\text{if } s_j > s_{\max} \text{ then}$$

$$s_{\max} = s_j \tag{3.3}$$

$$\hat{c} = p_j$$

• The node takes the mean of the received positions as the computed position.

• The mean of the received positions is weighted with the signal strength, and the estimated position is:

$$\hat{c}_i = \frac{\sum_j s_j p_j}{\sum_j p_j} \tag{3.4}$$

• The median of the received positions as the estimate:

$$\hat{c}_i = \text{median}(p_{1...j}) \tag{3.5}$$

• Each received position is a constraint on the node position which is considered to lie within the rectangular region Q. At each step, the node is constrained to lie in the intersection of its current region, Q_{i-1}, and a square of side length $2d$ centered on the GPS transmission. The position estimate of the node is taken as the centroid of the region Q_i:

$$Q_i = Q_{i-1} \bigcap [x_i - d, x_i + d] \times [y_i - d, y_i + d] \tag{3.6}$$

The parameter d should reflect the size of the radio communication region.

These algorithms do not require inter-node communications, so power consumption and network congestion is reduced. However, the localization accuracy is higher when the inter-node communication is used.

3.5.5.2. *Coverage algorithms*

Some applications of WSNs with mobile nodes require them to disperse throughout their environment. Exploration, surveillance and security applications all require coverage of large areas. In this section, an algorithm [164] for dispersing a number of mobile nodes into an enclosed space or an open environment is presented. Those algorithms must take into account several aspects: maintaining network connectivity, allowing for mobile node and communications failures, and providing an infrastructure for the mobile nodes to maintain their battery charge.

Uniform dispersion algorithm: this algorithm disperses mobile nodes uniformly throughout their environment. A thorough treatment of this technique is presented in [165]. Physical walls and a maximum dispersion distance between any two mobile nodes of r_{safe} are used as boundary conditions to help prevent the nodes from spreading too thin and fracturing into multiple disconnected components.

The algorithms works by moving each mobile node away from the vector sum of the positions $p = p_1, \ldots, p_c$ of their c closest neighbors $\mathrm{nbr} = \mathrm{nbr}_1, \ldots, \mathrm{nbr}_c$. The magnitude of the velocity vector that is given to the motor controller is:

$$v = \begin{cases} -\frac{v_{\max}}{cr_{safe}} \sum_{i=1}^c p_i & |p_i| \le r_{safe} \\ 0 & |p_i| > r_{safe} \end{cases} \tag{3.7}$$

where $v_{\max}$ is the maximum allowable velocity output. This vector directs the active mobile node away from its c nearest neighbors. The drive velocities are:

$$v_{rot} = v \cos(\mathrm{nbr}_i.\mathrm{bearing}), \quad v_{trans} = v \sin(\mathrm{nbr}_i.\mathrm{bearing}) \tag{3.8}$$

where $\mathrm{nbr}_i.\mathrm{bearing}$ is the bearing to nbr_i.

This is a relaxation algorithm. Imagine replacing a graph G with its Delaunay triangulation G', and then placing compressed springs between connected mobile nodes. This will tend to expand the network to fill the available space, but once the space is occupied, mobile nodes will position themselves to minimize the energy in the springs. Total group energy is minimized by minimizing local contributions, which happens when all the internode distances are equal. In practice, using the two closest neighbors works very well.

3.5.5.3. *MAC algorithms*

One of the tightest constraints in a wireless sensor network is the energy consumption, and communication is the highest energy consumption in a sensor node. The

MAC algorithms deal with the mobile nodes in a special manner with regard to the static nodes. In a sensor network composed by static nodes, after the configuration of the communication in the data link layer, the waste of energy in messages that are not information messages is going to be very low because the topology of the network will be the same most of the time. Only when a sensor fails will the network need to be configured again. However, when there are mobile nodes, the density of configuration messages grows because the topology of the network changes.

One protocol that deals with mobile nodes is the SMACS-EAR (Self-Organizing Medium Access Control for Sensor Networks, Eavesdrop And Register) [1, 2]. SMACS is a protocol that configures the nodes of a sensor network in a distributed way and has a flat topology for static nodes. However, EAR is an extension of the protocol for mobile nodes that attempts to offer continuous service to the mobile nodes under both mobile and static constraints. The key to the EAR protocol is to configure the communication channel between a static node and a mobile node with the least number of messages transmitted by the static node. Therefore, most of the energy is consumed by the mobile nodes. This policy is based on the fact that the mobile nodes can have better autonomy or the chance to recharge their batteries in special places or with the use of alternative energy sources.

With the use of the EAR protocol, the topology is no longer flat and the mobile nodes assume full responsibility for making and breaking connections with the static nodes. The algorithm consists of different steps:

1. The static node invites the other nodes to open a new connection (this step is done periodically and whether there are mobile nodes or not).

2. The mobile node responds to the invitation of the static node when it is going to create a connection with the static node.

3. The static node sends a reply accepting or denying the connection.

4. To break the link, the mobile node only has to send a disconnection message.

Finally, the mobile nodes need to have a registry of neighbors in order to keep a constant record of neighboring activity.

3.5.5.4. *Routing algorithms*

Because the mobile nodes interact with the network, it is possible that they become involved in the routing paths calculated at the network layer. For mobile nodes working as information sources, such as data collectors, routing is not an issue since the only goal is to place the information on the network, transmit it to the static nodes and allow them to route the information to the required destinations. If the mobile node is a sink of the sensor network, but its speed is low enough, the routing trees could be calculated with the movement of the mobile node. To save some energy, if the mobile node moves a short distance, only the nodes around the mobile node will recompute the routing trees. However, when the mobile node moves large distances, all the nodes will recompute the routing trees.

There are also some routing algorithms which are specially designed for sensor networks with mobile nodes [166]. Relevant algorithms are SPIN and TTDD, already presented in section 3.4.4 and briefly reported in the following, for the reader's convenience:

• SPIN: the SPIN family of protocols uses data negotiation and resource-adaptive algorithms. Nodes running SPIN assign a high-level name to completely describe their collected data (called metadata) and perform metadata negotiations before any data is transmitted. SPIN is a three stage protocol as sensor nodes use three types of messages (ADV, REQ and DATA) to communicate. ADV is used to advertise new data, REQ to request data and DATA is the actual message itself. The protocol starts when a SPIN node obtains new data that it is willing to share. It does so by broadcasting an ADV message containing metadata. If a neighbor is interested in the data, it sends a REQ message for the DATA and finally the DATA is sent to this neighbor node. The neighbor sensor node then repeats this process with its neighbors. As a result, the entire sensor area (interested in that information) will receive a copy of the data. On the other hand, these protocols are well-suited for an environment where the sensors are mobile because their forwarding decisions are based on local neighborhood information.

• TTDD: this protocol provides data delivery to multiple mobile nodes working as base stations. In TTDD, each data source proactively builds a grid structure which is used to disseminate data to the mobile sinks by assuming that sensor nodes are stationary and location-aware, whereas sinks may change their locations dynamically. Using the grid, a base station can flood a query, which will be forwarded to the nearest dissemination point (a node that stores the source information) in the local cell to receive data. Then, the query is forwarded along other dissemination points upstream to the source. The requested data then flows down in the reverse path to the sink. Trajectory forwarding is employed as the base station moves in the sensor field. Although TTDD is an efficient routing approach, there are some concerns about how the algorithm obtains location information, which is required to set up the grid structure.

3.5.5.5. *Mobile nodes planning algorithms*

In the following section, the predictable and controlled ad hoc motion of the mobile nodes is considered. Different techniques can be applied to obtain the required paths and trajectories to collect and store the information of the static nodes of the network. Path planning techniques used in robotics or well known techniques from the operational research field can be applied.

This section will be mainly devoted to the algorithms that compute the optimal path to visit a group of static nodes by using a team of mobile nodes (this problem is known as the *multiple Traveling Salesmen Problem*, often referred to as *m-TSP*).

It is an instance of the *Optimal Assignment Problem* (OAP) [167], which is a well-known problem originally studied in game theory, and then in operations research in the context of personnel assignment.

Problem statement. Given m mobile nodes, each visiting one static node (a more general case will be studied later) and n prioritized static nodes, each require one mobile node to visit it for a given purpose (collect information, share energy, etc.). Also, given for each mobile node a non-negative estimate of the utility of visiting a mobile node, if a mobile node is incapable of undertaking a task (visiting a static node or performing a required operation after reaching a static node), then the mobile node is assigned a rating of zero for that task. The goal is to assign mobile nodes to static nodes so as to maximize overall expected performance, taking into account the priorities of the static nodes and the skill ratings of the mobile nodes.

This problem can be cast as an ILP [167]: find mn non-negative integers α_{ij} that maximize:

$$U = \sum_{i=1}^{m} \sum_{j=1}^{n} \alpha_{ij} U_{ij} w_j \tag{3.9}$$

subject to:

$$\sum_{i=1}^{m} \alpha_{ij} = 1, \quad 1 \le j \le n$$
$$\tag{3.10}$$
$$\sum_{j=1}^{n} \alpha_{ij} = 1, \quad 1 \le i \le m$$

The sum (3.9) is the overall system utility (note that since α_{ij} are integers, they must all be either 0 or 1). Given an optimal solution to this problem (i.e. a set of integers α_{ij} that maximizes (3.9) subject to (3.10)), an optimal assignment is constructed by assigning mobile node i to static node j only when $\alpha_{ij} = 1$.

If the mobile nodes' utilities can be collected at one machine, then a centralized linear programming approach (e.g. Kuhn's Hungarian method [168]) will find the optimal allocation in $\mathcal{O}(mn^2)$ time. Alternatively, a distributed auction-based approach (e.g. Bertsekas's auction algorithm [169]) will find the optimal allocation, usually requiring time to be proportional to the maximum utility and inversely proportional to the minimum bidding increment. The two approaches (i.e. centralized and distributed) represent a trade-off between solution time and communication overhead. To implement a centralized assignment algorithm, n^2 messages are required to transmit the utility of each mobile node for each visit; an auction-based solution usually requires far fewer (sometimes fewer than n) messages to reach equilibrium. Moreover, the time required to transmit a message cannot be ignored, especially in wireless networks, which can cause significant delay.

In the previous problem, it has been assumed that at most one static node should be assigned to each mobile node. When the system consists of more static nodes than mobile nodes, this problem is one of building a time-extended *schedule* of tasks for each robot, with the goal of minimizing total weighted cost. Using Brucker's terminology [170], this problem is an instance of the class of scheduling problems

$$M||\sum w_j C_j \qquad (3.11)$$

that is, the mobile nodes execute tasks in parallel (M) and the optimization criterion is the weighted sum of execution costs ($\sum w_j C_j$). Problems in this class are strongly $\mathcal{NP}$-hard [171]. Even for relatively small problems, the exponential space of possible schedules precludes enumerative solutions.

A way of treating these problems is to ignore the time-extended component. For example, given m mobile nodes and n static nodes, the following approximation algorithm can be used:

1. Optimally solve the initial $m \times n$ assignment problem.

2. Use the greedy algorithm to assign the remaining tasks in an online fashion, as the robots become available.

The performance of this algorithm is bounded below by the normal greedy algorithm, which is 3-competitive for online assignment. The more nodes that are assigned in the first step, the better this algorithm will perform. As the difference between the number of mobile and static nodes that are initially presented decreases, performance approaches optimality.

Another way to approach this problem is to employ an iterative task allocation system, such as [172] price-based market. The mobile nodes would opportunistically exchange static nodes to visit over time, thereby modifying their schedules.

3.5.5.6. *Mobile nodes reactive algorithms*

It could also be possible to apply reactive techniques used in robotics to guide the mobile node. In this case, the mobile platform moves reacting to sensorial stimulus of the environment. There are several techniques that have been used in WSNs:

• The diffusion-based path planning [173] that is applied when it is known that the quantities of interest in the system are generated via a diffusion process. This algorithm assumes that a network of mobile sensors can be commanded to collect samples of the distribution of interest. These samples are then used as constraints for a predictive model of the process. The predicted distribution from the model is then used to determine new sampling locations.

• The random walk algorithm using gradient descent [160] is applied to guide the mobile node to a focus of interest. The algorithm proceeds as follows:

    1. Record the current sense value reading from the sensor $P(i)$.

    2. Move along a straight line for a constant distance.

    3. Record again the current sensor intensity reading from the sensor $P(i+1)$.

    4. If the difference $P(i+1) - P(i) > 0$, move along the same direction.

    5. If the goal is reached then stop.

    6. If the goal is not reached then rotate by a random angle and go back to step 1.

3.5.5.7. *Network repairing algorithm*

As described before, the use of mobile nodes makes it possible to repair the network connectivity of a sensor network [162]. The use of mobile nodes could be important to increase the fault tolerance in a WSN.

Assuming that a manual or automatic network deployment is executed, the connectivity repair algorithm could consist of two phases:

• The mobile node measures the connection topology of the deployed network and compares it to the desired topology. If they match, no deployment is done. This phase can be run at any time, with the objective detecting the potential failure of sensor nodes and ensuring sustained connectivity.

• Otherwise, the measured connectivity graph is used to compute new deployment locations that will repair the desired topology. These deployment locations can be represented as a set of waypoints.

In a real-life implementation [162], a simplified version of this procedure has been applied. The algorithm computes deployment locations whose connectivity graph is one connected component. Therefore, the task of the connectivity repair algorithm is to determine the number of connected components in the deployed network. The algorithm that determines the number of connected components works as follows:

    1. Each static sensor broadcasts its identification number and forwards the identification numbers that it hears.

    2. Each sensor keeps the largest value it has heard. The number of different values is the number of connected components in the graph.

    3. The mobile node can collect this information and determines how many components there are.

    4. If the network has at least two connected components, it can compute the separation region and determines how to cover it with waypoints in a way that connects the two components.

3.5.6. *Critical issues and future research*

In WSNs with mobile nodes, it is not necessary (relying on direct or multi-hop connections) to directly transmit information from the static nodes to the sink. Indeed, mobile nodes can relay the information among the static nodes and also the sinks; they can even relay information in the case of temporary connectivity holes and network partitioning. The static nodes can be organized in clusters and the mobile nodes can approach the clusters and transmit the information to others sinks or to the main sink. There are advantages in the collection of the information, and advantages for the diffusion of the information because the path of the information is the same but in the opposite direction. It should be pointed out that in this architecture there could be some delay problems. These problems could be solved by using more mobile nodes or mobile nodes with larger communication ranges. Higher energy consumption is the drawback in this case, but as has been mentioned before, a mobile node can always recharge its batteries. Furthermore, the use of mobile nodes allows energy-savings in the static nodes. For example, a mobile node can move near the locations of the static nodes, reducing their energy consumption due to communication.

Mobile nodes can improve or repair the communication in the sensor network. Furthermore, they can be used to calibrate the sensors of the static nodes. Mobile nodes can offer a better coverage among subnetworks of static nodes. A mobile node with better communication capabilities can even be used as a gateway between the sensor network and another type of network such as the Internet, cell phone, etc.

Several main trends have been identified in the literature about WSNs with mobile nodes:

• Energy constraints: in relation to WSNs, energy constraints is one of the most relevant issues. In WSNs with mobile nodes, this topic includes, for example, the study of the impact of mobile nodes on energy consumption of the entire network.

• Motion planning: an important trend is related to the study of how the sensor network can compute, in a distributed way, the path that the mobile node must follow. Also, this path can be updated depending on changes of the environment or by using new data collected by the sensor network.

• Cooperative perception: the sensor network can be considered an extension of the sensory capabilities of a mobile node, and therefore an improved model of the environment can be built with the information from the WSN for navigation purposes.

Finally, it should be pointed out that WSNs with mobile nodes is a relatively new topic and it is still a developing field. The following topics have been found in the literature about WSNs with mobile nodes: localization, communications (MAC and routing) and planning and reactivity. However, more algorithms and theoretical studies are needed in those fields. Furthermore, there are some topics that have been poorly addressed, such as self-hierarchical organization and/or clustering techniques, fault tolerance and coverage techniques. Therefore, more effort should be devoted to those aspects and their study should be promoted in the future. The following points can be suggested as relevant research trends in the future:

• Energy considerations in: static and mobile trade-off, computation of optimal parameters (speed, delay, etc.) and the relationship between data collection and data diffusion.

• Motion planning, taking into account communication constraints and the relation with self-hierarchical organization and/or clustering techniques.

• Cooperative perception algorithms to exploit sensory information from the nodes of the network.

• New reliability and fault tolerant concepts.

• Integrate and exploit heterogenity in mobile nodes.

• More field experiments with WSNs and mobile nodes.

3.6. Autonomous robotic teams for surveillance and monitoring

3.6.1. *Introduction*

Although most mobile robotic systems involve a single robot operating alone in its environment, a number of works have considered the problems and potential advantages involved in having an environment inhabited by a group of robots cooperating in order to complete a task. There are several reasons which could lead to the development of a multi-robot system. The main reason may be that it is possible to build more robust and reliable systems by combining unreliable but redundant components. Performance is another important advantage (many hands make light work) which becomes critical in dangerous environments over a broad area such as forest fires, contaminated areas, etc. In such an environment, a robotic team could provide several services: surveillance, searching, detection, tracking, monitoring, measurement, etc. For example, the *COMETS Project* of the European Commission (IST-2001-34304) on multiple heterogenous unmanned aerial vehicles considers the coordination and control of multiple heterogenous UAVs for applications such as detection and monitoring (see Figure 3.3). Figure 3.4 shows photographs of a demonstration of fire detection and extinguishing in the CROMAT project (see Chapter 2) on the cooperation of aerial and ground robots. Successes in the field over the last decade seem to have shown the feasibility, effectiveness and advantages of a multi-robot system for some specific robotic tasks: exploring an unknown planet [174], pushing objects [175, 176], or cleaning up toxic waste [177].

In other works, there are not many formal models for multi-robot coordination, because research has mainly covered the construction and validation of working systems. In the next section, several taxonomies (see [178, 179]) that categorize the

Figure 3.3. *Aerial robotics team*

bulk of existing multi-robot systems along various requirements are presented. Those requirements include team organization (e.g. centralized vs. distributed), communication topology (e.g. broadcast vs. unicast), and team composition (e.g. homogenous vs. heterogenous).

3.6.2. *A taxonomy of multi-robot systems*

A key difficulty in the design of multi-robot systems is the size and complexity of the space of possible designs. In order to make principled design decisions, an understanding of the many possible system configurations is essential. Several taxonomies have been proposed which provide the following aspects:

 • defining key features that help identify different multi-robot systems in a precise and complete way;

 • making it possible to state relations, analysis and formal proofs for the groups derived from the taxonomy;

 • making it easier to compare different systems in a common and simple framework;

 • describing the extent of the space of possible designs for a multi-robot system; in this way, during the development of a new system, it is possible to identify problems previously solved.

Different taxonomies are possible depending on the selection of the main requirements and the subset of systems studied (i.e. a team of mobile ground robots). In the following, several taxonomies that categorize the bulk of existing multi-robot systems are presented:

Yuta and Premvuti [180] The following classification for multiple autonomous robot systems is proposed:

> • the degenerate case – a single robot system;

> • there is common objective or mission to be performed, and the robots work toward their purpose together – a common objective system;

> • each robot has its own objective or mission, but there is some interference between robots during the execution of missions – an individual objective system.

In the case of a system with a single robot, the robot does not need to consider the idea of others or have any thoughts for a community. If there are some movable objects traveling in the robot's world, then these objects are merely obstacles, each with its own purpose.

Cao [178] This taxonomy, presented in [178] in the framework of cooperative mobile robotics, classifies different works taking into account the *cooperative behavior* of the system, which is a subclass of collective behavior that is characterized by cooperation. The main requirements in this survey are group architecture, resource conflicts, origin of cooperation, learning and geometric problems. Within group architecture, issues such as centralization/decentralization, differentiation and communication structures are considered. Resource conflict can arise due to physical objects, communication channels or space sharing. Spatial coordination problems have been traditionally solved by using multi-robot motion planning techniques, which take into account critical issues such as deadlock and collision avoidance. Within geometric problems, formation and marching problems are also considered.

Balch [181] There are many cases, especially in multi-robot systems, where task and reward should be treated separately in the framework of reinforcement learning. Autonomous agents embedded in their environment are not always able to accurately access their performance, and overall performance may also depend on other agents over which the learner has no direct control. Taxonomies of task and reward are presented, providing a framework for investigating the impact of differences in the performance metric and reward on multi-robot system performance. Taxonomies of multi-robot tasks are based on six requirements (time, criteria, subject of action, resource limits, group movement and platform capabilities), whereas five requirements are provided for rewards (source of reward, relation to performance, time, continuity and locality). In both cases, only certain values are valid for each requirement.

Ali [182] This research compares the performance of different classes of mobile behavior-based multi-agent telerobotics systems in relation to the kinds of tasks they are performing. The systems are compared in terms of safety, effectiveness and ease-of-use, for applications representing classes of tasks in a newly-developed taxonomy of mobile multiagent tasks. This taxonomy categorizes the tasks in terms of the relative motion of the agents. Four different task classifications from this taxonomy were studied.

Todt [183] This taxonomy is restricted to the framework of robot motion coordination and provides five requirements for the classification: coupled/decoupled coordination, coordination time, existence of coordination priorities, coordination cost evaluation and workspace representation. In addition to mobile robots, systems composed of several manipulators are considered. Several conclusions and trends are derived from the taxonomy. For example, the coordination problem in the time domain is further abstracted. The first work deals with the problem in the physical space, and then the formulation of the coordination problem moves to a more appropriate form. A simplification in the complexity of the methods and a trend towards decoupled methods is also identified.

Dudek [179] This classification can be applied to any multi-robot system and the *natural dimensions* form its basis. Those parameters are related to the properties of the group of robots as a collective. Seven main requirements are defined, covering several aspects of the system: the size of the collective, different parameters related to the communication system (bandwidth, range, topology, etc.), collective reconfigurability, processing capabilities of each element, and the composition of the group (in terms of hardware and software homogenity). Within each requirement, a bounded set of values is considered. By using this taxonomy, the authors classify some classical problems in multi-robot systems and several existing architectures. Furthermore, for certain tasks, formal proofs are provided to show higher performances for a collective when comparing to individual robots.

In Table 3.2, a summary of those multi-robot taxonomies is presented.

It is clear that many classifications are possible, with many different features, depending on the domain and the purpose of the system. However, there are several aspects that are common in many taxonomies:

• Differentiation between the robots in the system. A group of robots is defined as *homogenous* if the capabilities of the individual robots are identical and *heterogenous* otherwise. In general, heterogenity introduces complexity since task allocation becomes more difficult, and robots have a greater need to model other individuals in the group. In general, the realities of individual robot design, construction and experience will inevitably cause a multi-robot system to drift to heterogenity over time. This

Taxonomy	Domain	Number of requirements	Description
Yuta and Premvuti	Multi-robot	2	Derived from objectives and decision mechanisms.
Cao	Cooperative mobile robotics	5	Based on problems and solutions related to the cooperation mechanisms.
Balch	Tasks and rewards	6/5	Useful in systems with reinforcement learning.
Ali	Mobile multi-agent telerobotics systems	3	Based on the relative motion of the robots.
Todt	Multi-robot	5	Limited to multi-robot motion planning.
Dudek	Multi-robot	7	Based on characteristics of the group of robots.

Table 3.2. *Multi-robot taxonomies*

(a) HERO2 helicopter and Romeo 4R (University of Seville)

(b) the HERO2 helicopter landing in the AURIGA robot (University of Malaga)

Figure 3.4. *Aerial and ground robotic teams in the CROMAT project experiments (see description in Chapter 2)*

tendency has been recognized by experienced roboticists who have seen that several copies of the same model of robot can vary widely in capabilities due to the differences in sensor tuning, calibration, etc. This means that to employ robot teams effectively, it is important to understand diversity and predict how it will impact performance. In fact, differentiation can also be treated as an advantage in terms of complementarity between different components in order to complete a given task or mission.

• Centralized/decentralized approach in the design (see section 3.6.3.2).

• Allocation: in multi-robot systems, each robot is assumed to be able to perform tasks in response to tasks requests. The issue is to decide which robot should be endowed with each given task to be performed. This requires the capability to assess the advantage of providing a given robot with a given task. This is a difficult issue when the decision has to take into account the current individual plans of the robot as well as the tasks left to be assigned. Information needed to implement the optimal choice includes the models, each robot's plan and the current states of tasks to be executed for each robot.

• Communication between components (from different points of view). Some aspects are noted in section 3.6.3.3.

• Motion of the components in a common environment (see section 3.6.3.4).

It seems that these aspects are relevant and should be taken into account in the design and classification of a multi-robot system.

3.6.3. *Paradigms for coordination and cooperation*

Coordination is a process that arises within a system when given (either internal or external) resources are simultaneously required by several components of this system. In the case of a multi-robot system, there are two classic coordination issues to deal with:

• Temporal coordination: this can be achieved relying on robot synchronization. Several schemes to enable incremental negotiations related to possible time interval synchronization can be defined and implemented. As a result, a group of robots acknowledge a common time interval in which the synchronization should occur. Temporal coordination can be necessary in a wide spectrum of applications. For instance, in the case of event monitoring, several synchronized perceptions of the event are required.

• Spatial coordination: the sharing of space between the different robots to ensure that each robot will be able to perform its plan safely and coherently with respect to the plans of the other robots. Interaction models can be considered in order to reason about the interaction requirements within the joint tasks. Afterwards, during plan execution, collision avoidance can be safely achieved by applying several algorithms on the planned trajectories of robots in a neighborhood.

There are several explicit definitions of cooperation in robotics literature. Cooperation is defined as a "joint collaborative behavior that is directed toward some goal in which there is a common interest or reward". Furthermore, in [178] a definition for *cooperative behavior* can be found: given some task specified by a designer, a multiple-robot system displays *cooperative behavior* if, due to some underlying mechanism (i.e. the "mechanism of cooperation"), there is an increase in the total utility of the system.

The amount of research in the field of cooperative mobile robotics has grown substantially in recent years. This work can be broadly categorized into two groups [177]: swarm-type cooperation and intentional cooperation. The swarm-type approach to multi-robot cooperation deals with large numbers of homogenous robots. This approach is useful for non-time-critical applications involving numerous repetitions of the same activity over a relatively large area. The approach to cooperative control taken in these systems is derived from the fields of neurobiology, ethology, psychophysics and sociology. The second primary area of research in cooperative control deals with achieving intentional cooperation among a limited number of typically heterogenous robots performing several distinct tasks. In this type of cooperative system, the robots often have to deal with some sort of efficiency constraint that requires a more directed type of cooperation than is found in the swarm approach described above. Furthermore, this second type of mobile robotic mission usually requires several distinct tasks to be performed. These missions thus usually require a smaller number of possibly heterogenous mobile robots involved in more purposeful cooperation. Although individual robots in this approach are typically able to perform some useful task on their own, groups of such robots are often able to accomplish missions that no individual robot can accomplish on its own. Key issues in these systems include robustly determining which robot should perform which task (*task allocation*) so as to maximize the efficiency of the team and ensure the proper coordination among team members to allow them to successfully complete their mission.

3.6.3.1. *Paradigms in the architecture of multi-robot systems*

Robots individually have their own hardware/software structure, which is usually called robot architecture. Those architectures should take into account the particular characteristics of a multi-robot system, because a robot architecture designed for a single robot is not necessarily valid when this robot has to interact with other robots.

Only one of the taxonomies mentioned in section 3.6.2 ([178]) proposes a definition of a multi-robot system *group architecture*. It is an element which provides the infrastructure upon which collective behaviors are implemented, and determines the capabilities and limitations of the system. This is just a functional definition, so components of the architecture or integration aspects are not mentioned. In the literature, relatively little work has covered those aspects. Research in multi-robot systems has focused primarily on construction and validation of working systems, as opposed to more general analysis. As a result, we can find many architectures for multi-robot coordination, but relatively few formal models.

A list of representative working architectures developed during the last few years could include SWARM [184], ACTRESS [185], CEBOT [186], GOFER [187], ALLIANCE [188], MARTHA [189, 190] and COMETS among others.

3.6.3.2. *Centralized/decentralized architecture*

This specific section is devoted to this issue because the most fundamental decision that is made when defining a group architecture is whether the system is centralized or decentralized. A centralized decision configuration (with a minimal distributed supervision) is compatible (at least) and even complementary with a configuration endowed with fully distributed decision capabilities. Aspects related to the decision can be developed either within a central decisional component or between several distributed components (e.g. possibly the different robots within the system). However, several trade-offs should be considered regarding the decision:

• **Knowledge's scope and accessibility**: a preliminary requirement, to enable decisional aspects within a central component, is to permanently ensure the availability of (relevant) up-to-date knowledge within this central component. It requires the centralization of any decisional-related knowledge from any component of the system, and to have them updated permanently. However, assuming that this requirement can be fulfilled, it can perform more informed decisions, and hence manage the mission operations in a more efficient way.

Regarding distributed decision, the local scope of the available knowledge is a double-edged issue: as the only available knowledge is the knowledge related to the considered component (or close to the considered component), this knowledge is usually far more easy to access and to refresh. On the other hand, local and partial knowledge leads to decisions that may turn out to be incoherent regarding the whole system.

• **Computational power and scalability**: in a multi-robot system, the amount of data to process is quite huge: the processing of this knowledge in a centralized way obviously requires powerful computational means. Moreover, such a centralized computation reaches its limits when the number of robots increases: a centralized system cannot be scalable to any number of robots.

In contrast, a distributed approach within a multi-robot system can stay available when the number of robots increases, since the complexity remains bounded: each robot still only deals with a local partial knowledge of the system that leads us to manipulate local information.

The respective disadvantages of each approach can be mitigated by constraining or extending their framework. A centralized approach will be relevant if:

• the computational capabilities are compatible with the amount of information to process;

• the exchange of data meets both the requirements of speed (up-to-date data) and expressivity (quality of information enabling well-informed decision-making).

As a consequence, one way to help to satisfy these two points is to reduce the complexity of the data exchanged to a minimal level, still satisfying the decision-making but without overloading communications. This can be achieved by designing a task communication protocol meeting this minimal expressivity need, and then fitting this protocol to the particular relevant field of application.

On the other hand, a decentralized approach will be relevant if:

• the available knowledge within each distributed component is sufficient to perform coherent decisions;

• this required amount of knowledge does not endow the distributed components with the problems of a centralized system (in terms of computation power and communication bandwidth requirements).

One way to ensure that a minimal global coherence will be satisfied within the whole system is to enable communication between the robots of the system up to a level that will guarantee that the decision is globally coherent.

However, instead of definitely choosing one of these extreme configurations, an alternative possibility lies in hybrid solutions that may fit at best the requirements of a heterogenous system.

3.6.3.3. *Communication between components*

Network and data link layers in the communication of an autonomous robot team are similar to a MANET (Mobile Ad hoc NETwork). A MANET can be described as a peer-to-peer network which usually comprises tens to hundreds of meters, and aims to form and maintain a connected multi-hop network capable of transporting a large amount of data between nodes. The main goal for a MANET, like other conventional wireless networks, is to provide high QoS and high bandwidth efficiency when mobility exists. In contrast, the main goal in a sensor network is to extend the network lifetime by reducing as much as possible the energy consumption while providing the services that the WSN was designed for. As a consequence, a lower performance in other aspects of operation such as QoS and bandwidth usage is assumed.

Two examples of multi-hop routing algorithms for MANET are Ad Hoc On Demand Distance Vector (AODV) routing and Temporally Ordered Routing Algorithm (TORA). Both are examples of demand-driven systems that eliminate most of the overhead associated with table updating in high mobility scenarios. However, the energy cost during route set up (path discovery) is high, so they are not used in WSNs. Another algorithm, called Power-Aware Routing, finds the minimum metric path on two different power metrics: minimum energy per packet and minimum cost per packet. The first metric is intuitive and produces substantial energy-saving while the network retains full connectivity. However, performance degradation due to node/link failure is not accounted for. The minimum cost metric is obtained by weighting the energy consumption by the energy reserve on each node. It has the nice property of delaying failures by steering traffic away from low-energy nodes, but overhead for path maintenance could be high. As a conclusion, it is obvious that different protocols are needed for autonomous robotic teams and WSNs.

3.6.3.4. *Path planning for multiple robot systems*

As mentioned before, one of the main issues in multi-robot coordination is the *spatial coordination*. In this section, a formal statement of this problem is presented, and the main approaches to solve it are summarized.

Consider multiple robots that share the same world, $\mathcal{W}$. A path must be computed for each one that avoids collisions with obstacles and with other robots. Superscripts will be used in this section to denote different robots. The ith robot will be denoted by $\mathcal{A}^i$. Suppose there are m robots, $\mathcal{A}^1, \mathcal{A}^2, \ldots, \mathcal{A}^m$. Each robot, $\mathcal{A}^i$, has its associated configuration space, $\mathcal{C}^i$, and its initial and goal configurations, q^i_{init} and q^i_{goal}.

A state space can be defined that considers the configurations of all of the robots simultaneously,

$$X = \mathcal{C}^1 \times \mathcal{C}^2 \times \cdots \times \mathcal{C}^m \qquad (3.12)$$

A state $x \in X$ specifies all robot configurations, and may be expressed as $x = (q^1, q^2, \ldots, q^m)$. Let N denote the dimension of X, which is given by

$$\sum_{i=1}^{m} \dim(\mathcal{C}^i). \qquad (3.13)$$

There are two sources of obstacle regions in the state space: 1) *robot-obstacle* collisions, and 2) *robot-robot* collisions. For each i such that $1 \leq i \leq m$, the subset of X that corresponds to robot $\mathcal{A}^i$ in collision with the obstacle region, $\mathcal{O}$, is defined as

$$X^i_{obs} = \{x \in X \mid \mathcal{A}^i(q^i) \cap \mathcal{O} \neq \emptyset\} \qquad (3.14)$$

This models the robot-obstacle collisions.

For each pair, $\mathcal{A}^i$ and $\mathcal{A}^j$, of robots, the subset of X that corresponds to $\mathcal{A}^i$ in collision with $\mathcal{A}^j$ is given by

$$X^{ij}_{obs} = \{x \in X \mid \mathcal{A}^i(q^i) \cap \mathcal{A}^j(q^j) \neq \emptyset\} \qquad (3.15)$$

Both (3.14) and (3.15) will be combined in (3.17) to yield X_{obs} as it will be described now.

Formulation (Multiple-Robot Motion Planning)
1. There are m robots, $\mathcal{A}^1, \ldots, \mathcal{A}^m$, which may consist of one or more moving bodies.

2. Each robot, $\mathcal{A}^i$, for $1 \leq i \leq m$ has an associated *configuration space*, $\mathcal{C}^i$.

3. The state space, X, is defined as the Cartesian product

$$X = \mathcal{C}^1 \times \mathcal{C}^2 \times \cdots \times \mathcal{C}^m \qquad (3.16)$$

The obstacle region in X is

$$X_{obs} = \left(\bigcup_{i=1}^{m} X_{obs}^i \right) \cup \left(\bigcup_{ij, i \neq j}^{m} X_{obs}^{ij} \right) \qquad (3.17)$$

in which X_{obs}^i and X_{obs}^{ij} are the robot-obstacle and robot-robot collision states from (3.14) and (3.15), respectively.

4. A state $x_I \in X_{free}$ is designated as the *initial state*, in which $x_I = (q_I^1, \ldots, q_I^m)$. For each i such that $1 \leq i \leq m$, q_I^i specifies the initial configuration of $\mathcal{A}^i$.

5. A subset $x_G \in X_{free}$ is designated as the *goal state*, in which $x_G = (q_G^1, \ldots, q_G^m)$.

6. The task is to compute a continuous path, $\tau : [0, 1] \rightarrow X_{free}$ such that $\tau(0) = x_{init}$ and $\tau(1) \in x_{goal}$.

X can be considered as an ordinary configuration space. Classical planning algorithms for a single robot with multiple bodies [191] may be applied without adaptation in the case of centralized planning (planning that takes into account all robots). The main concern, however, is that the dimension of X grows linearly with the number of robots. Complete algorithms require time that is exponential in dimension, which makes them unlikely candidates for such problems. Sampling-based algorithms are more likely to scale well in practice when there many robots, but the resulting dimension might still be too high.

The motions of the robots may be decoupled in many interesting ways. This leads to several interesting methods which first develop some kind of partial plan for the robots independently, and then consider the plan interactions to produce a solution. This idea is referred to as decoupled planning. In [191], two approaches are given: (i) *prioritized planning* considers one robot at a time according to a global priority, while (ii) the *path coordination method* essentially plans paths by scheduling the configuration space-time resource.

3.6.4. *Robots using WSNs*

A robot can use a sensor network to expand its capabilities, for example sensing at inaccessible locations using the information from the network. In this case, the sensors of the robot are distributed in the environment and are not centralized onboard the robot itself. Therefore, the robot has multiple inputs from the same event, so its reactivity and performance is improved. Furthermore, a sensor network can be useful in the robot localization, navigation and tracking.

A sensor network can also be used for the guidance of robots, and as has been explained previously (see section 3.5.3), the same algorithms can be applied for other mobile nodes (such as people) with a suitable interface. This guidance can be based on the following principles:

- follow the movement of the source to be sensed;
- extract motion direction from the sensor network;
    - safe path from the type of danger detected by sensors (temperature, contaminants, etc.),
    - path to improve sensing of the source (source localization).

In particular, gradient-based methods can be applied for local reactive navigation. Two different situations can be considered:

- Motion to increase the gradient (increasing perception of the event). In this case, the objective is to detect and sense a source, such as pollution.

- Motion to decrease the gradient (decreasing perception of the event). This case deals with a situation in which a escape path is needed.

Two examples of the use of a sensor network by a robot can be found in [163, 192]. In the first paper, a robot without a GPS or a map can follow a path and reach a goal position, merely using the information received from the sensor network. In the second paper, an autonomous UAV follows a path determined by the sensor network. Furthermore, the sensor network can change the path dynamically according to several environment events.

3.6.5. *Algorithms for navigation of autonomous robots using WSNs*

The navigation of autonomous robots has been studied from a centralized approach: the autonomous vehicle has different sensors (GPS, compass, ultrasonics, etc.) and can localize itself and follow a path using the information provided by those sensors. However, the use of sensor networks by robots opens a new research area which considers the navigation of autonomous robots using distributed information. In this way, navigation is possible even without the robot carrying any sensors, exclusively relying on communication with the WSN.

3.6.5.1. *Potential field guiding algorithm*

A moving object, such as a robot or a person with a suitable interface, is guided across the network along a safe path, away from the type of danger that can be detected by the sensors [193]. Each sensor can sense the presence or absence of such types of danger. A danger configuration protocol runs across all the nodes of the network generating a danger map. In this map, the dangerous areas detected by the sensor network are represented as obstacles. Those obstacles will have negative values and the destination will have a positive value according to some metric.

The danger map is generated by a potential field protocol that works as follows [193]:

• Each node whose sensor triggers "danger" diffuses the information about the danger to its neighbors in a message that includes its source node id, the potential value and the number of hops from the source of the message to the current node.

• When a node receives multiple messages from the same source node, it keeps only the message with the smallest number of hops (the message with the least hops is kept because that message is likely to travel along the shortest path).

• The current node computes the new potential value from the source node. The node then broadcasts a message with its potential value and number of hops to its neighbors.

• After this configuration procedure, nodes may have several potentials from multiple sources. To compute its current danger level information, each node adds all the potentials.

On the other hand, the potential field information stored at each node can be used to guide an object equipped with a sensor that can talk to the network in an online fashion. The following algorithm [193] can compute the safest path to the destination:

• The goal node broadcasts a message with the danger degree of the path, which is zero for the goal.

• When a sensor node receives a message, it adds its own potential value to the potential value provided in the message, and broadcasts a message updated with this new potential to its neighbors.

• If the node receives multiple messages, it selects the message with the smallest potential (corresponding to the least danger) and records the sender of the message.

Finally, there is another algorithm [193] with the navigation guiding protocol. In this algorithm, the user asks the network where to go next. The neighboring nodes reply with their current values. The user's sensor chooses the best possibility from the returned values. Note that this algorithm requires the "integrated" potential computed by the first two algorithms in order to avoid getting stuck in local minima.

3.6.5.2. *Path computation and following algorithm*

When considering a sensor network deployed with localized nodes, the algorithm presented in [163] can compute the nodes of the sensor network that are within a certain *pathwidth* distance from a path defined by a list of coordinates provided in a broadcast message. Nodes on the path will store the path segment, will rebroadcast the path message and will be activated for robot guidance. The rest of the nodes use the knowledge of their location and the location of the sender (contained in the message) to determine if they are close enough to the direction vector pointing to where the path starts. If they are, they forward the message, otherwise they remain silent.

It is interesting to note that multiple paths can be computed, stored and updated by the network to match multiple robots and destinations. Furthermore, a map computation algorithm could be implemented, where the map could be constructed incrementally and adaptively such as an artificial potential field (see section 3.6.5.1) using hop-by-hop communication. The "obstacles" could correspond to events and will have negative values whereas the destination will have a positive value. Finally, joining the two algorithms, a distributed motion planning protocol could be computed by the sensor network where different path computation algorithms could be run as distributed protocols on top of the distributed map, updating the path dynamically according to different events in the environment.

On the other hand, this path stored in the sensor network can be used for the navigation of a robot (which has communication with the nodes of the sensor network). In the same way as the path message is propagated, the process has two phases: first, the robot has to reach the location where the path starts, and then the robot has to be guided along the path. The first phase contains the following steps:

1. One (or all) of the sensors aware that they are near to the start of the path send out a message that contains the location of the start of the path.

2. That message is forwarded throughout the sensor network.

3. The robot sends its location by a message in three different directions (120° dispersal angle).

4. The sensor node that receives the message from the robot knows the start location, the robot location and the direction which the message came from. Using this information, the node can send a directional message to the robot that directs it to the start point.

After the initialization phase which places the robot on the path, the navigation guidance algorithm is used to control the motion direction of the robot. This algorithm can be summarized as follows [163]:

1. The robot starts by sending out a `QueryOnPath` message which includes the sender's identification and location.

2. If it is received by a sensor on the path, this sensor replies with a `QueryAck` message which includes the path section, some consecutive way points, and an indication of where these way points fit into the path.

3. By gathering lists of segments from multiple sensors, the entire path can be assembled piece by piece as the robot moves following the way points in order.

Figure 3.5. *Autonomous helicopter carrying a sensor node communicating with ground sensor nodes (University of Seville)*

3.6.5.3. *Probabilistic navigation*

With this algorithm [192], assuming neither a map nor a GPS are available, the robot can navigate through the environment from point A to point B communicating with a sensor network.

The algorithm has two stages:

• *Planning*: when the navigation goal is specified (either the robot requests to be guided to a certain place, or a sensor node requires the robot's assistance), the node that is closest to the destination triggers the navigation field computation. During this computation, every node probabilistically determines the optimal direction in which the robot should move when in its vicinity. The computed optimal directions of all nodes in conjunction compose the navigation field. The navigation field provides the robot with the "best possible" direction that has to be followed in order to reach the destination. The navigation field is computed based on a value iteration algorithm which considers the deployed sensor network as a graph, where the sensor nodes are vertices. Assume that a finite set of vertices S in the deployed network graph and a finite set of actions A the robot can take at each node. Given a subset of actions $A(s) \subseteq A$, for every two vertices $s, s' \in S$ in the deployed network graph, and an action $a \in A(s)$, the transition probabilities $P(s' \mid s, a)$ (probability of arriving at vertex s' given that the robot started at vertex s and commanded an action a) for all vertices are determined. The general idea behind the value iteration is to compute the utilities for every state and then pick the actions that yield a path towards the destination with maximum expected utility. The utility is incrementally computed [192]:

$$U_{t+1}(s) = C(s,a) + \max_{a \in A(s)} \sum_{s' \in S-s} P(s' \mid s,a) \times U_t(s') \qquad (3.18)$$

where $C(s,a)$ is the cost associated with moving to the next vertex. Initially, the utility of the destination state is set to 1 (0 for the other states). Given the utilities, an action policy for every state s will be as follows [192]:

$$\pi(s) = \arg \max_{a \in A(s)} \sum_{s' \in S-s} P(s' \mid s,a) \times U(s') \qquad (3.19)$$

Finally, the robot maintains a probabilistic transition model for the deployed network graph, and can compute the action policy at each node for any destination point.

However, a more attractive solution is to compute the action policy distributively in the deployed network. The idea is that every node in the network updates its utility and computes the optimal navigation action (for a robot in its vicinity) on its own. When the navigation destination is determined (either a robot requiring to be guided to a certain node, or a node requiring a robot's assistance), the node that is closest to the destination triggers the computation by injecting a *Start Computation* packet into the network containing its *id*. Every node redirects this packet to its neighbors using flooding, and updates the utilities according to equation (3.18). After the utilities are computed, every node computes an optimal policy for itself according to equation (3.19). Neighboring nodes are queried once again for the final utility values. The computed optimal action is stored at each node and is sent as part of a suggestion packet that the robot would receive if it is in the vicinity of the node.

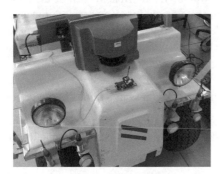

(a) Romeo 4R with a node

(b) the Romeo navigation interacting with the nodes

Figure 3.6. *Navigation of Romeo 4R interacting with the wireless sensor nodes (University of Seville)*

- *Navigation*: the algorithm explained above allows the robot to navigate through the environment between any two nodes of the deployed network. Initially, the current node is set to the node closest to the robot. Using the algorithm above, the node suggests a direction to the robot, and the robot takes that direction. Afterwards, using the signal strength, the robot can know that its closest node has changed, so it takes the new direction as the direction suggested by the new closest node. In this way, the robot is able to navigate without using neither a map nor a GPS.

Figures 3.5 and 3.6 shows the navigation of an autonomous helicopter with a sensor node onboard (Figure 3.5) and a ground autonomous vehicle (Figure 3.6) also with an onboard node interacting with the static nodes deployed in the environment.

3.6.6. Critical issues and future trends

One of the main reasons to develop a multi-robot system may be the fact that it is possible to build more robust and reliable systems by combining unreliable but redundant components. As has been mentioned before, research in multi-robot systems has focused primarily on the construction and validation of working systems. Therefore, in other works there are few general analyses of problems and solutions. Furthermore, regarding the fundamental decision between a centralized or a decentralized multi-robot architecture, we are not aware of any published theoretical comparison. Such a comparison would be of interest, particularly in scenarios where the team of robots is relatively small, and it is not clear whether the scaling properties of decentralization offsets the coordinative advantage of centralized systems.

On the other hand, a combination of multi-robot systems with WSNs can improve the overall reliability and performance of the whole system. Benefits can be identified in both directions: for example, a multi-robot system could provide network repairing services to a WSN, and a WSN could provide extended sensory capabilities to a multi-robot team. Unfortunately, the interaction between a WSN and a robotic team has not been addressed in the literature.

The following issues have been identified as relevant research trends in the future:

- Robotic teams networked in the environment: integration with WSNs and other cooperating objects should be studied in order to exploit many potential benefits and complementarities.

- Multi-robot planning with reliability constraints has not been properly solved, especially in a distributed way.

- Centralized/decentralized architecture trade-off: a set of rules should be provided to allow optimal design decisions.

- Hard real-time interaction with the environment (transportation, etc.).

- Development of a set of metrics for the performance of a robotic team.

- Increase heterogenity in the robotic team: complementarities, synergies, etc. can be exploited, but heterogenity implies complexity in the design and the algorithms.

- More experimental field studies are definitely needed.

3.7. Inter-vehicle communication networks

Urban areas will have dense network coverage in the near future with a large number of deployed COs to sense and act in the environment. Various vehicle and traffic management applications including road safety are expected to be supported by the major technological penetration of Intelligent Traffic Systems (ITS).

Cooperation among vehicles is a promising approach to address critical road safety and efficiency. For instance, coordinated collision avoidance systems could significantly reduce road accidents. Road safety becomes rather problematic in countries which heavily rely on the road networks for the transportation of goods.

Various research initiatives aim at developing and applying advanced technological systems to increase highway capacity and safety, and to reduce road congestion and air pollution. The PATH project is a US collaboration between the California state government, the University of California, and private industry [194] who have been working towards this objective. Projects in Europe [195, 196, 197] and Japan [198] have similar long-term research goals.

Despite the fact that promising results have been achieved in some of these projects, major issues remain to be resolved. This might explain the large list of proposed traffic applications with no real-life deployment.

This section intends to outline the other works and discuss the issues in the area of inter-vehicle and road-vehicle communication.

3.7.1. Road-vehicle communication (RVC)

This system supports data exchange between vehicles and devices deployed along the roadside. The simplest form of RVC is to sense the vehicle flow at a junction and forward the raw traffic data directly (or through third vehicles) to a remote unit for human assisted data processing. A more intrusive but accurate approach is to place sensors in the vehicle [199]. In this case, critical fine-grained information such as the average/instantaneous speed and flow statistics could be computed. Recommendations to the drivers, including countermeasures, would be transmitted to vehicles on the road via a tunable radio station or wireless network. In addition, the driver could receive important information (e.g. speed limits) about a road section.

This centralized approach provides more control over the data processing. However, it may introduce issues for applications that cannot tolerate high system processing time.

3.7.2. Inter-vehicle communication (IVC)

Inter-vehicle communication (IVC) enables vehicles to communicate with other vehicles without line-of-sight (LOS) using wireless communications. We classify the IVC approaches identified in the literature with respect to the infrastructure they are built upon. When the communication among and between vehicles relies on a deployed system infrastructure, say a cellular network, the IVC is termed *infrastructure-based*.

In contrast, vehicles may communicate where essentially no pre-planned infrastructure is present. They usually form a wireless MANET in order to deliver data among themselves. We call this an *infrastructure-less* or *ad hoc* network set-up.

IVC can be supported by any of these two approaches. RVC is often an infrastructure-based scheme as ground support including pre-deployed road sensors needs to be in place.

3.7.3. Communication scenario

IVC networks exhibit characteristics that are specific to scenarios of *high node mobility* [200]. We list some of these properties:

• Low power consumption and small physical size have been some of the fundamental issues in designing software and hardware for low-cost wireless sensor nodes. However, vehicles taking up large amounts of physical space and energy consumption is not an issue. Unlike typical WSN deployment scenarios, vehicles can be equipped with powerful sensors and a radio to achieve long transmission ranges and high quality sensor data.

• Contraints on mobility may impact the coverage of the network in cases where no ground infrastructure (e.g. base stations) to support the communication between vehicles is present. Adding more nodes to the network may not improve coverage but rather negatively affect the use of available bandwidth.

• Despite the constraints on the movement of vehicles (i.e. they must stay on the lanes), communication can be severely impaired by the high relative velocities of vehicles, even when moving in the same direction. In ad hoc short-range communication, the network will tend to experience very rapid changes in topology.

• The mobility pattern, driver behavior and vehicle responses are difficult to model in a simulator. Also, the current lack of information on these aspects poses a challenge in validating models created for real-life applications.

• Vehicle density varies from one area to another. Urban areas tend to have higher density than, for instance, remote rural areas.

• The adoption rate of IVC/RVC technology in vehicles is expected to be low in the near future.

Although these characteristics have some impact on infrastructure-based IVNs, the major issues arise from ad hoc network set-ups. Blum *et al.* [200] pointed out that IVC networks have mobility characteristics different from the typical MANETs considered in the literature. Specifically, these characteristics cause rapid topology changes, frequent fragmentation of the network, a small effective network diameter, and limited utility from network redundancy. Such issues must be tackled from the outset since they directly influence the IVC system design at all communication layers.

3.7.4. *IVN applications*

In this section, we introduce categories of IVC applications.

3.7.4.1. *Safety*

To make roads safer for passengers, data among vehicles can be exchanged to determine and avoid dangerous situations ahead of time. Proactive safety systems are at the top of the list of priorities in the current IVC research.

Cooperative assistance systems coordinate vehicles at critical points such as in highway entries, roundabouts and crossings without traffic lights. In the case of a traffic jam, event messages could be broadcasted to vehicles following behind just in time for a turn onto the next exit (alternative path discovery).

Also, sophisticated vehicle onboard warning systems can disseminate real-time information about roadway and environmental hazards to near-by vehicles. The driver's behavior (steering, braking and acceleration) could be determined by inspecting the controller area networks (CAN) of the vehicle. This information may be associated with data exchanged through the IVC network in order to give early warning of dangerous driving situations [201].

Such information includes weather and road surface conditions (e.g. slippery, icy), accidents events immediately detected by sensing the air bags ignition. Zones of relevance for such events could be defined in order to make a focused and efficient hazard warning data dissemination [202].

Also, it would be useful for vehicles to receive information about cars preceding the immediate vehicle. For instance, consider a lorry driving in front of a vehicle on a two-way road. It would help if the visual field of the lorry driver could be "passed" to the driver following behind. The lorry could send real-time video images from a camera installed in its front side. This would extend the visibility of the vehicle driver and therefore could potentially avoid frontal crashes in take-overs. This is even more relevant when the weather conditions affect the driver's visibility.

Intelligent cruise control is another safety and comfort application area. A common scenario is when the vehicle attempts to automatically keep a safe distance margin from the vehicle ahead. The distance can be measured using an accurate range finder system (e.g. milimeter radar or infrared laser). Real-time adaptation algorithms would use IVC in order to properly control the speed and acceleration of vehicles.

In addition, the vehicle can use an IVC network to exchange messages between the vehicle ahead and vehicles in adjacent lanes to perform control maneuvers such as smoothly braking when the adjacent vehicle changes lane. This application scenario is discussed in [203].

The safety applications discussed in the literature rarely include other parties that may be involved in road accidents such as motorcyclists and pedestrians. For instance, speed limit control based on the location-awareness of vehicles requires some local knowledge of speed constraints and the density of the pedestrians in the section a vehicle is entering. COs deployed on the roadside could communicate this information to the vehicle.

3.7.4.2. Traffic management

A route finder system avoids an otherwise convenient-looking route in favor of another that is less congested. Cooperating vehicles could monitor areas in order to identify traffic congestion ahead of time by communicating dynamic traffic flow information such as the density of nearby vehicles.

The envisaged system would give enough time for vehicles to opt for alternative routes and therefore avoid traffic jams. This real-time congestion information would be frequently forwarded from one vehicle to another as long as there is still traffic congestion.

In addition, efficient traffic flow can be achieved with platooning, which is a technique that arranges two or more vehicles at a regular distance. Inside a platoon all the vehicles follow the leader with a small intra-platoon separation of usually 1 meter. The inter-platoon spacing is assumed to be large so as to isolate the platoons from each other [204]. The available capacity of the road network is expected to increase as the overall vehicle headway is significantly reduced.

GPS-based mapping and guidance are other areas of application. Current GPS-based navigational systems use limited information on the current traffic conditions since real-time data collection through satellite is currently expensive to deploy on a large scale. This can be achieved with information exchange among vehicles through wireless communications.

3.7.4.3. *Environmental protection*

The Kyoto Protocol was established to set specific targets for reductions in greenhouse gas emissions [205]. Emission restriction targets were made in some countries, including Member States of the EU.

Appropriate means for monitoring and reducing gas emissions will be required in the near future. To support this, vehicle-based sensor networks may be deployed where vehicles not only gather dynamic traffic and in-vehicle information but also sense the air quality around the car (nitric oxide, carbon monoxide, etc.).

The sentient car project [206] proposed a sensor-based vehicle to collect real-time information about the car's performance and its surrounding air quality, and overlays this data on a map of the area the vehicle is visiting. In this application, multiple cars locally sense levels of pollution (nitric oxide, carbon monoxide, noise level) measured with sophisticated sensing equipment placed in the tailpipe. The sensor data can be communicated to roadside devices or forwarded to a remote processing unit through a hybrid wireless infrastructure, such as a cellular network and IEEE 802.11.

3.7.4.4. *Traffic and vehicle information for billing*

Congestion-based charging has gained interest over the past years as a means to mitigate traffic congestion in urban areas. There are different models in use today, ranging from area to time-based charging. Central London and Singapore are two cities which have implemented this charging scheme.

Other approaches discussed in the research literature rely on dynamic traffic information in order to implement a dynamic congestion-based charging scheme. The charge to use a particular route varies over time according to the levels of measured congestion. This scheme has been regarded as more efficient than time or area-based charging.

To put this dynamic charging scheme into practice, an efficient distributed traffic monitoring system needs to be in place. IVC networks could assist vehicles in cooperatively detecting congested areas. Such a charging scheme could offer choices of routes to vehicles entering a congested region. The driver could opt, for instance, between inexpensive but longer alternative routes or pay more to use the current congested route.

A slightly related application is the dynamic insurance policy as pointed out in [201]. The premium should change to indicate the vehicle usage and driver's behavior on the road. Information that could be taken into account include the common routes selected, time of day, and acceleration patterns.

3.7.4.5. *Data communication using delay-tolerant networks*

The problem of providing data communications to remote and rural areas is discussed in [207]. The authors consider the approach of asynchronous messaging in order to greatly reduce the cost of connectivity.

The Wizzy Digital Courier service provides asynchronous Internet access to schools in remote areas of South Africa [208]. A courier on a motorbike or bicycle, equipped with a USB flash storage device, travels from a village school to a city carrying all the outbound email and web requests for the day. The courier may forward the data collected from distant schools to another courier using wireless communication.

A similar project in Lapland [209] aims at providing intermittent Internet connectivity to the Saami population, who live in widely dispersed communities in remote areas, and are not well served by either wired, fixed wireless or satellite Internet service. As they travel on snow vehicles from community to community, the data can be stored and forwarded through opportunistic communication between devices on these vehicles supplemented by a few solar-powered base-stations positioned on tracks in the wilderness [210].

Because the data communication is asynchronous and it relies on mobile routers to collect, forward and deliver messages between static nodes (sensor networks, and central servers) [211], this type of application is best supported by a delay-tolerant network framework (DTN) [212]. Such networks are assumed to experience frequent, long duration partitioning and may never have an end-to-end path.

This is a niche scenario since few applications will tolerate high delay for web access and other instant-based services.

3.7.4.6. *Added-value services*

We observe that there is a potential for using vehicle-to-vehicle communication to leverage a class of opportunistic communication services that, although non-safety critical, could be realized with an IVC network. We include in this list services such as data look-up for file sharing among moving vehicles.

The IVC/RVC would be used as an ad hoc distributed storage mechanism which would be accessed by the in-vehicle entertainment system. Gerla *et al.* [213] put forward a proposal for a file sharing system between cars based on the bit torrent peer-to-peer system. To be legally acceptable, however, such an application would require appropriate incentives to avoid the unauthorized distribution of copyrighted material.

Interactive communication among vehicles is another type of service that drivers and passengers may find useful. This service could establish a voice connection to other vehicles driving in the same direction for the purposes of traffic information exchange. Interactive applications including instant messaging, multiplayer games and onboard Internet access would make the journey smoother for families driving with children.

3.7.4.7. *Important aspects*

We consider in this section a non-exhaustive list of design issues that should be taken into account when engineering an IVC network:

• Time constraints: establish the application's relative tolerance to overall delay including network and processing delays.

• Reliability: the application's tolerance to errors made visible to it. A zero tolerance threshold indicates that the application requires a 100% guaranteed data delivery.

• Scale: refers to the number of destination vehicles intended to receive a particular data item.

• Levels of infrastructure: indicates to what extent a pre-planned infrastructure is required to support the deployment of the application. In ad hoc network set-ups, no infrastructure is necessary and vehicle devices self-organize into an IVN.

• Security and privacy: some applications require authentication schemes for communication and degrees of anonymity of vehicle and driver information.

According to these aspects, an IVC service may experience severe network delays and still be reliable. We envisage, however, that a significant number of services including the dissemination of warning messages require low delay and highly guaranteed data delivery. Such requirements pose interesting design challenges in pure ad hoc inter-vehicle communications. An open question here is whether any level of infrastructure will be necessary to support reliable and timely IVC services.

In Table 3.3 we intersect each application category with the design issues discussed above.

These applications have been discussed in the research literature along with proposed schemes for their implementation either using infrastructure or ad hoc network set-ups. The following sections present an overview of these schemes and issues.

	Safety	Environmental protection	Charging	Delay-tolerant	Added-value services
Time con-straints	Real-time (bounded low latency)	High tolerance to delay	May tolerate high communication delay	Asynchronous communication, may tolerate extreme delays – e.g. hours or days.	Interactive systems may not tolerate high delay
Reliability	High	Medium – tolerate loss of data (e.g. multiple sensor sources in a region)	High – e.g. data communication for "pay as you drive" insurance must be reliable and accurate	High reliable communication. Similar to the email system – guaranteed delivery and variable delay	Medium-high. For instance, voice calls between vehicles require reliable communication with guaranteed bounds on data loss.
Scale	Source of event towards multiple vehicles	Multiple sources to individual destinations	Multiple sources to multiple vehicles	Single source-single destination	Single source-multiple destination; single source-single destination
Security	Source authentication	Source authentication	Source authentication, high degree of privacy	Authentication and privacy	Authentication and privacy

Table 3.3. *Important aspects to IVC applications*

Our approach is to break down the issues at various levels of system and communication protocol design. The lowest level we discuss is the MAC layer in the next section. We then describe unicast and multicast routing issues and approaches.

3.7.5. MAC layer

This communication layer should strive to maximize the packet throughput of the IVC/RVC network by minimizing the delay and packet loss rates. Thus, congestion control becomes a key design issue for MAC layer protocols.

The scale of the IVC network can significantly complicate the engineering and validation of a suitable MAC protocol. We expect that on a busy highway, thousands of cars travel within a road segment. This does not necessarily mean that thousands of nodes will communicate. However, hundreds may require some form of IVC.

Luo *et al.* [214] suggest that there are two general approaches for designing MAC layers for IVC networks. One scheme uses the MAC functionality of existing wireless LAN systems including the medium-range radio IEEE 802.11. In contrast, the other approach extends the MAC layer of 3G cellular radio systems. The next sections discuss these two approaches.

3.7.5.1. *Wireless LAN*

Usually, WLAN MAC protocols have limited support for ad hoc distributed coordination to the medium access. The 802.11a is the MAC layer protocol chosen as the basis of the Dedicated Short-Range Communication (DSRC) standards, which has been allocated 75MHz (5.85–5.925 GHz) of the spectrum in the USA by the FCC for any type of vehicular communication.

The motivation for this technical choice is to take advantage of the distributed coordination in an ad hoc mode that is currently built-in in the PHY and MAC standards specifications of these systems [215].

The difficulty arises from the high mobility characteristic of an IVC/RVC scenario which significantly increases the probability of network partioning. Thus, a MAC mechanism that explicitly allocates resources (timeslots, frequency spectrum or codes) introduces a major problem [200]. Contentions created on congested roads, for instance, may require dynamic allocation and deallocation of codes in order to optimize the network transmission throughput. Such a system overhead could potentially add an extra delay to the system. This is acceptable for non-safety applications but it is likely to be problematic for real-time data service delivery of safety-critical messages.

3.7.5.2. *Cellular networks*

An alternative to the previous approach is the use of an unmodified MAC layer of current proposed cellular networks. This scheme relies on aggregation points (cells) to forward the data packets from one vehicle to another. The advantage of this is the availability of pre-deployed cellular network infrastructure. To what extent, however, the system delay (network and processing) and the limited transmission rate would impact on the applications remains to be investigated.

Today, reasonable cellular network coverage may have a positive aspect when providing reliable and timely IVC service. The problem for some applications would be the limited transmission data rate of a 3G cellular network system which is up to 144 Kbps for users in high-speed motor vehicles.

Another approach extends 3G cellular systems in order to add functionality for decentralized medium access [197]. This scheme has potential for a better control of the radio resource because of the Code Division Multiple Access (CDMA) subsystem. Consequently, this gives flexibility to adjust the data transmission rate when compared with WLAN extensions. The design challenge is how the decentralized ad hoc coordination could be engineered.

The discussion of which approach should be used in IVC networks is tightly coupled with the question of whether to have an infrastructure-based or an ad hoc system deployment. The extended WLAN approach would certainly be more suitable for ad hoc than the 3G cellular network.

However, an infrastructure-based 3G cellular system can offer the appropriate functionality to design the MAC layer for an ad hoc vehicular network. The ad hoc network could be *overlayed* on the 3G network.

A hybrid of these two approaches is being explored in some research projects [206]. In the sentient car, the researchers equipped an experimental vehicle, a Ford Transit, with two dashboard LCD displays, one for the driver and a larger one for the navigator. A PC placed in the back of the van integrating all the sensing and communication capabilities. A GSM link (9.6 Kb/s) provided a low-speed data connection but with a good network coverage. The 802.11b network interface provided much higher data rate but at the expense of very limited coverage.

In hybrid approaches, an issue that arises is how the system seamlessly handles the transition from one type of wireless medium to another (vertical handover). This may introduce packet loss and delay which can severely damage the data communication. To minimize these adverse effects, network handover approaches have been discussed in the literature.

Although handover for homogenous wireless networks is a well-understood research topic, there are ongoing research efforts in developing mechanisms for hetegenous networks. An approach is to leave the handover decisions to the network. The mobile node reports the received signal strength from various base stations to the network which then decides when to switch the node to another base station [216]. Usually, buffering schemes can be used to mitigate the effects of handovers.

The problem with network-controlled handover is the lack of information on the current status of the mobile node including factors such as the applications running, processor load, physical context and so forth. This complicates the process of deciding precisely the appropriate time to handoff. Recently, practical results obtained from an IPv6-based testbed composed of GPRS, WLAN (802.11b) and LAN network systems [217, 218] show that the major issue in vertical handovers is the delay. The average handover delay of a WLAN to GPRS TCP connection is 3.8 sec in upward handovers (maximum of 4.4 sec) and 6.8 sec in downward handovers (maximum of 8.8 sec).

Network coverage can be significantly improved with a hybrid IVC network (infrastructured and ad hoc). However, these results suggest that vertical handover issues in heterogenous cellular-based IVC systems should be addressed in order to offer reliable communication services to the applications.

3.7.5.3. *Approaches*

The issue of designing a MAC layer for ad hoc IVC networks has been addressed in [219]. The authors discuss an extension to the reservation ALOHA to efficiently deal with distributed slot reservation. Vehicles rely on their neighbors to determine if their request for a slot has succeeded. As pointed out in [200], the high mobility of vehicles results in varying sets of neighbors. It is unclear, therefore, whether the proposed distributed reservation scheme can efficiently deal with high mobile nodes by keeping low the number of packet collisions in the network.

There are other proposals based on traditional LAN technologies such as the non- or p-persistent CSMA used by DOLPHIN [220]. The contribution of this work is to show that the non-persistent CSMA outperforms the p-persistent scheme regarding packet loss in those cases usually involved in IVC. As a result, the non-persistent CSMA is adopted as the IVC protocol of the DEMO 2000 cooperative driving application [198].

FleetNet Design

The FleetNet project have chosen the UMTS Terrestrial Radio Access Time Division Duplex (UTRA TDD) [202, 221] as the MAC layer. The original specification of this layer relied on a centralized coordination scheme between mobile nodes. Such a design choice may introduce a problem when extensive ad hoc communication capabilities are needed. For instance, in cooperative driver assistance, high speed vehicles may only be required to communicate with vehicles in the closest vicinity and usually for a short period of time.

To overcome the centralized control, several changes to the UTRA TDD layer have been proposed. The structure of the protocol framing was excluded from the original specifications. The frame duration is 10ms and 14 slots are available in each frame in the aspired Low Chip-Rate mode (LCR).

The MAC is organized in a decentralized manner where each station is individually responsible for managing the available resources. The Reservation ALOHA (R-ALOHA) scheme is used to coordinate the distributed access to the medium. Reserved slots are used in subsequent frames as long as there are packets to be transmitted. The reservation of the next slot is indicated by piggy-back signaling.

To offer high priority services, an adjustable portion of the transmission capacity in terms of number of slots per frame can be constantly reserved. The remaining part can be dynamically assigned and temporarily reserved by different stations for services with lower priority. In order to avoid the problem of near-far-probe and to keep the power control scheme simple, the MAC protocol states that only one station transmits in a slot at a time.

The simulation model assumed in this work that all nodes were within the same radio range and the interference was measured as a binary value of "yes or no". The physical layer was entirely based on the UTRA-TDD PHY layer. The simulated MAC layer assumed that nodes have at least one reserved time-slot. In practice, this implies that they need to establish a circuit-switched broadcast connection which they never quit. The maintenance of a long-term connection is an issue.

An important result obtained is that as long as the traffic load is below saturation, the mean delay is almost constant. For messages of small length, the constant delay was 25 ms. Although this was constant, it is reasonably high for the direct communication model of the simulation.

A great weakness of this protocol validation is the simulation model used. First, only direct communication among nodes within the same radio range has been considered. Secondly, there is no discussion on the mobility model simulated. Thirdly, the authors intend to include detailed channel models with respect to path loss and interference conditions in future simulations. Thus, it is unclear whether the presented results can be representative of a real application scenario.

WRTP

The Wireless Token Ring Protocol (WTRP) [222] addresses the distributed coordination issue by proposing the design of the MAC layer protocol based on the IEEE 802.4 standards. This protocol is commonly referred to as the token ring and it was extensively used in the late 1980s and early 1990s.

In particular, WTRP addresses the single points of failure of centralized coordination mechanisms in order to support dynamic changing topologies, where nodes can be partially disconnected from other nodes. The authors made a case for using a token ring protocol for platooning applications because of the high spatial reuse that can be achieved.

A connectivity manager component in each node maintains a connectivity table that is an ordered list of stations in the ring. This table scales linearly with the length of the ring. Similarly to the original token ring protocol, the WTRP protocol offers a ring recovery mechanism that is triggered when the monitoring node decides that its successor is unreachable.

In such a case, the station recovers from the failure by reforming the ring. When looking up in the connectivity table, a node can find the next connected node in the transmission order to send a SET-PREDECESSOR message. This mechanism allows nodes to dynamically join and leave the ring.

It is unclear, however, how the system copes with frequent joins/leaves, in which case the ring needs to be frequently recreated. This process may render the system a high maintenance overhead and *livelock* where the communication overhead (control token messages) is so high that the system cannot make any useful progress, i.e. communicate data among stations.

The results presented cannot be taken as representative of a realistic IVC network scenario. Only four nodes were used in the simulation model where far more moving nodes would be expected to simulated increased topology changes with announced joins and leaves.

Also, the protocol is implemented on top of an 802.11/DCF mode. Thus, it is difficult to assimilate the real performance of this protocol when compared to the 802.11 based approaches. However, a token ring is more robust to collisions compared to the 802.11. The cost to pay is the maintenance of the ring.

MAC for DSRC

Qing Xu *et al.* [223] argue that MAC protocols that rely on centralized schemes for dynamic allocation of resources such as TDMA (time slots), FDMA (channels) or CDMA (codes) are difficult to use in highly dynamic IVC/RVC networks. Also, the authors rule out protocols that use synchronous control communication schemes such as the RTS and CTS schemes.

This paper discusses the design of various random access protocols for IVC and RVC in the DSRC. The design is optimized for safety applications, which require strict quality of service guarantees. The communication should be reliable and with few delays.

The authors pointed out that the distributed coordinated function (DCF) for MAC protocols cannot provide QoS required by safety applications. For instance, the enhanced DCF protocol addresses QoS by packet prioritization. The performance, however, degrades when the number of packets of equal priority increases. This is the case in safety applications where differentiation of packets may not be suitable since it is difficult to establish the priorities among different types of safety messages.

In this scenario, receivers are specified as a geographic region (geo-cast zone) relative to the transmitter. The sender broadcasts messages to all the receivers in its communication range. The receiver applies a multicast filter to determine whether its is in the geo-cast zone of the message. Each message has an associated lifetime which is regarded as the usefulness of the message. In the case where the lifetime expires, more messages are transmitted if those sent could not be delivered.

In this work, the following protocols have been studied:

• Asynchronous fixed repetition (AFR): the radio does not listen to the channel prior to communication and the protocol randomly selects k distinct slots out of the n available slots constituting the lifetime.

- Asynchronous p-persistent repetition (APR): the node transmits a packet in each of n available slots in a lifetime with probability $p = k/n$, while with probability $1 - (k/n)$, it delays the message transmission to the next time slot. In this case, the radio does not listen to the channel before it sends out a packet. The positive integer $k \leq n$ is a design parameter of the protocol.

- Synchronous fixed repetition (SFR): the same as the AFR protocol except that all nodes are synchronized to a global clock similar to the slotted ALOHA scheme. The generation and transmission of messages happen at the beginning of a slot. This partially avoids overlap between packets. It does require, however, a global clock.

- Synchronous p-persistent repetition (SPR): similar to the APR except for the global synchronization among the nodes of packet generation and transmission.

- Asynchronous fixed repetition with carrier sensing (AFR-CS): if a node with a new message to transmit senses that the channel is busy, the message is regarded as backlogged and the node will attempt the message transmission in the next empty slot. The node repeats the attempt until the channel becomes idle.

- Asynchronous p-persistent reception with carrier sensing (APR-CS): similar to the AFR-CS protocol with the repetition slots selected to follow the p-persistent scheme.

These protocols were studied analytically through simulation in NS2 (network simulator) and SHIFT (highway vehicle traffic simulator). Messages were generated based on events including on-board sensor measurements. The message is passed down to the MAC layer, which attempts to send a message only within the lifetime of the message.

The preliminary results presented validated their simulation. There is an optimal probability for p-persistence protocols when transmitting packets beyond this point increases the number of congestion in the channel. Also, the idea of sensing the channel before transmitting improves the performance as observed in this work, but introduces issues that have not been fully discussed such as synchronization of colliding stations.

The main conclusion is that AFR-CS showed the lowest probability of reception failure to be around 0.0008, less than one-tenth of the packet failure rate obtained for IEEE 802.11 under the same conditions. This indicates that repetitions with equal probability (fixed repetitions) help to overcome interference by giving a transmitter more chances to transmit. The synchronous and asynchronous versions of this protocol have shown the same benefit. However, it is more practical for the radios to sense the channel before transmitting than to synchronize the transmission of all nodes. This is the reason why the AFR-CS protocol was the preferred choice by these authors.

The MAC protocols discussed here have been evaluated in simulations with limited real traffic data and mobility models. This raises the question of their efficacy to address the communication needs at link layer of real IVC/RVC applications.

3.7.6. *Routing*

Application messages need to be transmitted from a source to a destination point. When these nodes are within the same radio range (closest vicinity), then the MAC layer protocol is sufficient to directly forward these messages.

In other cases, however, the radio device of a source node might not be sufficiently powerful to reach a far away node. To overcome this, we can extend the communication range by routing packets through third nodes. This creates a multi-hop path between any two points.

The source node transmits its packets to destination nodes through its neighboring nodes, which decide the next hop to forward the packets using static or dynamic routing policy information with local or global scopes. The scalability of the routing protocol depends on various factors including the average amount of status information each node must store about other nodes and the update frequency of the network status data. In large scale ad hoc deployments, the maintenance of global network status in each node introduces scalability issues. Ideally, the routing protocol should rely on local status information (neighboring nodes) in order to make global routing decisions.

3.7.6.1. *Traditional MANET protocols*

Traditional Internet routing protocols such as RIP and OSPF are both *proactive* routing protocols. Periodic broadcasts of network topology updates (e.g. distance vector or link state information) are used in order to compute the shortest path from the source to every destination. This operation consumes a reasonable amount of network bandwidth.

Although they are widely used in the Internet backbone, they cannot be directly used in MANET because of the differences between these two types of networks including the limited bandwidth in MANET (up to 54 Mbps with 802.11 g compared to gigabits in Internet backbones). Thus, the routing control overhead cannot be ignored in MANET. The network topology in MANET is dynamic, changing at rates that depend on how nodes move around. The topology in this case for proactive protocols needs to be updated at a higher frequency than a fixed network.

These are the main reasons why several proposed proactive MANET protocols stemmed from Internet routing protocols had major issues when dealing with the high mobility and limited bandwidth of MANETs. The issue to address is how we can decrease the amount of routing control traffic. Protocols in this category include the Destination-Sequenced Distance Vector routing protocol (DSDV) [36].

To keep the amount of routing control traffic low, reactive protocols have been proposed to notify the network on how packets should be treated by forwarding nodes in a multi-hop scenario. This may range from pre-defined routing paths that packets should follow to specific policies implemented in the forwarding nodes, for instance, to QoS-based packet scheduling. Typical MANET reactive protocols include the Ad hoc On Demand Distance Vector (AODV) and the Dynamic Source Routing (DSR). The reader is referred to [224] for further information on these protocols.

Reactive routing follows two steps. In the route discovery, the source node broadcasts a packet throughout the network to find the route between itself and the destination. In the route maintenance phase, the established routes are checked for validity since the nodes along the path are free to move arbitrarily. When any failure is found along the path, the source will be notified and it may decide to restart the route discovery process.

The high mobility characteristics of IVC networks create a challenge for any type of reactive and proactive protocols. The rapid topology changes and the high probability of disconnection in some parts of the network are issues that need to be addressed in the design of routing protocols. Blum *et al.* [200] showed using simulations that current MANET routing protocols fail to address these issues. Proactive protocols will be overwhelmed in maintaining fresh routing status with rapid topological changes. In contrast, reactive protocols attempt to discover routing paths before sending a message. However, the short lifetime of routing paths and the fragmentation of the networks make reactive routing in IVC/RVC networks extremely difficult.

3.7.6.2. *Location-based routing*

Communication messages will be likely to have an *area of relevance* which defines the subset of the traveling vehicles that would be interested in receiving such messages. In intelligent cruise control, for instance, the vehicle attempts to maintain a safe distance to the other vehicles in a highway. The geographical area of interest for message exchange can be geometrically represented as a circumference with the centre being the vehicle which generates events.

Location-based routing protocols have been proposed for IVC/RVC networks. Although some of these protocols were originally proposed to MANETs with low mobility, they were adapted to address the issues that arise in high mobility scenarios. In this mechanism, the location of the vehicles (e.g. GPS position) is used to route messages from a source to a destination point through neighboring nodes. An intermediate node upon receiving a message decides locally whether there are neighbors closest to the destination than the node itself. This strategy is called *greedy forwarding*.

This approach can fail when there is no neighbor available that is closer to the destination than the current forwarding node. To address this, several recovery mechanisms have been discussed in the literature including the Perimeter Mode in Greedy Perimeter Stateless Routing (GPSR) [225]. The perimeter mode constructs a planar graph from the local connectivity graph where the void space has been identified. The planar graph eliminates any redundant edge and packets are routed using the paths of this constructed graph using the *right hand rule*. This rule states that when arriving at node x from node y, the next edge traversed is the next one sequentially counter-clockwise about x from edge (x;y). The rule traverses an exterior region, in this case, the region outside the same triangle, in counter-clockwise edge order.

To address the deficiencies of the perimeter mode approach, Lochert *et al.* [226] proposed a strategy to deal with the high mobility of nodes that use information about the specific topological structure of a city. With this information, the routing can be assisted with data to decide ahead of time probable trajectories that vehicles may take. This has been shown to overcome the major problem of "void spaces".

The proposed position-based routing is called Geographic Source Routing (GSR) and uses maps of a small part of Berlin. The authors argue that the availability of maps is a valid assumption as vehicles are often equipped with onboard navigational systems.

Similar to other position-based routing protocols, the GSR strategy relies on position information in order to make forwarding decisions. One of the requirements for this protocol is the availability of a location service that can provide the current position of a node. To send a packet to a destination node, the sender needs the fresh position of the destination so that it can be appended to the message header. The GSR proposes a location service called reactive location service (RLS) to fulfill this requirement. Location queries are flooded on the network and the reply message will contain the current geographical position of a given vehicle.

By means of simulation, this research work compares the GSR approach with non-position-based ad hoc routing strategies such as the Dynamic Source Routing (DSR) and Ad Hoc On demand Distance Vector Routing (AODV) [224]. While the DSR's performance is severely affected by issues such as scalability and high mobility (e.g. short route lifetimes), both AODV and the GSR position-based approaches have good performance, with the position-based approach outperforming AODV.

The DSR protocol has the highest network load among all three protocols as it generates a large amount of signaling traffic. This protocol sends large packets due to the routes appended to the packet headers, especially during the route discovery phase, which leads to a significant network overload. The second cause of failure for DSR is that mobility of vehicles cause frequent route breaks. This is more severe in highway scenarios which have higher mobility than city scenarios.

On the other hand, the GSR position-based approach shows a slightly higher packet delivery rate when compared to DSR and AODV. The GSR and DSR show that there is an overhead for the first packet of a connection. As these two protocols are source-based routing, a route establishment for the DSR or the location discovery for the GSR are two similar processes that need to be carried out. However, AODV presents the highest delay as it uses an expanding ring search technique.

The authors recognize the ad hoc routing approach as a feasible mechanism compared to cellular network-based telematics. The benefits discussed are the low delay for data transport in emergency warning systems, the greater reliability of the ad hoc paradigm thanks to the grid structure of the network, and the low cost deriving from the use of unlicensed frequency bands. However, it remains to be investigated whether these benefits will happen in real-life deployment of this system. A review of position-based unicast protocols can be found in [227, 228].

3.7.7. *Multicast networking in the context of wireless inter-vehicle and road networks*

3.7.7.1. *Multicast addressing and delivery*

Many of the envisaged applications for vehicle-vehicle and road-vehicle communication require communication with destinations that are groups of vehicles. The group may be defined by:

• *geocast:* group membership defined by the current locations of the destination vehicles;

• *property-based:* group membership defined by other properties of the vehicles (e.g. vehicle type: private car, taxi, heavy truck, light truck, PSV);

• *explicit groups:* a group may consist of vehicles that have explicitly joined a named grouping, such as the subscribers to a commercial traffic alert system.

From the application point of view, the purpose of multicast communication is to address messages to a group of destination nodes without any need for the application to be concerned with the following:

• Group membership: the identities of the nodes in the group and dynamic changes in its membership.

• Reliable delivery: guaranteed delivery to all members of the group. Approaches to this include the use of negative acknowledgements and "gossip" protocols in which nodes keep neighboring nodes informed about the messages they have received.

• Ordered delivery: several messages sent to a group may arrive at different nodes in the group in a different order. If the messages are all from the same source, then the problem is called source ordering and can be achieved by the use of timestamps and delayed delivery of out-of-order messages. However, for messages from different source nodes whose clocks may not be perfectly synchronized, the problem is non-trivial and has at least two sub-classes: total ordering and causal ordering.

Not all applications require reliable or ordered delivery, but it seems likely that some vehicle-related applications will. For example, safety warnings, traffic control commands and speed limit information need to be delivered reliably. Traffic control commands must be received in the same order by all vehicles and speed limit end messages must be delivered after the corresponding begin message.

The above discussion is based on established distributed system techniques that can be found in several sources such as [229], but the standard techniques for reliable and ordered delivery often incur significant delivery delays and many vehicle-based applications are sensitive to delay.

3.7.7.2. Multicast routing

Single-hop

Since wireless networking is inherently a multiple-access technology, all messages sent are received by all the nodes in the same cell as the source node. However, multicast messages are delivered to the application layer only in nodes that are members of the destination group identified in the message header. This is achieved by filtering on the destination address in the network layer or in the network interface hardware.

Multi-hop

If the members of a multicast group are in more than one cell, then the network layer must ensure that the message addressed to the group is received by all members of the group. Some nodes must act as relays to transmit messages between cells. This can be done in one of the following ways:

• Flooding: the multicast messages are forwarded to all cells. Nodes in each cell select messages by filtering. This is clearly extremely inefficient and non-scalable.

• Simulated multicast: a list of group members' network addresses is supplied to every sending node and multicast messages are actually sent by unicasting to each member of the group.

• Multicast-aware relaying: relay nodes maintain sufficient knowledge of the multicast tree to enable them to efficiently transmit messages to all of the cells in which members of the destination group reside. Note that this can be viewed as an optimized form of flooding in which the tree of cells to which messages are sent is pruned to cover only cells that are likely to contain group members. The pruning needs not be perfect, just good enough to achieve adequate efficiency, but the maintenance of the multicast tree in the case of rapidly-changing groups such as those that are used in geocasting may be unacceptably expensive.

3.7.7.3. Geocasting

The message recipients in a multicast data communication are likely to exhibit some spatial correlation. To address this need, location-based multicast regarded as *geocasting* may be a suitable communication paradigm for group-based data dissemination that takes into account geographical areas.

Geocasting enables the transmission of a message from a set of sources to all nodes within a geographical destination area. Geocasting is a subclass of multicasting which can be implemented with a multicast membership defined by the current locations of the destination vehicles. Also, further multicast filtering can be applied in order to refine the received data and associate it with other groups of interest. For instance, this could be used to distinguish messages targeting vehicles and pedestrians.

In this section, we give an overview of the most important geocasting algorithms based on the descriptions presented in [230]. When appropriate, we also comment on the suitability of the algorithm for IVC/RVC applications.

3.7.7.4. *Flooding-based geocasting*

Pure flooding

The simplest form for implementing geocasting is to flood the network with packets targeted at a geographical destination region. The receiver checks whether its position is within this region upon the packet arrival. The major drawback of this scheme is the high number of generated packets in the network which significantly increases the overhead as the number of participating vehicles increases.

Location-based multicast (LBM)

An approach to minimize such an overhead is to give orientation to the flooding process by establishing a forwarding zone. Nodes discard packets when they are outside this zone and forward them otherwise. In [7, 8] the authors proposed two LBM schemes which differ in the way these zones are defined.

In one, a zone consists of at least the destination region and a path between the sender and this region. The zone size can be adjusted by a system parameter in order to increase the probability for message reception at the expense of an overhead increase.

The second scheme specifies the zone by the coordinates of the sender, the destination region, and the distance of a node to the center of this region. When a node receives a geocast packet, it determines whether it belongs to the forwarding zone specified in the packet by computing its own geographic distance to the center of the destination region. If such a distance is less than the one recorded in the packet, the node forwards this packet to its neighbors.

Voronoi diagrams

When the forwarding zone is empty or partitioned, the LBM protocol can perform poorly. To address this, neighbors that belong to a zone are determined using Voronoi diagrams, which partition the network in n regions, where n is the number of neighbors. The Voronoi region of a neighbour consists of all nodes that are closer to this neighbor than to any other neighbor.

When a node receives a geocast packet, it first computes the Voronoi diagram. The Voronoi partitions intersecting with the geocast destination region are used as the forwarding zone. Inside the destination region, flooding is often used.

Although Voronoi diagrams minimize the problem of empty forwarding zones through network partition, the overhead for flooding packets in the network is still high. Also, in scenarios of high mobility the computation of Voronoi regions can be a major issue as the node distribution in the network will vary over time.

Mesh

Mesh algorithms [13, 14] discover redundant paths between the source and destination regions to address issues that arise from node link failures. In such schemes, the forwarding zone approach of LBM is used to create the mesh network. Once this is established, packets are forwarded using the routing paths of the mesh. The overhead of this type of geocasting routing is mainly associated with the set-up of the mesh and its maintenance over time. This approach may work reasonably in low mobility scenarios but unlikely to succeed in high mobility IVC/RVC networks. It may be necessary to update or even reconstruct an established mesh because of the rapid rate of changes in topology.

3.7.7.5. *Routing without flooding*

GeoNode

Imielinski and Navas [20, 21, 22] considered the problem of geographic multi-point to multi-point routing in fixed networks. Their assumption is that the network has a cellular architecture with a GeoNode assigned to each cell. This results in two-level routing, one between a sender and the GeoNode and the other between a GeoNode and the destination region. This approach may be suitable for fixed networks but it seems unrealistic for IVC/RVC networks.

GeoTORA

GeoTORA [24, 25] is based on TORA (Temporally Ordered Routing Algorithm), which is a unicast routing protocol for ad hoc networks. In TORA a directed acyclic graph (DAG) is maintained for each destination. It represents for each node the direction to the destination node. Thus, this approach can be used for forwarding a packet to a destination starting at any node.

As nodes are moving in IVC/RVC scenarios, the DAG needs frequent updates which can lead to instability. It also introduces a maintenance problem since relevant information about the network topology needs to be gathered to construct each node's DAG.

3.7.7.6. *Summary of simulation results*

The authors in [230] carried out simulations with NS2. The simulated 802.11 network was configured with a 250 m wireless transmission range and the number of nodes varied between 100 and 1,000.

The first observation is that none of the protocols previously discussed guaranteed delivery. Redundancy, as in the flooding-based approaches, may not provide high delivery success rates as expected. One reason is because redundant routes in highly dynamic scenarios, for example vehicles traveling at high relative speeds, exhibit a very short lifetime [200]. The delivery success rates for the protocols described above were between 95% and 100% for an edge length between two nodes of 1,000 meters and this rate drops to between 10% and 30% with edges length of 3,000 meters.

3.7.8. *Time synchronization*

IVC/RVC applications need a mechanism to establish a common sense of time among the vehicles on the road. Generated events should be associated with a timestamp at the source in order to determine the reliability and accuracy of received travel and traffic information – a recent event data will usually be more accurate [199].

We list two time frameworks that could suit these applications:

• *Relative time:* the sense of time is established with respect to an agreed time reference which could be an elected vehicle among a specific group. This is applicable to scenarios where preserving the order of events is the only required function from a time synchronization service.

• *Absolute time:* in other cases, preserving the order of events is crucial but it is not the only functionality needed. Applications might require event data to be timestamped when they occurred with an absolute time value with respect to a true time standard such as the universal coordinated time (UTC).

MAC protocols based on a slotted TDMA structure require time synchronization to align nodes to the commonly used slot structure. The authors in [231] proposed and analyzed a decentralized time synchronization protocol. The protocol avoids systematic timing drift of the nodes, in a steady state. Results show that all nodes are kept synchronous within a time interval of twice the propagation delay between them. It is unclear, however, how the protocol deals with rapid changes in network topology which is a characteristic of IVC/RVC networks.

The question that remains to be explored is to what extent decentralized time synchronization protocols are really a requirement when we assume the majority (at least those which have wireless communication) will be equipped with GPS receivers. Depending on the type of receiver, a time with an error below 1 microsec can be achieved.

3.7.9. *Simulations: more real-life models*

Many applications require a model of the real world. The question is how such a model can be designed and validated. Traffic simulation for ITS can be classified in two categories: (a) microscopic modeling – suitable for group communication as the applications are often concerned with local behaviors of vehicles (e.g. instantaneous velocity and position); (b) macroscopy modeling – the mobility pattern is defined by four parameters: average car speed v (m/s), traffic density (in vehicles/km), traffic flow (in vehicles/second) and net time gap in seconds. Some of these parameters are assigned according to normal or uniform distribution.

The mobility patterns are quite different from the random waypoint model that is extensively used for ad hoc network simulations. There are a few simulators built and used for the simulations in the literature. We refer to PATH/CORSIM and GloMoSim that can simulate a vehicle network in addition to NS2 that is commonly used to simulate the network communication protocols. Other research have used MatLab and Simulink in order to simulate the vehicular network [203].

Although there is research work that has used realistic data such as the study discussed in [211], where GPS data collected from actual buses in San Francisco were used, there is a lack of realistic data in the majority of the simulated work that includes real road/traffic conditions and also realistic vehicle models. The research community should attempt to incorporate more GIS data with dynamic information of the traffic systems.

We argue that complete and realistic simulators can only be built when we take a multidisciplinary research approach that includes simulations from areas such as GIS data, network/protocol, road transport and health/safety.

3.8. Classification of the concepts

This section provides a taxonomy of the concepts that have been dealt with in this chapter, according to the application requirements and characteristics defined in Chapter 2 and summarized in the following table.

REQUIREMENTS AND CHARACTERISTICS	
Topology	Scalability
Fault tolerance	Localization
Data traffic characteristics	Networking infrastructure
Mobility	Node heterogenity
Power consumption	Real-time
Reliability	

Table 3.4. *Application requirements and characteristics as defined in Chapter 2*

3.8.1. *Classification of the thematic areas*

The four thematic areas have been selected to cover most of the aspects that concern the various types of CO-based systems. In the following, each thematic area is analyzed according to the list of aforementioned requirements.

3.8.1.1. *Wireless sensor networks for environmental monitoring (WSNEMs)*

A WSNEM is characterized by a large number of stationary sensor nodes, disseminated in a wide area, designated to collect information from the sensors and act accordingly. For example, a WSNEM can be used in greenhouses to monitor environmental conditions (air humidity/temperature, light intensity, fertilizer concentration in the soil and so on) and deliver the sensed data to a central controller that determines the actions to take (activating watering springs, sliding shatters, air humidifiers and so on). Also, WSNEMs may be used for monitoring the integrity of buildings, bridges or, more generally, structures. Sensors inserted in the structure can detect significant variations of pressure, positions of landmarks or relative positions of the surrounding sensors, and send data to a controller node, either spontaneously on in response to an explicit request from the controller. Another possible example is the smart environments. The environment is equipped with sensor nodes that allow interaction with people. In particular, a WSN can track the movement of objects or persons in a given (also wide) area, thus providing the basic functionalities for the development of surveillance applications:

• *Topology.* The topology can be: pre-planned, semi-random or random. In the first case, nodes are accurately placed in the field for different purposes, such as monitoring the deformation of rigid structures or the environment in a specific area (greenhouses). A semi-random topology is obtained when most of the nodes are randomly scattered over the area, but some specific nodes are placed in specific positions, for instance to guarantee connectivity or to act as beacons for the other nodes. Finally, a random topology is obtained when nodes are scattered in the area without any plan, as in the case of airplane dissemination of sensors over a forest or a contaminated area. Although nodes are static, changes in topology can still occur because of the on/off cycles that the nodes go through to conserve energy, and because of the dynamic nature of the wireless medium.

• *Scalability.* Such networks can be composed of thousands of nodes. Therefore, scalability is a primary issue that has to be dealt with.

• *Fault tolerance.* Nodes are prone to failure due to energy depletion or physical crashing.

• *Localization.* Localization can play an important role in cases of random or quasi-random topologies. Furthermore, WSNEM can also be used to trace the motion of (non-cooperative) objects in the area, such as the motion of enemy troops in a battlefield, or wild animals in a forest and so on. Notice that the case of cooperative moving objects can be referred to the WSNMN scenario.

• *Data traffic characteristics.* Traffic patterns are rather peculiar and differ from classic all-to-all paradigms considered in classic ad hoc networks. Flows are mostly unidirectional, from sensors to one or more sinks and vice versa (e.g. greenhouse). Traffic is usually light (very low average bit rate), though future applications may require the transmission of heavy data bursts (e.g. image transfer). Data may show strong spatial correlation. For instance, the readings of temperature sensors that are placed in close proximity each other will be strongly correlated.

• *Networking infrastructure.* WSNEMs do not rely upon any network infrastructure. However, interest is being devoted to connecting WSNEMs with other networks (e.g. WLANs).

• *Mobility.* Nodes are normally static. In some cases, a node's position may undergo small variations due to external causes (e.g. wind, quake, vibrations, and so on).

• *Node heterogenity.* Nodes can differ in their sensing, computational and transmission capabilities. However, the literature is mainly focused on homogenous networks.

• *Power consumption.* Since nodes are usually battery powered and not (easily) rechargeable, power consumption is a primary issue.

• *Real-time.* Real-time can be a requirement for WSNEM when the detection of some events (fire, intrusions and so on) has to be notified to the sink/controller in a strictly limited time period.

• *Reliability.* Reliability of data transport in WSNEM may or may not be a requirement, depending on the specific type of application. Clearly, in each scenario that involves safety, surveillance or, in general, monitoring of potentially hazardous events, data reliability is a primary issue. However, many other application scenarios for WSNEM may exploit data redundancy to increase their robustness to data corruptions (e.g. climate monitoring in greenhouses, temperature controllers, smart environments and so on).

3.8.1.2. *Wireless sensor networks with mobile node (WSNMNs)*

A WSNMN is characterized by the use of mobile nodes in a WSN, ranging from a network with only mobile nodes to a network with a combination of static and mobile nodes. The use of mobile nodes in sensor networks increases the capabilities of the network and allows dynamic adaptation to changes in the environment. Some applications of mobile nodes could be: *collecting and storing sensor data in sensor networks*, reducing the power consumption due to multi-hop data forwarding; *sensor calibration*, using mobile nodes with different and eventually more accurate sensors; *reprogramming nodes*, "by air" for a particular application; and *network repairing*, when the static nodes are failing to sense and/or to communicate:

• *Topology.* The topology can be pre-planned, semi-random or random. The first case appears when the static nodes are deployed in known places and mobile nodes have a controlled and predicted motion. The second case can be found when either the static nodes are randomly placed or the mobile nodes have uncontrolled and non-predicted motion. Finally, a random topology is obtained when the static nodes are placed without any plan and the mobile nodes move uncontrolled and with a non-predicted motion. The communication can be single-hop because mobile nodes can approach different static sensors to collect data. However, multi-hop communication can be used in other cases, for example, when the number of static nodes is much higher than the number of mobile nodes. Finally, it should be mentioned that the topology is dynamic due to the presence of mobile nodes.

• *Scalability.* As has been stated in the section about WSN with static nodes, scalability is a primary issue. Mobile nodes can improve and make easier the scalability of the network, because those nodes can manage different regions containing a fixed number of static nodes, allowing communications between them.

• *Fault tolerance.* The use of mobile nodes can increase the fault tolerance of a WSN, because a mobile node can be used to replace a non-working node or to calibrate the different sensors of the static node.

• *Localization.* This is an important issue due to the use of mobile nodes. Furthermore, the mobile nodes can be used to localize the static nodes with random or quasi-random topologies.

• *Data traffic characteristics.* In WSNMN, the mobile nodes can be used as sinks that send the information to a central station. Usually, the traffic between static and mobile nodes is low, whereas the traffic among mobile nodes and the central station is high. Finally, the mobile nodes can also process the information gathered from the static nodes in order to reduce the data traffic.

• *Networking infrastructure.* WSNMN does not need any networking infrastructure. However, it could be of interest to connect the WSNMNs (or at least some mobile nodes) to other networks, such as the Internet.

• *Mobility.* WSNMN considers a range of configurations, from WSN composed only by mobile nodes to WSN composed mainly by static nodes and only a few mobile nodes.

• *Node heterogenity.* Nodes can be static or mobile. Within these two groups, nodes can be heterogenous in terms of sensing capabilities, batteries, locomotion system, etc.

• *Power consumption.* Mobile nodes can reduce the power consumption due to multi-hop data forwarding. Furthermore, it is possible to have places where mobile nodes can recharge their batteries when a low energy level is detected. In any case, power consumption is also a primary issue in WSNMN.

• *Real-time.* When a mobile node has to move sequentially to approach different static nodes in distant locations, an important disadvantage is the delay required. Then, in general, this strategy can be applied only in delay tolerant scenarios.

- *Reliability.* Mobile nodes can increase the reliability of the network because they can be used to repair or increase the connectivity of the network by deploying new static nodes.

3.8.1.3. *Autonomous Robotics Team (ART)*

One of the main reasons to develop a multi-robot system may be that it is possible to build more robust and reliable systems by combining unreliable but redundant components. Research in multi-robot systems has focused primarily on construction and validation of working systems, rather than more general analysis of problems and solutions. As a result, many architectures can be found in the literature for multi-robot coordination, but relatively few formal models. Those models can only be found when addressing specific aspects of a multi-robot system, such as path planning or task allocation, but not for the whole architecture. On the other hand, a combination of multi-robot systems with WSNs can improve the overall reliability and performance of the whole system. Benefits can be identified in both directions. For example, a multi-robot system could provide network repairing services to a WSN and a WSN could provide extended sensory capabilities to a multi-robot team:

- *Topology.* In ART, topology can be pre-planned because the motion of the robots is always to some extent controlled and predictable.

- *Scalability.* In a multi-robot system, the amount of data to process is quite large: the processing of this knowledge in a centralized way obviously requires powerful computational means. Moreover, such a centralized computation reaches its limits when the number of robots increases (for example, in swarm robotics): a centralized system cannot be scalable to any number of robots. In contrast, a distributed approach can stay available when the number of robots increases, since each robot still only deals with a local partial knowledge of the system, so that it is required to manage only a limited set of information related to its location.

- *Fault tolerance.* By fault tolerance, we mean the ability of the robot team to respond to individual robot failures or failures in communication that may occur at any time during a mission.

- *Localization.* Many techniques have been developed in robotics in the last decades in the research field of localization. Each robot can localize itself by using a satellite GPS in outdoor scenarios, beacons or landmarks in the environment, etc. A perception system involving several sensors such as cameras (visual and infrared), ultrasonics, lasers, etc. can also be used in combination with odometry techniques and systems to localize the robot. In ART, new techniques can be applied to exploit the information from other robots in terms of localization improvement.

- *Data traffic characteristics.* The data traffic among the robots is usually higher (images, telemetry, etc.) than the traffic among the nodes of a WSN. Moreover, traffic patterns are more similar to the classic all-to-all paradigm considered in ad hoc networks.

• *Networking infrastructure.* In ART, a networking infrastructure is not usually required. However, if a standard wireless communication system is used (i.e. WiFi), some kind of infrastructure such as an access point will be needed. In any case, it is usually of interest to connect the robots to an external network such as the Internet for monitoring, remote control and supervision of the tasks execution.

• *Mobility.* In ART, mobility is implicit since the interactions with the environment usually requires mobile robots or robotic manipulators.

• *Node heterogenity.* In general, heterogenity introduces complexity since task allocation becomes more difficult, and robots have a greater need to model other individuals in the group. However, differentiation can also be treated as an advantage in terms of complementarity between different components in order to complete a given task or mission.

• *Power consumption.* Robots can autonomously recharge their batteries when a low level of energy is detected. Furthermore, mixed solutions including solar panels are also possible.

• *Real-time.* Real-time can be a requirement for ART depending on the particular application considered.

• *Reliability.* This requirement should be addressed during the design of the ART in order to allow mechanisms and algorithms to increase the reliability during the operation of the system, making it possible to detect and overcome various sources of system failure, like the reduction in the number of visible GPS satellites or communication breakdowns. Perception techniques and special planning functions can be used to detect and monitor a faulty robot making it possible to increase reliability in the execution of a mission.

3.8.1.4. *Inter-Vehicular Networks (IVN)*

Cooperation among vehicles is a promising approach to address critical road safety and efficiency in IVN scenarios. For instance, coordinated collision avoidance systems could significantly reduce road accidents. Road safety becomes rather problematic in countries which rely heavily on the road network for transportation of goods.

Despite the fact that promising results have been achieved in some of the research projects in this area, major issues remain to be resolved. This might explain the large list of proposed traffic applications with no real deployment. IVC exhibits characteristics that are specific to scenarios with *high node mobility*:

• *Topology.* Despite the constraints on the movement of vehicles (i.e. they must stay in the lanes), the network will tend to experience very rapid changes in topology.

• *Scalability.* Such networks can be formed by thousands of vehicles that can interact with each other. Thus, scalability is a major issue that needs to be addressed.

• *Fault tolerance.* It is important to understand how IVC/RVC networks can detect and recover from faults. Fault tolerance is more easily achieved with a hybrid of ad hoc approaches for direct IVC and infrastructure-based schemes. The latter could introduce fallback mechanisms through a pre-deployed network backbone.

• *Localization.* Location is a key characteristic in IVN. The design choices for localization schemes in this scenario range from centralized but inexpensive satellite positioning systems (e.g. GPS) to decentralized systems. An important aspect to consider is the accuracy of localization systems and its implications to the applications, especially safety applications. For instance, routing strategies need to consider an error region while making the decision to forward packets.

• *Data traffic characteristics.* Some of the IVN applications require communication with destinations that are groups of vehicles. Thus, the type of traffic should be predominantly multicast, in particular, geocasting with physical areas of coverage.

• *Networking infrastructure.* Some of the IVC approaches assume a pre-deployed infrastructure, say a cellular network. Others rely on ad hoc communication when vehicles may communicate where essentially no pre-planned infrastructure is present. IVC can be supported by any of these two approaches. However, road-to-vehicle communication is often an infrastructure-based scheme, as ground support including pre-deployed road sensors needs to be in place.

• *Mobility.* The high mobility of vehicles poses issues that need to be addressed. Communication can be severely impaired by the high relative velocities of vehicles, even when moving in the same direction. In ad hoc short-range communication, the network will tend to experience very rapid changes in topology. Also, constraints on mobility may impact the coverage of the network in cases where no ground infrastructure (e.g. base stations) to support the communication between vehicles is present. Adding more nodes to the network may not improve coverage but rather negatively affect the use of available bandwidth.

• *Node heterogenity.* Unlike typical WSNs deployment scenarios, vehicles can be instrumented with powerful sensors and radios to achieve long transmission ranges and high quality sensor data. Interoperability between different types of devices is a major issue.

• *Power consumption.* Low power consumption and small physical size have been some of the fundamental issues in designing software and hardware for low cost wireless sensor nodes. However, vehicles exibit more physical space and energy consumption is not an issue.

• *Real-time.* Safety applications will not tolerate high overall delay including network and processing delays.

• *Reliability.* In this scenario, reliability is the application's tolerance to errors made visible to it. A zero tolerance threshold indicates that the applications requires 100% guaranteed data delivery.

Although these characteristics have some impact on infrastructure-based IVNs, the major issues arise from ad hoc network set-ups. Blum *et al.* [200] pointed out that IVC networks have mobility characteristics different from the typical MANETs considered in the literature. Specifically, these characteristics cause rapid topology changes, frequent fragmentation of the network, a small effective network diameter and limited utility from network redundancy. Such issues must be tackled from the outset since they directly influence the IVC system design at all communication layers.

The thematic areas taxonomy is summarized in the following table.

Taxonomy of the thematic areas				
	WSNEM	WSNMN	ART	IVN
Topology	Typically multi-hop: slow dynamic due to on/off duty cycles	Single-hop: medium dynamic due to node mobility	Multi-hop: high dynamic due to node mobility	Multi-hop: high dynamic due to node mobility
Scalability	Primary issue	Primary issue	Primary issue	Primary issue
Fault tolerance	Prone to nodes failure	Partially resilient to nodes failure	Prone to nodes failure	Partially resilient to failures
Localization	Relevant	Primary issue	Primary issue	Primary issue
Data traffic characteristics	Generally from the sensors to one or few sinks	Generally from the sensor to the mobile nodes	Robot-to-robot and environment-to-robot	Road-to-vehicle or vehicle-to-vehicle
Networking infrastructure	None	None	Possible interaction with access points	Possible interaction with cellular network or fixed access points
Mobility	None	Medium and limited to mobile nodes	Medium-high	High
Node heterogenity	Possible	Always	Possible	Possible
Power consumption	Energy-efficient algorithms on/off duty cycles	Rechargeable mobile nodes	Rechargeable robots	Irrelevant
Real-time	Secondary issue (depending on application)	Secondary issue (depending on application)	Secondary issue (depending on application)	Primary issue
Reliability	Secondary issue (depending on application)	Secondary issue (depending on application)	Primary issue	Relevant issue (depending on application)

Table 3.5. *Taxonomy of the thematic areas*

3.8.2. *Classification of the algorithms*

In this section, we will mainly refer to the algorithms covered by this chapter. For further details on the algorithms described, please refer to the specific sections of the chapter.

3.8.2.1. *MAC algorithms*

● *Topology.* MAC algorithms can be classified on the basis of the topology information they need to operate. Topology-independent MAC algorithms, such as those based on CSMA (MACA, MACAW, PAMAS) or SMAC and DBMAC, do not require any knowledge of the network topology. Other protocols, like SIFT and STEM, require nodes to have local information only, i.e. information regarding the nodes in their proximity. Finally, topology-dependent protocols, such as TRAMA, assume nodes are aware of the entire network topology.

• *Scalability.* Generally, the performance of MAC protocols, in terms of medium access delay, is affected by the number of contending users. Contention-based access protocols, such as MACA, MACAW, PAMAS and so on, scale rather well with the number of nodes, when the traffic offered to the network is low. In contrast, with high traffic loads, random protocols performance (in terms of medium access delay) worsens rather rapidly as the number of nodes increases. Contention-free MAC algorithms (e.g. time-division based algorithms) scale better with high traffic loads, while for low traffic such solutions may incur in longer access delay than random algorithms.

• *Fault tolerance.* In general, MAC algorithms are not affected by node failure, even though a certain performance loss may be experienced in case of topology-dependent algorithms.

• *Localization.* MAC algorithms can be classified as location-aware and location-independent. Location-aware solutions usually follow a cross-layer approach, since the location information is used both to manage the access to the medium and the forwarding of the information towards the intended destination (see GeRaF, Smart Broadcast). Some algorithms assume only that each node is acquainted (in some way) with its own spatial coordinates, while others require the knowledge of the positions of the surrounding nodes only, or of all the nodes in the network. Pure medium access algorithms are generally location-independent. Location-aware algorithms are usually much more efficient that location-independent algorithms. However, they may turn out to be very sensitive to localization errors. These aspects have not been sufficiently covered in the literature yet.

• *Data traffic characteristics.* Traffic characteristics may have a strong impact on MAC algorithms performances. Contention-based protocols usually show better performance in the case of sporadic traffic bursts, while deterministic access mechanisms are more suited for handling periodic traffic generation patterns. Currently, MAC algorithms do not consider the traffic flow patterns, i.e. the set of nodes that exchange data. An exception is represented by cross-layer solutions that provide a integrated mechanism for both MAC and routing, and are sometimes designed according to the specific traffic flow pattern expected in the system.

• *Networking infrastructure.* Generally, the presence of a network infrastructure makes it possible to resort to contention-free MAC protocols based on polling strategies or resource reservation. However, most of the protocols covered by this chapter can be operated in absence of any networking infrastructure.

• *Mobility.* Mobility might represent an issue for MAC protocols for two reasons. First, mobility involves topology variations that may affect algorithms that need to adjust some parameters according to the density of nodes in the contention area (SIFT,

TRAMA, TSMA, MACAW). Second, MAC algorithms based on medium reservation mechanisms (MACA, MACAW) may fail in the event of mobility, since the reservation procedures usually assume static nodes. For instance, algorithms based on the RTS/CTS handshake to reserve the medium may fail because either the corresponding nodes move outside the mutual coverage range after the handshake or external nodes get into the contention area and start transmitting without being aware of the medium reservation.

Nevertheless, many MAC algorithms considered in this chapter are capable of self-adapting to the topology variations in the case of node mobility. Algorithms like TRAMA, TSMA and SMACS-EAR can still adapt to topology variations, but at the expense of energy-efficiency and access delay.

● *Node heterogenity.* MAC algorithms for heterogenous networks have not yet been investigated in the literature. Algorithms based on channel sensing (CSMA-based) provide some resilience to interference produced by other radio interfaces operating on the same frequency band and, thus, can be adopted in a heterogenous system. However, this solution would not leverage on the nodes diversity. This topic is, indeed, still to be investigated in the literature.

● *Power consumption.* Energy-efficiency is considered in several MAC protocols, particularly in the case of WSNs. A typical method to reduce energy consumption is to let nodes alternate periods of activity and sleeping. Notice that such on/off cycles may either be managed independently of the MAC protocol or as part of it. For instance, CSMA, MACA and MACAW protocols do not explicitly consider the presence of such on/off cycles. Nevertheless, CSMA behavior is not affected by on/off cycles, while MACA and MACAW may fail since they assume nodes are always notified of the channel state. Protocols like PAMAS and SMAC, on the other hand, take into account the sleeping periods of the nodes, thus allowing a more efficient power management of the system. Usually, this is obtained at the cost of a higher complexity of the MAC protocol.

● *Real-time.* Contention-based MAC protocols cannot usually provide any real-time guarantees. Conversely, contention-free algorithms, such as TSMA or TRAMA, are able to guarantee a given maximum access delay, which depends on the number of competing nodes.

● *Reliability.* Almost all the MAC algorithms considered in the chapter require explicit acknowledgment (ACK) of correct data reception from the receiver. Usually, in case of missing or negative ACK, the data link layer entity retransmits the data unit. However, the process is stopped when a given number of retransmissions is reached. In this case, the data unit is discarded. Hence, in general, MAC protocols can provide only limited reliability. Notice that contention-based MAC algorithms are prone to transmission errors due to collisions, events that, on the contrary, never occur in contention-free algorithms. Therefore, contention-based algorithms are typically less reliable that contention-free ones.

The MAC algorithms taxonomy is summarized in the following table:

Taxonomy of the MAC algorithms									
	CSMA	MACA MACAW PAMAS	SMACS-EAR	Sift	STEM	DB-MAC	TRAMA	TSMA	Energy-aware TDMA
Topology	Indep.	Indep.	Indep.	Only local	Only local	Indep.	Compl. top. knowledge	Only local	Compl. top. knowledge
Scalability	Partial (low traffic)	Partial (low traffic)	Partial (low traffic)	Medium (low traffic)	Medium (low traffic)	Partial (low traffic)	Good (high traffic)	Good (high traffic)	Good (high traffic)
Fault tolerance	Resilient	Resilient	Resilient	Resilient	Resilient	Resilient	Partially res.	Partially res.	Partially res.
Localiz.	Not required	Not required	Not required	Not required	Not required	Not required	Not required	Not required	Not required
Data traffic charac-teristics	Better for sporadic traffic	Better for sporadic traffic	Better for sporadic traffic	Better for sporadic traffic	Better for sporadic traffic	Better for sporadic traffic	Better for periodic traffic	Better for periodic traffic	Better for periodic traffic
Netw. infrastruct.	None	None	None	None	None	None	None	None	None
Mobility	High resilience	Medium resilience	Medium resilience	Medium resilience	Low resilience	Medium resilience	Low resilience	Low resilience	Low resilience
Node hetero-genity	None	None	None	None	None	None	None	None	None
Power con-sumption	High	High	Medium (on/off)	High	Low (sleep)	Medium (data aggr.)	Low (sleep)	Low (sleep)	Low (sleep)
Real-time	Partial	Partial	Partial	Partial	No	Partial	Yes	Yes	Yes
Reliab.	Partial	Partial	Partial	Partial	Partial	Medium	High	High	High

Table 3.6. *Taxonomy of the MAC algorithms*

3.8.2.2. *Routing algorithms*

• *Topology.* Routing algorithms can be differentiated on the basis of the routing topology they realize. Usually, table-based algorithms create tree topologies, so that each node is the root of a routing tree towards each other node in the network.

On-demand routing algorithms, on the contrary, realize point-to-point routing topologies, where a path from a node to its destination is created when needed (GedRaF, GAF, GEDIR). Cluster-based routing algorithms construct a hierarchical topology, where some nodes are elected as CHs (or coordinators) and forward data collected from their neighbors towards the final destination (LEACH, PEGASIS, TEEN). Request-driven routing algorithms aim at defining a path from possible multiple sources to the node that issues a specific data request. These algorithms lead to a star-shaped routing topology, where many paths originating from the source nodes converge to the destination node. Examples are RR and DD routing.

• *Scalability.* Scalability is a important issue for routing protocols. Table-based protocols usually show scalability problems when the number of nodes (and, consequently, routing table entries) grows, particularly for systems with limited storing and computational capabilities (typically sensors networks). To alleviate this problem, many protocols resort to clustering techniques that, in turn, bring forth some control overhead (LEACH, TEEN, PEGASIS). State-less algorithms have been introduced to cope with scarce storing capabilities, while maintaining good scaling properties. Typical examples are location-based algorithms, such as GeRaF, GAF, GEDIR, where nodes need to maintain the information regarding their own location and that of the destination node. Nevertheless, the literature does not consider in detail the issue of distributing and maintaining the location information over the network.

Algorithms that make use of broadcast packets to gather and/or diffuse topological information usually show scalability problems in large networks due to the broadcast storm problem, unless broadcasting is obtained by means of a specific broadcast-diffusion algorithm (DD, RR).

• *Fault tolerance.* Usually, routing algorithms can adapt to topology variations due to nodes failure. However, the reaction to a topology variation due to nodes failure may require some time and, hence, bring some performance degradation. During this time, data can be delayed, duplicated or lost.

• *Localization.* As seen for the MAC algorithms, also routing algorithms can be classified in location-aware and location-independent. Location-aware routing algorithms include the cross-layer solutions discussed in the classification of MAC algorithms, such as GeRaF, and other pure routing algorithms, such as GAF, GEDIR and GEAR. Usually, location-aware routing algorithms assume that each node is acquainted (in some way) with its own spatial coordinates and those of the intended destination node. Hence, the next hop is determined in order to move the packet towards the destination.

Data-centric routing algorithms, such as DD, RR and SPIN, make use of broadcast techniques to disseminate and gather routing information and, therefore, do not require any localization feature.

Hierarchical routing algorithms, in general, are based on topological information, though do not require exact node localization (LEACH, TEEN). Nevertheless, localization may help the process of creating the cluster structure, thus resulting in better performance (VGA, TTDD).

• *Data traffic characteristics.* Data traffic characteristics may affect routing algorithm design. In WSNs, for instance, spatial correlation among data generated by nodes in close proximity is exploited by cross-layer solutions that merge routing and data processing functionalities (TEEN, VGA, COUGAR). Specific routing algorithms have been proposed for centralized traffic patterns, where information flows to and from a single central node (e.g. a sink node in WSN) and several peripheral nodes. Some examples are SOP and MCFA.

• *Networking infrastructure.* Most of the routing algorithms for COs are designed according to an ad hoc paradigm. Therefore, solutions are completely distributed and do not require any backbone infrastructure.

• *Mobility.* Generally speaking, all the routing algorithms considered are able to cope with dynamic topologies due to node mobility. However, most of them react to topology variations by dropping the broken paths and computing new ones from scratch, thus incurring performance degradations. In particular, mobility may strongly affect cluster-based algorithms due to the cost of maintaining the cluster-architecture over a set of mobile nodes. Examples of routing algorithms specifically designed for networks with slow-mobile nodes are GAF and TTDD, which attempt to estimate the nodes trajectories. Other protocols that are well-suited for an environment where the sensors are mobile are the SPIN family of protocols because their forwarding decisions are based on local neighborhood information.

• *Node heterogenity.* Node heterogenity can be a winning feature to develop efficient routing algorithms, particularly for WSNs with mobile nodes. Notwithstanding, the literature still lacks solutions that provide leverage on node heterogenity to enhance the routing process.

• *Power consumption.* Power consumption is typically a very important issue in the design of routing protocols, since many CO systems involve battery-powered units. Accordingly, several energy-efficient routing algorithms have been presented in the literature, in particular for WSNs. The simplest method of reducing power consumption is to allow each node to schedule sleeping periods. Therefore, routing protocols have to be designed to work also in the presence of on/off duty cycles (GAF, ASCENT).

Other protocols reduce the amount of exchanged information in order to improve the energy-efficiency of the system. Moreover, other techniques such as data aggregation, overhead reduction, CH rotation and so on can be used to improve energy-efficiency (LEACH, GEAR, PEGASIS, TEEN and HPAR).

• *Real-time.* Routing algorithms that provide tight constraints on the packet delivery time are rather rare.

• *Reliability.* Most of the considered routing protocols cannot guarantee data reliability, especially when the network is rarely populated. Some routing algorithms, such as GeRaF, GAF, GEDIR and SPIN, may fail to discover a path in cases with connectivity holes within a connected network. Other algorithms can guarantee delivery if source and destination nodes are connected (GOAFR, SPAN, LEACH and PEGASIS). Broadcasting-based algorithms, such as RR and DD, generally offer high reliability due to the capillary diffusion of the routing control packets.

The routing-algorithm taxonomy is summarized in the following table.

Taxonomy of the routing algorithms							
	GeRaF	GAF SPAN	SPIN	LEACH PEGASIS	TEEN	HPAR	Rumor Direct-Diffusion
Topology	Point-to-point	Point-to-point	Star	Hierar.	Hierar.	Hierar.	Star
Scalability	Good	Good	Good	High	High	High	Low
Fault tolerance	High	High	High	Medium	Medium	Medium	High
Localization	Required	Required	Not required	May help	May help	May help	Not required
Data traffic characteristics	Irrelevant	Irrelevant	Relevant	Relevant	Relevant	Relevant	Relevant
Networking infrastructure	None	None	None	None	None	None	None
Mobility	High resilience	High resilience	High resilience	Medium resilience	Medium resilience	Medium resilience	High resilience
Node heterogenity	None	None	None	None	None	None	None
Power consumption	Low (on-off)	Low (sleep)	Medium (data aggr.)	Low (clust.)	Low (clust.)	Low (clust.)	High
Real-time	No	No	No	No	No	No	No
Reliability	Partial	Partial	Medium	Medium	Medium	Medium	High

Table 3.7. *Taxonomy of the routing algorithms*

3.8.2.3. *Localization algorithms*

• *Topology.* Localization algorithms are used to infer the geographical position of a node by elaborating the signals received from position-aware nodes (beacons/landmarks). The precision of the estimation is usually strictly dependent upon the placement of the beacons. Therefore, the network topology may have effects on the performance of most localization algorithms. For example, in the case of range-free approaches, inhomogenous node density may lead to an incorrect distance estimate (Centroid, DV-Hop).

• *Scalability.* Localization algorithms are usually scalable with the network population. However, if the geographical extension of the network increases, a higher number of beacons may have to be deployed (APIT). Even if a multilateration approach is adopted, relaxing the need for direct beacon visibility, an increase in the average number of hops from the beacons leads to localization error accumulation (DV-Host, DV-Dist). The complexity of the localization algorithms, as well as the preciseness of the estimation usually increases with the number of beacons (AHLoS). To conclude, the localization algorithms, in general, might show scalability problems with the number of nodes that populate the network.

- *Fault tolerance.* The localization algorithms are usually tolerant to the failure of some nodes, given that they are not beacons. Failure of beacons is particularly critical for localization algorithm performance (Centroid, DV-Hop, DV-DIST). Also, malfunctioning nodes, for instance, nodes with defective HW, may have an impact on localization errors and on localization error propagation (DV-Hop, N-Hop TERRAIN, AHLoS).

- *Localization.* Localization algorithms require, in general, a suitable disposition of the beacon nodes in the network area. Estimation may also be refined by using the positioning information estimated by the surrounding nodes.

- *Data traffic characteristics.* To estimate the position, sensor nodes use the control packets sent by beacons that contribute to the network load (DV-Hop, DVB-Distance, N-Hop Terrain). In the case of networks with mobile nodes, the position estimation might be improved by increasing the beacon frequency. Hence, a trade-off between localization accuracy and network load can arise. Furthermore, localization algorithms might make use of data packets sent by position-aware nodes to adjust their position estimation. In this case, regular or periodic data traffic exchange involving position-aware nodes can improve the performance of the localization mechanisms without introducing extra control traffic.

- *Networking infrastructure.* In general, localization algorithms make use of infrastructures. Specifically, satellite-based positioning mechanisms require a complex satellite network infrastructure. More generally, localization algorithms require the presence of a network infrastructure that hosts beacon nodes whose positioning information is disseminated over the network. Localization algorithms that aim at providing only relative positions of a node in a network, on the other hand, do not require any permanent infrastructure.

- *Mobility.* In general, node mobility increases the localization error. However, in some contexts, mobile nodes capable of accurate position estimation might be used to disseminate positioning information over a network of elementary static nodes.

- *Node heterogenity.* The use of satellite-based positioning systems is not always possible, for it increases the cost of the nodes and the power consumption. The heterogenity of nodes play a fundamental role in these scenarios, since a bunch of localization-enabled nodes might be exploited by the other nodes in a network to derive an estimation of their position (APIT).

- *Power consumption.* Localization schemes increase the power consumption. In particular, the use of satellite-based schemes is very expensive in terms of power consumption in some contexts (such as WSNs). This cost might be reduced by installing a limited number of devices in the network (e.g. beacons) and by using localization algorithms to estimate the position of the other nodes in the network. Clearly, in this case, the energy consumption is due to the control packets exchange (DV-Hop). Other localization strategies encompass the use of ultrasonic transceivers (Cricket, AHLos) that determine further energy consumption.

- *Real-time.* In general, satellite-based positioning systems are capable of providing quasi-real-time localization services. However, localization algorithms that are based on the elaboration of beacon signals are not suitable for strictly real-time applications, since, in general, they require the reception of several control packets to reduce the estimation error. On the other hand, the problem of whether a real-time application can be supported or not only becomes an issue for mobile networks of COs. In many application scenarios in which nodes are static, the localization process can be performed at the network set-up, reducing its costs and allowing the use of different types of algorithms independent of the real-time constraints of the application.

- *Reliability.* Reliability of the localization algorithms depends on the number and position of the beacon nodes, possibly the number of hops over which the localization error propagates, or the presence of malfunctioning or malicious nodes. Malicious nodes are nodes whose purpose is to compromise the correct operation of the network. Such nodes can provide, for example, incorrect ranging estimates or incorrect information on their own position to other nodes, affecting other nodes' localization accuracy or the ranging estimate accuracy (DV-Hop, N-Hop TERRAIN, AHLoS). Ways to detect and filter the information provided by malfunctioning or malicious nodes have to be provided.

The localization-algorithm taxonomy is summarized in the following table.

	Range-free				Range-based		
	Centroid	DV-Hop	APIT	Monte Carlo	AHLoS, N-Hop Multilat.	DV-DIST, HOP-TER.	AFL
Topology	Symmetric	Uniform	Generic	Generic	Generic	Generic	Generic
Scalability	Good	Limited	Good	Good	Good	Limited	Very Good
Fault tolerance	Partial	Partial	Good	Good	Partial	Partial	Good
Localization	Coarse	Good/coarse	Good/coarse	Good	Good	Coarse	Good
Data traffic characteristics	Local	Flooding	Local	Local	Local	Flooding	Local
Networking infr.	Required	Required	Required	Required	Required	Required	Not required
Mobility	Fragile	Fragile	Required	Robust	Fragile	Fragile	Partially robust
Node heterogenity	Landmarks	Landmarks	Landmarks	None	None	Landmarks	None
Power cons.	Low	High	Medium	Medium	High	High	Low
Real-time	No	No	No	Partial	No	No	No
Reliability	Partial	Partial	Partial	Partial	Partial	Partial	Medium

Table 3.8. *Taxonomy of localization algorithms*

3.8.2.4. Data processing

• *Topology.* The network topology might play an important role in the design of specific data processing. The perfect knowledge of the network topology, for instance, can be used to determine the position of better collector nodes. Moreover, if the topology is pre-planned, nodes with more computational capabilities can be placed in strategic positions. On the contrary, in the case of random topologies, the choice of more suitable aggregation points have to be made in a distributed manner and can be less efficient. In regards to the organization of the network structure, the data processing techniques can lie on different communication topologies. Most known algorithms (DD, LEACH, PEGASIS, TAG and TiNA) run over tree-based or hierarchical structures. In contrst, other schemes such as Synopsis Diffusion, and Tributaries and Deltas organize the network in a concentric ring structure.

• *Scalability.* Scalability is an important goal in the design of efficient data processing techniques, especially in large and dynamic networks. Existing data processing techniques based on the construction of an aggregation tree are less scalable than the multipath schemes due to the high cost of maintaining the organization of the network. This characteristic is accentuated in large or dynamic networks where adding or removing nodes from the tree structure heavily impacts on the performance of the algorithms. On the other hand, multipath solutions offer good scalability, due to the local and distributed functionalities.

• *Fault tolerance.* Data processing, in general, is performed in order to reduce the intrinsic data redundancy that might characterize CO scenarios (e.g. WSNEM). On the other hand, data redundancy may ensure a higher reliability in the cases of sensor failure, connectivity holes and so on. Hence, a trade-off between fault tolerance and redundancy reduction has to be cut. In the case of low packet loss probability, tree-based algorithms achieve better performance because they are able to minimize the number of transmissions required to deliver a given amount of data, so the redundancy is kept to a minimum. On the other hand, the multipath schemes preserve some data redundancy so it performs better in the case of high packet loss probability. There are also some hybrid approaches such as Tributaries and Deltas which are able to adjust their behavior according to the link conditions.

• *Localization.* In some cases, data aggregation techniques require information about the location of nodes. Nevertheless, almost all data processing techniques do not require any type of localization methods.

• *Data traffic characteristics.* The design of data processing techniques is strongly correlated to the specific considered application. In some cases, for instance in WSNEM, it may be useful to perform data aggregation as near as possible to the data sources due to the high redundancy among data collected in the same spatial region. In other cases, data processing can be performed along the path, for instance to merge information flows directed to a common destination.

• *Networking infrastructure.* Data processing can take advantages from the presence of networking infrastructure. For instance, access points can play the role of data collectors and perform any type of simple data processing before forwarding information to the end destination. However, at this time, none of the proposed algorithms make use of the existing network infrastructure.

• *Mobility.* Mobility might improve the efficiency of data aggregation techniques. For instance, nodes with controlled or predictable motion can be driven all over the network to collect, process and store data generated by static nodes. On the contrary, mobility can affect the performance of the data processing schemes based on the aggregation tree.

• *Node heterogenity.* Data processing may exploit objects with higher storage and computational capabilities as aggregation centers, in order to convey and process data from less-powerful objects displaced in a common region. Moreover, it is necessary to take into account the different node capabilities when the aggregation functions or the data structures are designed. For instance, the proposed Q-digest structure can be used to store data with a different degree of precision according to the storage capabilities of the nodes.

• *Power consumption.* Data processing techniques are, in general, implemented to limit the energy consumption by reducing the amount of transmitted data or the network overhead. Nevertheless, the required processing power contributes to depleting the energy resources of collector nodes. This aspect, however, is rarely considered in the literature and needs further investigation. At this time, aggregation functions implemented by algorithms such as Direct Diffusion, LEACH, TiNA, TAG are very simple (in general, they are a statistical function) and they do not require additional power consumption.

• *Real-time.* Data processing techniques usually involve time delay for gathering and processing many data units from different sources. Consequently, such techniques might not be guaranteed real-time requirements. This drawback is independent of the algorithm because it derives from the need to collect more than one packet before aggregating data and sending a new packet.

• *Reliability.* Data processing techniques are usually reliable, though, as mentioned, they might incur large delays that could affect the utility of the delivered data for the final node. Also in these cases, multipath strategies could guarantee a higher reliability than the tree-based schemes.

The data processing algorithm taxonomy is summarized in the following table.

	Direct diffusion	LEACH, PEGASIS	TAG, Cougar, TiNA	Synopsis diffusion	Tributaries and deltas
Topology	Tree-based	Tree-based	Tree-based	Ring	Hybrid
Scalability	Low	Low	Low	High	Medium
Fault tolerance	Low	Low	Low	High	High
Localization	None	None	None	None	None
Data traffic characteristics	Relevant	Relevant	Relevant	Relevant	Relevant
Networking infr.	None	None	None	None	None
Mobility	High resilience	High resilience	High resilience	Low resilience	Medium resilience
Node heterogenity	None	None	None	None	None
Power cons.	Low	Low	Low	High	Medium
Real-time	No	No	No	No	No
Reliability	Medium	Medium	Medium	High	High

Table 3.9. *Taxonomy of the data processing algorithm*

3.8.2.5. *Navigation algorithms*

• *Topology.* Information about the topology can be used to improve the accuracy of the localization algorithms, allowing for a better performance from the navigation algorithm. Localization has a more significant impact on the performance of the path computation and the following algorithm (PACFA) when comparing with others. Therefore, information about topology can improve the accuracy in the navigation in general, and especially with PACFA.

• *Scalability.* The navigation algorithms presented in this chapter use local information provided by the WSN and these algorithms have only been tested with one mobile node. Thus, scalability would depend on the capability of the communication protocol to support it. Furthermore, the use of a team of mobile nodes would involve other considerations such as the coordination among them for optimal covering of an area or collision avoidance. Those aspects could have a significant impact in the amount of information exchanged.

• *Fault tolerance.* The information provided by WSNs used in the navigation algorithms improve the fault tolerance with respect to mobile nodes that only use the information from sensors installed on-board.

• *Localization.* Localization of both the static nodes of the WSN and the mobile nodes is required for most of the navigation algorithms. The performance of the potential field guiding (POFA) and probabilistic navigation (PRONA) algorithms is more robust to localization errors than PACFA. If the localization of the nodes is not provided *a priori*, the navigation algorithm should also involve a position estimation.

• *Data traffic characteristics.* The navigation algorithms involve the exchange of a large amount of data among the static and the mobile nodes. These data should be updated at a rate which depends on the speed of the mobile node. PRONA does not require an explicit computation of the path and therefore a lower amount of data is involved. Finally, the increase in the information exchanged due to the navigation algorithm cannot exceed the capacity of the sensor network.

• *Networking infrastructure.* The navigation algorithms found in the literature only use local information from the nodes close to the mobile node. Then, there is not any special requirement regarding the networking infrastructure.

• *Mobility.* This is an intrinsic characteristic of these algorithms that can be applied to guide a robot or a person with a suitable interface in a given environment.

• *Node heterogenity.* This is also an intrinsic characteristic of these algorithms due to the fact that both static and mobile nodes are present. Even among the mobile nodes, different characteristics, such as the locomotion system, are possible.

• *Power consumption.* Power consumption of the nodes is increased due to the higher information exchange rate required during navigation. PACFA and POFA involve a first stage to compute the path, so a higher power consumption is required.

• *Real-time.* Real-time requirements mainly depend on the speed of the mobile node. On the other hand, the navigation algorithms found in the literature are designed to consider negligible delays in the information exchange among the nodes.

• *Reliability.* Besides the general reliability issues in WSNs, the reliability of the mobile platform itself must also be considered.

The navigation algorithm taxonomy is summarized in the following table.

Taxonomy of the navigation algorithms			
	POFA	PACFA	PRONA
Topology	Non-relevant	Relevant	Non-relevant
Scalability	Not considered	Not considered	Not considered
Fault tolerance	High	High	High
Localization	Less required	More required	Less required
Data traffic characteristics	Very high rate	Very high rate	High rate
Networking infrastructure	None	None	None
Mobility	Intrinsic	Intrinsic	Intrinsic
Node heterogenity	Intrinsic	Intrinsic	Intrinsic
Power consumption	Very high	Very high	High
Real-time	Relevant	Relevant	Relevant
Reliability	High	High	High

Table 3.10. *Taxonomy of the navigation algorithms*

3.8.2.6. *Timetable of the literature on the subject*

The following figure presents the different concepts that have been classified in this book according to the year of their appearance in the literature. The figure offers an overview of the trend's evolution.

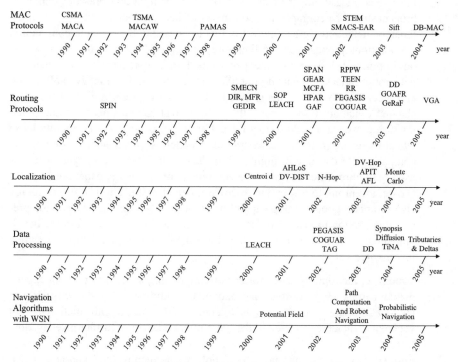

Figure 3.7. *Timetable of the literature*

3.9. Critical issues and research gaps

In this section, we point out the aspects that, according to the survey on the algorithms and paradigms presented in the previous sections, deserve further investigation in the future. We first discuss the research gaps that are transversal to the thematic areas and then we present the gaps that, on the contrary, are confined to a specific topic.

3.9.1. *Gaps with general scope*

Localization. Provisioning of localization information in distributed networks has been, and still is, an interesting research topic. Indeed, localization can be both a service offered by the network to the applications (object tracking, mapping, automatic driving, and so on) and a technique for improving the network operation itself. Many localization methods have been proposed, including the distance estimation based on the strength of the radio signals received by surrounding nodes, measurement of the time of flight of light and pressure impulses, broadcasting of geographical coordinates of beacon nodes, and so on. Unfortunately, despite the great interest in this topic within the research community, the solutions proposed in the literature often reveal their limits when carried out in practice. Indeed, initial experiments on this topic have revealed that signal strength measurements do not allow for a fine localization, even after long and precise calibration. This is mainly due to the unpredictability of the radio channel dynamic. Furthermore, the energy level has an impact on the transmission power of a node so that the calibration process becomes rapidly loose when nodes progressively discharge their batteries. Mechanisms based on the comparison of the time of flight of impulses with different propagation speeds usually require either cumbersome equipment, or are energy expending or, otherwise, are prone to errors due to environmental noise. The UWB-technology also promises good ranging capabilities that may be exploited to provide precise localization information. However, such a technology is still in the developmental phase and its performance in terms of ranging has not been widely investigated yet.

Therefore, the topic requires further investigation, particularly through experimental campaigns in various environmental conditions. Also, the sensitivity of the location-based algorithm to the tolerance in the location information has not yet been sufficiently covered in the literature and deserves further investigation.

QoS support. Several current and future applications require QoS guarantees. Although a relevant amount of work has been devoted to QoS-provisioning over wired and wireless networks, the topic has not been sufficiently covered in other contexts, such as that of WSNs or vehicular networks. In the perspective of seamless integration of hybrid systems, mechanisms to provide QoS over heterogenous platforms will become a primary issue.

Realistic models. Realistic channel, environment and object models would allow more accurate design and testing of algorithms. Actual solutions and analysis are based upon simplistic models of radio propagation as well as node capabilities. Therefore, the feasibility of the proposed solutions is still limited upon ideal models.

Evaluation of data processing costs. Most of the described algorithms make use of some data processing techniques to perform data fusion, data aggregation, routing and so on, in order to maximize the efficiency of the network functionalities and reduce the power consumption spent in the transmission procedure.

However, the computational and energy cost of this processing is rarely considered. Since many cross-layer approaches gain advantages from data processing, the cost of these utilities and their feasibility in sensor networks should be addressed.

Experimentation. Generally, solutions proposed in the literature hardly work when deployed in the real world. Solutions proposed are mostly tested by means of theoretical analysis or computer simulations under very simplistic hypotheses. There are only a few works that present empirical results. In the future, the research in WSN should be more experiment-oriented, in order to verify the limits of the theoretical analysis and to reveal the issues that inevitably arise when a system is actually deployed in the field. This knowledge can, thus, be used to refine existing solutions and develop novel approaches.

Heterogenity. The cooperation among heterogenous objects have the potential to drastically enhance the capabilities of the system. Up to now, heterogenity has been mainly considered in terms of different transmission ranges, computational capabilities, energy and storing capacity, and mobility. Several other possible synergies, however, have not been investigated yet. For instance, the integration of Radio-frequency Identifier (RFID) technology with WSNs provides a symbiotic solution that leads to improved performance of the system. Also, the cooperation among heterogenous autonomous objects, such as robots and WSNs, or WNS and IVN, as well as the integration of such systems with other communication systems (e.g. cellular), may result in a dramatic growth of the set of applications that might be provided to the users. Although this subject has often been addressed as a primary research topic in the recent past, the investigation is in a preliminary stage.

3.9.2. *Gaps in WSNs*

WSN with multiple sinks. Almost all the WSN solutions considered in this book take into account networks with only one sink. However, several application scenarios suggest the use of WSNs with multiple sinks. In this context, it is necessary, for example, to design and develop new routing protocols, new addressing mechanisms, more efficient localization methods and so on. These aspects need further investigation, from both theoretical and practical perspectives.

Energy and hardware constraints. The energy and hardware constraints are relevant issues in WSNs. The energy constraint limits the network lifetime, while hardware constraints impact on the complexity of the algorithms run by the network. Currently, energy-saving is mostly realized by allowing nodes to alternate between active and sleeping phases. As far as the hardware constraints are concerned, the literature usually considers very basic models, such as limited energy and storage capacity and limited computational power, without better specifying the actual bounds of such values. Following this approach, solutions are tailored to minimize the algorithms' complexity in terms of operations performed and/or memory occupied or, on the contrary, to minimize the energy consumption by limiting the number of transmissions, independently of the complexity of the algorithms used (see, for instance, data aggregation techniques). The result is that most of the solutions provided in the literature are not feasible or hardly effective when deployed in real-life devices. Therefore, further research is needed to better characterize the energy and hardware constraints of the devices and to develop more realistic models of the hardware architectures.

3.9.3. *Gaps in wireless sensor networks with mobile nodes*

WSN with mobile nodes. It should be pointed out that WSN with mobile nodes is a relatively new topic and it is still a developing field. The following topics have been found in the literature about WSN with mobile nodes: localization, communications (MAC and routing), planning and reactivity. However, more algorithms and theoretical studies are needed in those fields. Furthermore, there are some topics that have been poorly addressed, such as self-hierarchical organization and/or clustering techniques, fault tolerance and coverage techniques. Therefore, more efforts should be devoted to those aspects and their study should be promoted in future.

Motion planning. Mobile nodes can be introduced in a static WSN to collect and store the information of the static nodes of the network. An important trend is related to the study of how the sensor network can compute, in a distribute way, the path that the mobile node must follow. This path can be updated depending on changes of the environment or using new data collected by the sensor network. Also, we should consider communication constraints and the relation with self-hierarchical organization and/or clustering techniques. More algorithms and theoretical studies are needed in this topic.

3.9.4. *Gaps in autonomous robotics team*

Architecture. The system architecture affects the entire design space. Several trade-offs arise when comparing centralized and decentralized architectures. However, the literature lacks comparisons between these types of architecture. Therefore, more analysis is necessary in the future.

Networking with environment. A combination of multi-robot systems with WSNs can improve the overall reliability and performance of the whole system. The interaction between a WSN and a robotic team has not been addressed in the literature yet. A lot of work still has to be done in this field. In particular, there is a lack in distributed planning techniques and in the cooperative perception.

3.9.5. *Gaps in inter-vehicular networks*

Architecture. The literature published so far considers IVN with centralized and decentralized architecture. Both structures offer several advantages but it is not clear which is the best solution for this field. The future is probably in the use of hybrid architectures, but the topic deserves further investigation.

PHY and MAC layers. Most of the literature concerning IVN refers to either UMTS-based or 802.11-like PHY layers. However, the effectiveness of such transmission technologies in the context of IVN has not been thoroughly investigated yet. The IEEE organization has recognized this gap and is promoting the study of such aspects through the IEEE 802.11p working group, which is intended to investigate the wireless communication between fast moving vehicles or between a fast moving vehicle and a stationary object. However, the topic still leaves much room for further research.

Mobility. Clearly, routing is one of the most critical issues in IVN. Several algorithms proposed for vehicular networks are, in fact, derived by algorithms for ad hoc networks and, therefore, do not take advantage of the specific features of IVN. Furthermore, most of the solutions do not consider realistic mobility models. Therefore, this topic requires further investigation in order to define routing algorithms that are not only robust to the high topology dynamic that characterizes this scenario, but that are also able to exploit the specific features of the IVN context in order to improve the performance.

3.10. Conclusions

This last section is intended to wrap up the study by making some considerations on the definition of a unified framework for the design of algorithms for the different CO systems.

This chapter was conceived to provide an in-depth analysis of the literature regarding CO systems and, hence, to identify a set of algorithms and architectures that could form a common framework for the next generation of CO systems. Unfortunately, the CO umbrella encompasses systems with static and energy-limited nodes, very low duty cycles and very flexible delay constraints, as well as systems with autonomous mobile nodes, no energy supply problems, and strict requirements in terms of communication delay and reliability, which clearly arises from the analysis of the reference thematic areas considered in the study.

Therefore, the first conclusion that can be drawn from this chapter is that COs may present irreconcilable discrepancies, which hardly make feasible the definition of a unified approach for the design of algorithms and solutions for this type of system. Nonetheless, it is possible to identify some common *trends* that are transversal to the plethora of different design approaches.

Cross-layer. There are two main approaches that can be followed in the definition of algorithms for embedded systems: the *layered approach* and the *cross-layered approach*. The layered approach keeps a clear separation among the layers that carries out the different network functionalities, thus simplifying the implementation and updates of the algorithms. On the contrary, a cross-layered approach requires a joined design of the various network layers. This method usually makes it possible to define more effective algorithms by taking into consideration the interaction among the mechanisms that act at the different layers. In the case of WSN, for example, some physical parameters or application information can be exploited to define energy-efficient MAC or routing protocols. This performance improvement, though, is paid for in terms of flexibility of the software structure, which becomes more difficult to update and maintain. Therefore, there is a trade-off between the two approaches that has to be cut, weighting the performance advantage obtained by adopting a cross-layer approach against the difficulty in developing and maintaining the software stack. The general trend is to prefer a cross-layer approach, since the gain in performance that can be potentially achieved with a cross-layer approach largely compensates for the drawbacks of a more complex protocol structure.

Dynamic hierarchical architecture. From an architectural point of view, we can identify two general approaches: *distributed* and *centralized*. The first approach seems to be more suitable for WSNs due to the high number of nodes, the geographical extension of the network and the unpredictable network topology, which makes the realization of a centralized structure rather difficult. Nevertheless, the literature shows that, while preserving a basic distributed approach, the realization of dynamic clustering structures allows for a more efficient handling of the sensor resources. The trend is, hence, to follow a distributed approach in the design of algorithms for WSN, with the possibility of organizing the nodes in a dynamic hierarchical architecture, which allows for a more efficient management of the network resources. This advantage is particularly evident in the case of heterogenous networks, i.e. networks with some more powerful nodes that are capable of longer battery autonomy, longer transmission ranges, and larger storing and computational capabilities.

Location-based solutions. Many algorithms for WSN assume the availability of location information on the nodes. Some algorithms assume only that each node is acquainted (in some way) with its own spatial coordinates, while others only require the knowledge of the positions of the surrounding nodes, or of all the nodes in the network. Algorithms that make use of localization information for managing the medium access or the routing are usually much more efficient than location-independent algorithms, though they may turn out to be excessively sensitive to localization errors.

Asynchronous communication paradigms. Algorithms that are designed to work upon a network with synchronous nodes are usually more efficient that asynchronous algorithms. However, keeping the synchronization in a large network of nodes is a demanding task, both in terms of energy and control overhead. Indeed, cheap hardware leads to large clock drifts among the nodes, which have to exchange periodic information to maintain the synchronization. Furthermore, synchronous networks are fragile with respect to long-term topology variations (addition/elimination of nodes), which can be considered as fairly seldom events in static WSNs. On the other hand, asynchronous algorithms are much more flexible and easy to realize. If the cost to keep synchronization could be neglected, synchronous approaches would definitely overcome asynchronous solutions. However, the general trend, at least when considering protocols that are implemented in real networks, is to go for asynchronous solutions.

Wireless communications. Although the topic covered by this chapter refers exclusively to wireless networks, it is worth bearing in mind that many existing sensor networks make use of wired communication media. However, the trend is definitely towards the wireless solution, which opens the way to a large variety of new applications, most of which are in the safety, surveillance and monitoring area.

heterogenity. As stated, the cooperation among heterogenous objects may dramatically enlarge the potential of a communication system. Examples of hybrid solutions encompass the integration of RFID and WSN, or the cooperation among robots and WSN, or WSN and IVN. Although the investigation of this topic is still in a preliminary stage, this subject is considered a major research topic for the future.

3.11. Bibliography

[1] I.F. Akyildiz, W. Su, Y. Sanakarasubramaniam and E. Cayirci, "Wireless sensor networks: a survey", pp. 393–422, March 2002.

[2] J. E. H. Callaway, *Wireless Sensor Networks: Architectures and Protocols*, Auerbach Publications, Boca Raton, FL, August 2003.

[3] A.K. Sumit Khurana and A.P. Jayasumana, "Effect of hidden terminals on the performance of IEEE 802.11 MAC protocol", in *Proceedings of the 23rd Annual Conference on Local Computer Networks*, Institute of Electrical and Electronics Engineers, 1998.

[4] F. Tobagi and L. Kleinrock, "Packet switching in radio channels: Part II – the hidden terminal problem in carrier sense multiple access modes and the busy-tone solution", *IEEE Trans. on Communications*, vol. 23, pp. 1417–1433, December 1975.

[5] C.-K. Toh, "Ad hoc mobile wireless networks", *ACM Computer Communications Review*, July 1998.

[6] X. Kaixin and M. Gerla, "Effectiveness of RTS/CTS handshake in IEEE 802.11 based ad hoc networks", *Ad Hoc Networks*, vol. 1, no. 1, pp. 107–123, September 2003.

[7] H.S. Chhaya and S. Gupta, "Performance modeling of asynchronous data transfer methods of IEEE 802.11 MAC protocol", *Wireless Networks*, vol. 3, no. 3, pp. 217–234, 1997.

[8] M. Borgo, A. Zanella, P. Bisaglia, and S. Merlin, "Analysis of the hidden terminal effect in multi-rate IEEE 802.11b networks", in *Proceedings of WPMC04*, vol. 3, Abano Terme (Padova), Italy, 12-15 September 2004, pp. 6–10.

[9] P. Karn, "MACA – A New Channel Access Method for Packet Radio", *AARRL/CRRL Amateur Radio 9th Computer Networking Conference*, September 22 1990.

[10] V. Bharghavan, A. Demers, S. Shenker and L. Zhang, "MACAW: a media access protocol for wireless LANs", *ACM SIGCOMM Computer Communication Review*, pp. 212–225, August 1994.

[11] S. Singh and C. Raghavendra, "PAMAS: power aware multi-access protocol with signaling for ad hoc networks", *ACM SIGCOMM Computer Communication Review*, vol. 28, no. 3, pp. 5–26, July 1998.

[12] W. Ye, J. Heidemann and D. Estrin, "An energy-efficient MAC protocol for wireless sensor networks", *Proc. 21st Int'l. Joint Conf. IEEE Comp. Commun. Soc. (Infocom 2002)*, vol. 3, pp. 1567–1576, June 2002.

[13] K. Jamieson, H. Balakrishnan and Y.C. Tay, "Sift: a MAC protocol for event driven wireless sensor networks", MIT, Tech. Rep. LCS-TR-894, May 2003.

[14] C. Schurgers, V. Tsiatsis, S. Ganeriwal and M. Srivastava, "Optimizing Sensor Networks in the Energy-Latency-Density Design Space", *IEEE Transactions on Mobile Computing*, vol. 1, no. 1, pp. 70–80, January 2002.

[15] G. di Bacco, T. Melodia and F. Cuomo, "A MAC protocol for delay-bounded applications in wireless sensor networks", *Proc. 3rd Annual Mediterranean Ad Hoc Networking Workshop (Med Hoc Net)*, pp. 208–220, June 2004.

[16] K. Arisha, M. Youssef and M. Younis, "Energy-aware tdma-based MAC for sensor networks", in *IMPACCT 2002*, May 2002.

[17] V. Rajendran, K. Obraczka and J.J. Garcia-Luna-Aceves, "Energy-efficient collision-free medium access control for wireless sensor networks", *Proceedings of the 1st ACM International Conference on Embedded Networked Sensor Systems*, pp. 181–192, 2003.

[18] A.F. Imrich Chlamtac, "Making transmission schedules immune to topology changes in multi-hop packet radio networks", *IEEE/ACM Transactions on Networking*, vol. 2, no. 1, pp. 23–29, February 1994.

[19] H.Z. Imrich Chlamtac and András Faragó, "Time-spread multiple-access (TSMA) protocols for multi-hop mobile radio networks", *IEEE/ACM Transactions on Networking*, vol. 5, no. 6, pp. 804–812, December 1997.

[20] J.N. Al-Karaki and A.E. Kamal, "Routing techniques in wireless sensor networks: a survey", *Wireless Communications, IEEE*, vol. 11, no. 6, pp. 6–28, December 2004.

[21] F. Ye, H. Luo, J. Cheng, S. Lu and L. Zhang, "A two-tier data dissemination model for large-scale wireless sensor networks", *Proc. 8th ACM Int'l. Conf. Mob. Comp. and Net. (MOBICOM)*, pp. 148–159, September 2002.

[22] Y. Xu, J. Heidemann and D. Estrin, "Geography-informed energy conservation for ad hoc routing", *Proc. 7th ACM Int'l. Conf. Mob. Comp. and Net. (MOBICOM)*, pp. 70–84, 2001.

[23] B. Chen, K. Jamieson, H. Balakrishnan and R. Morris, "Span: an energy-efficient coordination algorithm for topology maintenance in ad hoc wireless networks", *Proc. 7th Int'l. Conf. Mob. Comp. Net. (MOBICOM)*, pp. 85–96, July 2001.

[24] I. Stojmenovic and X. Lin, "GEDIR: loop-free location based routing in wireless networks", *Int'l. Conf. Parallel and Distrib. Comp. and Net.*, November 1999.

[25] M. Zorzi and R.R. Rao, "Geographic random forwarding (GeRaF) for ad hoc and sensor networks: energy and latency performance", *IEEE Trans. on Mobile Computing*, vol. 2, no. 4, pp. 349–365, 2003.

[26] M. Zorzi and R. R. Rao, "Geographic random forwarding (GeRaF) for ad hoc and sensor networks: multihop performance", *IEEE Trans. on Mobile Computing*, vol. 2, no. 4, pp. 337–348, 2003.

[27] M. Zorzi, "A new contention-based MAC protocol for geographic forwarding in ad hoc and sensor networks", *Proc. IEEE Int'l. Conf. Commun. (ICC)*, vol. 6, pp. 3481–3485, June 2004.

[28] Y. Yu, D. Estrin and R. Govindan, "Geographical and energy-aware routing: a recursive data dissemination protocol for wireless sensor networks", UCLA Comp. Sci. Dept., Tech. Rep. 010023, May 2001.

[29] C. Intanagonwiwat, R. Govindan, D. Estrin, J. Heidemann and F. Silva, "Directed diffusion for wireless sensor networking", *ACM/IEEE Transactions on Networking*, vol. 11, no. 1, pp. 2–16, February 2003.

[30] W.R. Heinzelman, J. Kulik and H. Balakrishnan, "Adaptive protocols for information dissemination in wireless sensor networks", *Proc. 5th ACM/IEEE Int'l. Conf. Mob. Comp. and Net. (MOBICOM)*, pp. 174–185, August 1999.

[31] J. Kulik, W.R. Heinzelman and H. Balakrishnan, "Negotiation based protocols for disseminating information in wireless sensor networks", *Wireless Networks*, vol. 8, pp. 169–185, 2002.

[32] D. Braginsky and D. Estrin, "Rumor Routing Algorithm for Sensor Networks", *Proc. of the 1st ACM International Workshop on Wireless Sensor Networks and Applications*, 2002.

[33] F. Ye, A. Chen, L. Songwu and Z. Lixia, "A scalable solution to minimum cost forwarding in large sensor networks", *Proceedings of the 10th International Conference on Computer Communications and Networks*, vol. 8, pp. 304–309, October 2001.

[34] M. Chu, H. Haussecker and F. Zhao, "Scalable information-driven sensor querying and routing for ad hoc heterogenous sensor networks", *Int'l. J. High Performance Computing Applications*, vol. 8, Fall 2002.

[35] Y. Yao and J. Gehrke, "The COUGAR approach internetwrok query processing in sensor networks", *SIGMOD Record*, September 2002.

[36] S. Servetto and G. Barrenechea, "Constrained random walks on random graphs: routing algorithms for large scale wireless sensor networks", *Proc. 1th ACM Int'l. Wksp. Wireless Sensor Networks and Apps.*, 2002.

[37] W.R. Heinzelman, A. Chandrakasan and H. Balakrishnan, "Energy-efficient communication protocol for wireless microsensor networks", *Proc. 33rd Hawaii Int'l. Conf. Sys. Sci.*, pp. 3005–3014, January 2000.

[38] S. Lindsey and C. Raghavendra, "PEGASIS: power-efficient gatering in sensor information systems", *IEEE Aerospace Conference Proceedings*, vol. 3, pp. 1125–1130, July 2002.

[39] A. Manjeshwar and D. Agrawal, "TEEN: a routing protocol for enhanced efficiency in wireless sensor networks", *Proc. of the 15th International Parallel and Distributed Processing Symposium*, pp. 2009–2015, April 2001.

[40] A. Manjeshwar and D.P. Agrawal, "APTEEN: a hybrid protocol for efficient routing and comprehensive information retrieval in wireless sensor networks", *Proc. 16th Int'l. Parallel and Distrib. Proc. Symp.*, pp. 195–202, 2002.

[41] V. Rodoplu and T.H. Meng, "Minimum energy mobile wireless networks", *IEEE Journal on Selected Areas in Communications*, vol. 17, no. 8, pp. 1333–1344, August 1999.

[42] L. Li and J.Y. Halpern, "Minimum energy mobile wireless networks revisited", *Proc. IEEE Int'l. Conf. Commun. (ICC)*, vol. 1, pp. 278–283, June 2001.

[43] L. Subramanian and R.H. Katz, "An architecture for building self-configurable systems", *Proc. 1st ACM/IEEE Wksp. Mob. Comp. and Ad Hoc Net. (MOBIHOC)*, pp. 63–73, August 2000.

[44] J.N. Al-Karaki, R. Ul-Mustafa and A.E. Kamal, "Data aggregation in wireless sensor networks – exact and approximate algorithms", *IEEE Wksp. on High Perf. Switching and Routing*, pp. 241–245, April 2004.

[45] A. Savvides, C. Han and M.B. Strivastava, "Dynamic fine-grained localization in ad hoc networks of sensors", *Proc. 7th ACM Int'l. Conf. Mob. Comp. and Net.*, pp. 166–179, July 2001.

[46] J.N. Al-Karaki and A.E. Kamal, "On the correlated data gathering problem in wireless sensor networks", *Proc. 9th IEEE Symp. Comp. and Commun.*, vol. 1, pp. 226–231, July 2004.

[47] Q. Li, J. Aslam and D. Rus, "Hierarchical power-aware routing in sensor networks", *Proc. DIMACS Wksp. Pervasive Networking*, May 2001.

[48] B. Krishnamachari, D. Estrin and S. Wicker, "The impact of data aggregation in wireless sensor networks", *International Workshop on Distributed Event-Based Systems*, 2002.

[49] M. Lotfinezhad and B. Liang, "Effect of partially correlated data on clustering in wireless sensor networks", in *IEEE SECON 2004*, Santa Clara, CA, US, October 2004.

[50] S. Madden, M. Franklin, J. Hellerstein and W. Hong, "TAG: a tiny aggregation service for ad hoc sensor networks", in *OSDI 2003*, 2003.

[51] A. Sharaf, J. Beaver, A. Labrinidis and K. Chrysanthis, "Balancing energy efficiency and quality of aggregate data in sensor networks", *The VLDB Journal, The International Journal on Very Large Data Bases*, vol. 13, no. 4, December 2004.

[52] S. Nath, P.B. Gibbons, S. Seshan and Z.R. Anderson, "Aggregation: Synopsis diffusion for robust aggregation in sensor networks", in *ACM SenSys 2004*, Baltimore, Maryland, US, November 2004.

[53] A. Manjhi, S. Nath and P.B. Gibbons, "Tributaries and deltas: efficient and robust aggregation in sensor network streams", in *ACM SIGMOD 2005*, Baltimore, Maryland, US, June 2005.

[54] D. Baker, A. Ephremides and J. Flynn, "The design and simulation of a mobile radio network with distributed control", *IEEE Journal on Selected Areas in Communications SAC-2*, pp. 226–237, January 1984.

[55] A. Ephremides, J.E. Wieselthier and D.J. Baker, "A design concept for reliable mobile radio networks with frequency hopping signaling", in *Proceedings of the IEEE 75*, vol. 1, January 1987, pp. 56–73.

[56] E.M. Belding-Royer, "Multi-level hierarchies for scalable ad hoc routing", *ACM/Kluwer Wireless Networks 9*, vol. 5, pp. 461–478, September 2003.

[57] M. Gerla and C.R. Lin., "Multimedia transport in multi-hop dynamic packet radio networks", in *Proceedings of International Conference on Network Protocols*, Tokyo, Japan, 7–10 November 1995, pp. 209–216.

[58] M. Gerla and J.T.-C. Tsai, *Multicluster, Mobile, Multimedia Radio Networks*, Conference Proceedings, 1995.

[59] C.R. Lin and M. Gerla, "Adaptive clustering for mobile wireless networks", *Journal on Selected Areas in Communications 15*, pp. 1265–1275, September 1997.

[60] I. Chlamtac and A. Faragó, "A new approach to the design and analysis of peer-to-peer mobile networks", *Wireless Networks 5*, pp. 149–156, May 1999.

[61] C. Bettstetter and R. Krauser, "Scenario-based stability analysis of the distributed mobility-adaptive clustering (DMAC) algorithm", in *Proceedings of the 2nd ACM International Symposium on Mobile Ad Hoc Networking and Computing, MobiHoc 2001*, Long Beach, CA, October 4–5 2001, pp. 232–241.

[62] M. Chatterjee, S.K. Das and D. Turgut, "An on-demand weighted clustering algorithm (WCA) for ad hoc networks", in *Proceedings of IEEE Globecom 2000*, vol. 3, San Francisco, CA, November 27–December 1 2000, pp. 1697–1701.

[63] G. Chen, F. Nocetti, J. Gonzalez and I. Stojmenovic, "Connectivity-based k-hop clustering in wireless networks", in *HICSS '02: Proceedings of the 35th Annual Hawaii International Conference on System Sciences (HICSS' 02)*: Volume 7. Washington, DC, USA: IEEE Computer Society, 2002, p. 188.3.

[64] S. Basagni, "Distributed clustering for ad hoc networks", in *Proceedings of the 1999 International Symposium on Parallel Architectures, Algorithms, and Networks (I-SPAN'99)*, Perth/Fremantle, Australia, June 23–25 1999, pp. 310–315.

[65] S. Basagni, "Distributed and mobility-adaptive clustering for multimedia support in multi-hop wireless networks", in *Proceedings of the IEEE 50th International Vehicular Technology Conference, VTC 1999-Fall*, vol. 2, Amsterdam, The Netherlands, September 19–22 1999, pp. 889–893.

[66] C. Bettstetter, "The cluster density of a distributed clutering algorithm in ad hoc networks", in *Proceedings of the IEEE International Conference on Communications, ICC 2004*, vol. 7, Paris, France, June 20–24 2004, pp. 4336–4340.

[67] C. Bettstetter and B. Friedrich, "Time and message complexity of the generalized distributed mobility adaptive clustering (GDMAC) algorithm in wireless multi-hop networks", in *Proceedings of the 57th IEEE Semiannual Vehicular Technology Conference, VTC 2003-Spring*, vol. 1, Jeju, Korea, April 22–25 2003, pp. 176–180.

[68] P. Wan, K.M. Alzoubi and O. Frieder, "Distributed construction of connected dominating sets in wireless ad hoc networks", *ACM/Kluwer Mobile Networks and Applications, MONET 9*, pp. 141–149, April 2004.

[69] U.C. Kozat, G. Kondylis, B. Ryu and M.K. Marina, "Virtual dynamic backbone for mobile ad hoc networks", in *Proceedings of the IEEE International Conference on Communications (ICC 2001)*, vol. 1, Helsinki, Finland, June 11–14 2001, pp. 250–255.

[70] T. Nieberg, P. Havinga and J. Hurink, "Size-controlled dynamic clutering in mobile wireless sensor networks", in *Proceedings of the 2004 Communication Networks and Distributed Systems Modeling and Simulation Conference, (CSDN'04)*, San Diego, CA, January 18–21 2004.

[71] S. Basagni, D. Turgut and S.K. Das, "Mobility-adaptive protocols for managing large ad hoc networks", in *Proceedings of the IEEE International Conference on Communications (ICC 2001)*, vol. 5, Helsinki, Finland, June 11–14 2001, pp. 1539–1543.

[72] A.B. McDonald and T. Znati, "A mobility-based framework for adaptive clustering in wireless ad hoc networks", *IEEE Journal on Selected Areas in Communications, Special Issue on Wireless Ad Hoc Networks 17*, pp. 1466–1487, August 1999.

[73] L. Wang and S. Olariu, "A unifying look at clustering in mobile ad hoc networks", *Wireless Communications and Mobile Computing 4*, pp. 623–637, September 2004.

[74] B. Das and V. Bharghavan, "Routing in ad hoc networks using minimum connected dominating sets", *IEEE International Conference on Communications (ICC'97)*, pp. 376–380, June 8–12 1997.

[75] P. Sinha, R. Sivakumar and V. Bharghavan, "Enhancing ad hoc routing with dynamic virtual infrastructures", in *Proceedings of IEEE Infocom 2001*, vol. 3, Anchorage, AK, April 22-26 2001, pp. 1763–1762.

[76] R. Sivakumar, B. Das, V. Bharghavan, "The clade vertebrata: spines and routing in ad hoc networks", in *Prooceedings of the IEEE Symposium on Computer Communications (ISCC'98)*, Athens, Greece, June 30–July 2 1998.

[77] R. Sivakumar, B. Das and V Bharghavan, "Spine-based routing in ad hoc networks", November 1998, pp. 237–248.

[78] R. Sivakumar, P. Sinha and V. Bharghavan, "CEDAR: A core-extraction distributed ad hoc routing algorithm", *IEEE Journal on Selected Areas in Communications*, vol. 17, pp. 1454–1465, August 1999.

[79] B. Liang and Z. Haas, "Virtual backbone generation and maintenance in ad hoc network mobility management", in *Proceedings of the 19th IEEE Infocom*, vol. 3, Tel Aviv, Israel, March 26–30 2000, pp. 1293–1302.

[80] J. Wu and H. Li, "On calculating connected dominating sets for efficient routing in ad hoc wireless networks", *Telecommunication Systems, Special Issue on Mobile Computing and Wireless Networks*, vol. 18, no. 1/3, pp. 13–26, September 2001.

[81] J. Wu, F. Dai, M. Gao and I. Stojmenovic, "On calculating power-aware connected dominating sets for efficient routing in ad hoc wireless networks", *Journal of Communications and Networks 4*, pp. 1–12, March 2002.

[82] F. Dai and J. Wu, "An extended localized algorithms for connected dominating set formation in ad hoc wireless networks", *IEEE Transactions on Parallel and Distributed Systems 15*, October 2004.

[83] I. Stojmenovic, M. Seddigh and J. Zunic, "Dominating sets and neighbors elimination-based broadcasting algorithms in wireless networks", *IEEE Transactions on Parallel and Distributed Systems 13*, pp. 14–25, January 2002.

[84] P. Krishna, N.H. Vaidya, M. Chatterjee and D.K. Pradhan, "A cluster-based approach for routing in dynamic networks", *ACM SIGCOMM Computer Communication Review 27*, pp. 49–64, 2 April 1997.

[85] A.D. Amis, R. Prakash, T.H.P. Vuong and D.T. Huynh, "Max-min d-cluster formation in wireless ad hoc networks", in *Proceedings of IEEE Infocom 2000*, vol. 1, Tel Aviv, Israel, March 26–30 2000, pp. 32–41.

[86] K. Xu, X. Hong and M. Gerla, "An ad hoc network with mobile backbones", in *Proceedings of the IEEE International Conference on Communications (ICC 2002)*, vol. 5, New York, NY, April 28–May 2 2002, pp. 3138–3143.

[87] L. Bao and J.J. Garcia-Luna-Aceves, "Topology management in ad hoc networks", in *Proceedings of the 4th ACM International Symposium on Mobile Ad Hoc Networking and Computing, MobiHoc 2003*, Annapolis, MD, June 1–3 2003, pp. 129–140.

[88] S. Basagni, A. Carosi and C. Petrioli, "Sensor-DMAC: dynamic topology control for wireless sensor network", in *Proceedings of the 60th IEEE Vehicular Technology Conference, VTC 2004 Fall*, Los Angeles, CA, September 26–29 2004.

[89] S. Basagni, M. Elia and R. Ghosh, "ViBES: Virtual backbone for energy saving in wireless sensor networks", in *Proceedings of the IEEE Military Communication Conference, MILCOM 2004*, Monterey, CA, October 31–November 3 2004.

[90] A. Marcucci, M. Nati, C. Petrioli and A. Vitaletti, "Directed diffusion light: low overhead data dissemination in wireless sensor networks", in *Proceedings of IEEE VTC 2005 Spring*, Stockholm, Sweden, May 29-June 1 2005.

[91] W.R. Heinzelman, A. Chandrakasan and H. Balakrishnan, "Energy-efficient communication protocol for wireless microsensor networks", in *Proceedings of the 33rd Annual Hawaii International Conference on System Sciences, HICSS 2000*, Maui, HA, January 4–7 2000, pp. 3005–3014.

[92] S. Bandyopadhyay and E.J. Coyle, "An energy efficient hierarchical clustering algorithm for wireless sensor networks", in *Proceedings of the 22nd IEEE Infocom 2003*, vol. 3, San Francisco, March 31–April 3 2003, pp. 1713–1723.

[93] S. Bandyopadhyay and E.J. Coyle, "Minimizing communication costs in hierarchically clustered networks of wireless sensors", in *Proceedings of the IEEE Wireless Communications and Networking Conference (WCNC 2003)*, vol. 2, New Orleans, LA, March 16–20 2003, pp. 1274–1279.

[94] J. Tillett, R. Rao and F. Sahin, "Cluster head identification in ad hoc sensor networks using particle swarm optimization", in *Proceedings of the IEEE International Conference on Personal Wireless Communications, ICPWC 2002*, New Delhi, India, December 15–17 2002, pp. 201–205.

[95] H. Chan and A. Perrig, "Ace: An emergent algorithm for highly uniform cluster formation", in *Proceedings of the 1st IEEE European Workshop on Wireless Sensor Networks, EWSN 2004*, H. Karl, A. Willig and A. Wolisz, Eds., Berlin, Germany, January 19–21 2004, pp. 154–171.

[96] S. Banerjee and S. Khuller, "A clustering scheme for hierarchical control in multi-hop wireless networks", in *Proceedings of the 20th IEEE Infocom 2001*, vol. 2, Anchorage, AK, April 22–26 2001, pp. 1028–1037.

[97] G. Gupta and M. Younis, "Fault tolerant clustering of wireless sensor networks", in *Proceedings of the IEEE Wireless Communications and Networking Conference (WCNC 2003)*, vol. 3, New Orleans, LA, March 16–20 2003, pp. 1579–1584.

[98] G. Gupta and M. Younis, "Performance evaluation of load-balanced clustering of wireless sensor networks", in *Proceedings of the 10th International Conference on Telecommunications (ICT 2003)*, vol. 2, Papeete, French Polynesia, February 23–March 1 2003, pp. 1577–1583.

[99] G. Jolly, M.C. Kuscu, P. Kokate and M. Younis, "A low-energy key management protocol for wireless sensor networks", in *Proceedings of the 8th IEEE International Symposium on Computers and Communications (ISCC 2003)*, Kemer-Antalya, Turkey, June 30–July 3 2003, pp. 335–340.

[100] G. Gupta and M. Younis, "Load-balanced clustering of wireless sensor networks", in *Proceedings of the IEEE International Conference on Communications (ICC 2003)*, vol. 3, Anchorage, AK, May 11–15 2003, pp. 1848–1852.

[101] M. Younis, K. Akkaya and A. Kunjithapatham, "Optimization of task allocation in cluster-based sensor networks", in *Proceedings of the 8th IEEE International Symposium on Computers and Communications (ISCC 2003)*, Kemer-Antalya, Turkey, June 30–July 3 2003, pp. 329–334.

[102] M. Younis, M. Youssef and K. Arisha, "Energy-aware routing in cluster-based sensor networks", in *Proceedings of the 10th IEEE International Symposium on Modeling, Analysis and Simulations of Computer and Telecommunication Systems (MASCOTS 2002)*, Fort Worth, TX, October 11–16 2002, pp. 129–136.

[103] K. Akkaya and M. Younis, "An energy-aware qos routing protocol for wireless sensor networks", in *Proceedings of the 23rd International Conference on Distributed Computing Systems Workshop, (ICDCSW 2003)*, Providence, RI, May 19–22 2003, pp. 710–715.

[104] M. Gerla and K.Xu, "Multimedia streaming in large-scale sensor networks with mobile swarms", *SIGMOD Rec.*, vol. 32, no. 4, pp. 72–76, 2003.

[105] W.-P. Chen, J.C. Hou and L. Sha, "Dynamic clustering for acoustic target tracking in wireless sensor networks", *IEEE Transactions on Mobile Computing 3*, vol. 3, pp. 258–271, July 2004.

[106] V. Mhatre and C. Rosenberg, "Homogenous vs. heterogenous clustered sensor networks: a comparative study", in *Proceedings of the 2004 IEEE International Conference on Communications, ICC 2004*, vol. 6, Paris, France, June 20–24 2004, pp. 3646–3651.

[107] V. Mhatre and C. Rosenberg, "Design guidelines for wireless sensor networks: communication, clustering and aggregation", *Ad Hoc Networks*, vol. 2, no. 1, pp. 45–63, 2004.

[108] P. Santi and J. Simon, "Silence is golden with high probability: maintaining a connected backbone in wireless sensor networks", in *Proceedings of the 1st IEEE European Workshop on Wireless Sensor Networks, EWSN 2004*, H. Karl, A. Willig and A. Wolisz, Eds., Berlin, Germany, January 19–21 2004, pp. 106–212.

[109] Y. Xu, J. Heidemann and D. Estrin, "Geography-informed energy conservation for ad hoc routing", in *Proceedings of the 7th ACM Annual International Conference on Mobile Computing and Networking*, Rome, Italy, July 16–21 2001, pp. 70–84.

[110] A.T.J. Blum, M. Ding and X. Cheng, *Connected Dominating Sets in Sensor Networks and MANETs*, vol. 1, D.-Z. Du and P.M. Pardalos, Eds. Kluwer Academic Publishers, 2004.

[111] N. Bulusu, J. Heidemann and D. Estrin, "GPS-less low cost outdoor localization for very small devices", University of Southern California, Tech. Rep. Technical Report 00-729, April 2000 [Online]. Available: URL: http://citeseer.nj.nec.com/bulusu00gpsless.html

[112] D. Niculescu and B. Nath, "Dv based positioing in ad hoc networks", *Telecommunication Systems*, vol. 22, no. 1–4, pp. 267–280, 2003 [Online]. Available: URL: http://paul.rutgers .edu/ dnicules/research/aps/aps-jrn.pdf

[113] T. He, C. Huang, B.M. Blum, J.A. Stankovic and T. Abdelzaher, "Range-free localization schemes for large scale sensor networks", in *Proc. of 9th Annual ACM/IEEE International Conference on Mobile Computing and Networking (MobiCom 2003)*. ACM, Sept 2003 [Online]. Available: URL: http://citeseer.nj.nec.com/he03rangefree.html

[114] L. Hu and D. Evans, "Localization for mobile sensor networks", in *MobiCom '04: Proceedings of the 10th annual international conference on Mobile computing and networking*, ACM Press, 2004, pp. 45–57.

[115] S. Thrun, D. Fox, W. Burgard and F. Dellaert, "Robust Monte Carlo localization for mobile robots", *Artificial Intelligence*, vol. 128, no. 1-2, pp. 99–141, 2001.

[116] L. Doherty, L. Ghaoui and K. Pister, "Convex position estimation in wireless sensor networks", in *Proc. of IEEE INFOCOM '01*, vol. 3, IEEE, April 2001, pp. 1655–1663.

[117] S. Simic and S. Sastry, "Distributed localization in wireless ad hoc networks", UC Berkeley, Tech. Rep. UCB/ERL M02/26, 2002.

[118] G. Stupp and M. Sidi, "The expected uncertainty of range free localization protocols in sensor networks", in *Algorithmic Aspects of Wireless Sensor Networks: First International Workshop, ALGOSENSORS 2004, Turku, Finland, July 16, 2004. Proceedings*, ser. Lecture Notes in Computer Science Series, vol. 3121, Springer, 2004.

[119] N. B. Priyantha, A. Chakraborty and H. Balakrishnan, "The cricket location-support system", in *Proceedings of the 6th Annual International Conference on Mobile Computing and Networking*, ACM Press, 2000, pp. 32–43.

[120] N.B. Priyantha, A.K. Miu, H. Balakrishnan and S. Teller, "The cricket compass for context-aware mobile applications", in *Proceedings of the 7th Annual International Conference on Mobile Computing and Networking*, ACM Press, 2001, pp. 1–14.

[121] D. Moore, J. Leonard, D. Rus and S. Teller, "Robust distributed network localization wit noisy range measurements", in *SenSys '04: Proceedings of the Second ACM Conference on Embedded Networked Sensor Systems*, Baltimore, MD, November 2004.

[122] P. Bahl and V. N. Padmanabhan, "Radar: an in-building rf-based user location and tracking system", in *INFOCOM (2)*, 2000, pp. 775–784 [Online]. Available: URL: http://citeseer .ist.psu.edu/bahl00radar.html

[123] A. Savvides, C.-C. Han and M.B. Strivastava, "Dynamic fine-grained localization in ad hoc networks of sensors", in *Proceedings of the 7th Annual International Conference on Mobile Computing and Networking*, ACM Press, July 2001, pp. 166–179 [Online]. Available: URL: http://doi.acm.org/10.1145/381677.381693

[124] A. Savvides, H. Park and M.B. Srivastava, "The bits and flops of the n-hop multilateration primitive for node localization problems", in *Proceedings of the 1st ACM International Workshop on Wireless Sensor Networks and Applications*, ACM Press, September 28 2002, pp. 112–121 [Online]. Available: URL: http://doi.acm.org/10.1145/570738.570755

[125] D. Niculescu and B. Nath, "Ad hoc positioning system (aps)", in *Proc. of IEEE Global Telecommunications Conference (GLOBECOM 2001)*, vol. 1. GLOBECOM, November 2001, pp. 2926–2931 [Online]. Available: URL: http://citeseer.nj.nec.com/519054.html

[126] D. Niculescu and B. Nath, "Ad hoc positioning system (aps) using aoa", in *Proc. of 22nd Annual Joint Conference of the IEEE Computer and Communications Societies (INFOCOM 2003)*, vol. 3. San Francisco, CA: IEEE, March 2003, pp. 1734–1743 [Online]. Available: URL: http://citeseer.nj.nec.com/niculescu03ad.html

[127] D. Niculescu and B. Nath, "Error characteristics of ad hoc positioning systems", in *Proceedings of the 5th ACM International Symposium on Mobile Ad Hoc Networking and Computing, MobiHoc 2004*, Roppongi Hills, Tokyo, Japan, 2004, pp. 20–30.

[128] C. Savarese, J.M. Rabaey and K. Langendoen, "Robust positioning algorithms for distributed ad hoc wireless sensor networks", in *Proceedings of the General Track: 2002 USENIX Annual Technical Conference*, USENIX Association, 2002, pp. 317–327 [Online]. Available: URL: http://rama.pds.twi.tudelft.nl/ koen/papers/robust-positioning.pdf

[129] C. Fretzagias and M. Papadopouli, "Cooperative location sensing for wireless networks", in *2nd IEEE International Conference on Pervasive Computing and Communications*, Orlando, Florida, March 2004 [Online]. Available: URL: http://www.cs.unc.edu/maria/percom04.pdf

[130] K.K. Chintalapudi, A. Dhariwal, R. Govindan and G. Sukhatme, "Ad hoc localization using ranging and sectoring", in *IEEE INFOCOM '04: The Conference on Computer Communications*, vol. 1, March 2004, pp. 2662–2672.

[131] N.B. Priyantha, H. Balakrishnan, E. Demaine and S. Teller, "Anchor-free distributed localization in sensor networks", in *in Proc. of the 1st International Conference on Embedded Networked Sensor Systems (SenSys 2003)*, November 5-7 2003, pp. 340–341.

[132] T. Fruchterman and E. Reingold, "Graph drwaing by force-directed placement", *Software – Practice and Experience (SPE)*, vol. 21, no. 11, pp. 1129–1164, November 1991.

[133] A. Howard, M. Mataric and G. Sukhatme, "Relaxation on a mesh: A formalism for generalized localization", in *Proceedings of IEEE/RSJ Intl. Conference on Intelligent Robots and Systems (IROS)*, Wailea, Hawaii, October 2001.

[134] F. Zhao, J. Shin and J. Reich, "Information-driven dynamic sensor collaboration", *IEEE Signal Processing Mag.*, vol. 19, pp. 61–72, March 2002.

[135] F. Zhao, J. Liu, J. Liu, L. Guibas and J. Reich, "Collaborative signal and information processing: an information-directed approach", *Proc. of the IEEE*, vol. 91, no. 8, pp. 1199–1209, August 2003.

[136] I. Chatzigiannakis, S. Nikoletseas and P. Spirakis, "An efficient communication strategy for ad hoc mobile networks", in *Proceedings of the 20th Annual ACM Symposium on Principles of Distributed Computing, PODC 2001, Newport, RI*, August 26–29 2001, pp. 320–332.

[137] I. Chatzigiannakis and S. Nikoletseas, "An adaptive compulsory protocol for basic communications in highly changing ad hoc mobile networks", in *Proceedings of the International Parallel and Distributed Processing Symposium, IPDPS 2002, Fort Lauderdale, FL*, April 15–19 2002, pp. 193–202.

[138] P. Juang, H. Oki, Y. Wang, M. Martonosi, L.-S. Peh and D. Rubenstein, "Energy-efficient computing for wildlife tracking: Design trade-offs and early experiences with zebranet", in *Proceedings of the 10th International Conference on Architecural Support for Programming Languages and Operating Systems, ASPLOS-X*, October 5–9 2002, pp. 96–107.

[139] Q. Li and D. Rus, "Sending messages to mobile users in disconnected ad hoc wireless networks", in *Proceedings of the 6th ACM Annual International Conference on Mobile Computing and Networking, MobiCom 2000*, Boston, MA, August 6–11 2000, pp. 44–55.

[140] M. Grossglauser and D.N.C. Tse, "Mobility increases the capacity of ad-hoc wireless networks", *IEEE/ACM Transactions on Networking*, vol. 10, no. 4, pp. 477–486, August 2002.

[141] W. Zhao, M. Ammar and E.W. Zegura, "A message ferrying approach for data delivery in sparse mobile ad hoc networks", in *Proceedings of the 5th ACM International Symposium on Mobile Ad Hoc Networking and Computing, MobiHoc 2004*, Roppongi Hills, Tokyo, Japan, May 24–26 2004, pp. 187–198.

[142] S. Jain, R.C. Shah, W. Brunete, G. Borriello and S. Roy, "Exploiting mobility of energy-efficient data collection in sensor networks", in *Proceedings of the IEEE Workshop on Modeling and Optimization in Mobile, Ad Hoc and Wireless Networks, WiOpt 2004*, Cambridge, UK, March 24–26 2004.

[143] R.C. Shah, S. Roy, S. Jain and W. Brunette, "Data MULEs: Modeling a three-tier architecture for sparse sensor networks", in *Proceedings of the 1st IEEE International Workshop on Sensor Network Protocols and Applications, SNPA 2003*, Anchorage, AK, May 11 2003, pp. 30–41.

[144] H.S. Kim, T.F. Abdelzaher and W.H. Kwon, "Minimum energy asynchronous dissemination to mobile sinks in wireless sensor networks", in *Proceedings of the 1st International Conference on Embedded Networked Sensor Systems, SenSys 2003*, Los Angeles, CA, November, 5–7 2003, pp. 193–204.

[145] C. Intanagonwiwat, R. Govindan, D. Estrin, J. Heidemann and F. Silva, "Directed diffusion for wireless sensor networking", *IEEE/ACM Transactions on Networking*, vol. 11, no. 1, pp. 2–16, February 2003.

[146] F. Ye, H. Luo, J. Cheng, S. Lu and L. Zhang, "A two-tier data dissemination model for large scale wireless sensor networks", in *Proceedings of the 8th ACM Annual International Conference on Mobile Computing and Networking, MobiCom 2002*, Atlanta, GA, September 23–28 2002, pp. 148–159.

[147] J.G. Jetcheva and D.B. Johnson, "Adaptive demand-driven multicast routing in multi-hop wireless ad hoc networks", in *Proceedings of the 2nd ACM International Symposium on Mobile Ad Hoc Networking & Computing, MobiHoc 2001*, Long Beach, CA, October 4–5 2001, pp. 33–44.

[148] L. Tong, Q. Zhao and S. Adireddy, "Sensor networks with mobile agents", in *Proceedings of the IEEE Military Communication Conference, MILCOM 2003*, vol. 1, Boston, MA, October 13–16 2003, pp. 705–710.

[149] P. Venkitasubramaniam, S. Adireddy and L. Tong, "Sensor networks with mobile agents: optimal random access and coding", *IEEE Journal on Selected Areas in Communications*, vol. 22, no. 6, pp. 1058–1068, August 2004.

[150] V. P. Mhatre, C. Rosenberg, D. Kofman, R. Mazumdar and N. Shroff, "A minimum cost heterogenous sensor network with a lifetime contraint", *IEEE Transactions on Mobile Computing*, vol. 4, no. 1, pp. 4–15, January/February 2005.

[151] A. Chakrabarti, A. Sabharwal and B. Aazhang, "Using predictable observer mobility for power efficient design of sensor networks", in *Proceedings of the Second International Workshop on Information Processing in Sensor Networks, IPSN 2003*, F. Zhao and L. Guibas, Eds., Palo Alto, CA, April 22–23 2003, pp. 129–145.

[152] P. Baruah, R. Urgaonkar and B. Krishnamachari, "Learning-enforced time domain routing to mobile sinks in wireless sensor fields", in *Proceedings of the 29th Annual IEEE International Conference on Local Computer Networks, LCN 2004*, Tampa, FL, November 16–18 2004, pp. 525–532.

[153] A. Kansal, A.A. Somasundara, D.D. Jea, M.B. Srivastava and D. Estrin, "Intelligent fluid infrastructure for embedded networks", in *Proceedings of the 2nd ACM/SIGMOBILE International Conference on Mobile Systems, Applications, and Services, MobySys 2004*, Boston, MA, June 6–9 2004, pp. 111–124.

[154] D.K. Goldenberg, J. Lin, A.S. Morse, B.E. Rosen and Y.R. Yang, "Towards mobility as a network control primitive", in *Proceedings of the 5th ACM International Symposium on Mobile Ad Hoc Networking and Computing, MobiHoc 2004*, Roppongi Hills, Tokyo, Japan, May 24–26 2004, pp. 163–174.

[155] S.R. Gandham, M. Dawande, R. Prakash and S. Venkatesan, "Energy efficient schemes for wireless sensor networks with multiple mobile base stations", in *Proceedings of IEEE Globecom 2003*, vol. 1, San Francisco, CA, December 1–5 2003, pp. 377–381.

[156] Z.M. Wang, S. Basagni, E. Melachrinoudis and C. Petrioli, "Exploiting sink mobility for maximizing sensor networks lifetime", in *Proceedings of the 38th Hawaii International Conference on System Sciences, Big Island, Hawaii*, January 3–6 2005.

[157] R. Peterson and D. Rus, "Interacting with sensor networks", in *Proc. of the 2004 IEEE Intl. Conference on Robotics & Automation*, April 2004, pp. 180–186.

[158] M.B. McMickell, B. Goodwine and L.A. Montestruque, "Micabot: A robotic platform for large-scale distributed robotics", in *Proc. of the 2003 IEEE, Intl. Conference on Robotics & Automation*, September 2003, pp. 1600–1605.

[159] S. Bergbreiter and K.S.J. Pister, "Costbots: An off-theshelf platform for distributed robotics", in *Proc. of the 2003 IEEE/RSJ Intl. Conference on Intelligent Robots and Systems*, October 2003, pp. 1632–1637.

[160] K. Dantu, M. Rahimi, H. Shah, S. Babel, A. Dhariwal and G. Sukhatme, "Robomote: enabling mobility in sensor networks", Dept. of Computer Science, University of Southern California, Los Angeles, California, Tech. Rep. CRES-04-006, 2004.

[161] L. Navarro-Serment, R. Grabowski, C.J. Paredis and P.K. Khosla, "Millibots", *IEEE Robotics & Automation Magazine*, pp. 31–40, December 2002.

[162] P. Corke, S. Hrabar, R. Peterson, D. Rus, S. Saripalli and G. Sukhatme, "Autonomous deployment and repair of a sensor network using an unmanned aerial vehicle", in *Proc. of the 2004 IEEE Intl. Conference on Robotics & Automation*, April 2004, pp. 3602–3608.

[163] P. Corke, R. Peterson and D. Rus, "Networked robots: Flying robot navigation using a sensor net", in *International Symposium of Robotic Research (ISRR)*, 2003.

[164] J. McLurkin and J. Smith, "Distributed algorithms for dispersion in indoor environments using a swarm of autonomous mobile robots", in *Proc. of the 7th International Symposium on Distributed Autonomous Robotic Systems*, Toulouse, France, June 2004, pp. 381–390.

[165] J. Cortes, S. Martinez, S. Karatas and F. Bullo, "Coverage control for mobile sensing networks", in *Proceedings of the IEEE International Conference on Robotics and Automation*, Arlington, VA: IEEE, 2002, pp. 1327–1332.

[166] J.N. Al-Karaki and A.E. Kamal, "Routing techniques in wireless sensor networks: a survey", *IEEE Wireless Communications Magazine*, pp. 6–28, December 2004.

[167] D. Gale, *The Theory of Linear Economic Models*, Nueva York, EEUU: McGraw-Hill Book Company, Inc., 1960.

[168] H.W. Kuhn, "The hungarian method for the assignment problem", *Naval Research Logistics Quarterly*, vol. 2(1), pp. 83–97, 1955.

[169] D.P. Bertsekas, "The auction algorithm for assignment and other network flow problems: a tutorial", *Interfaces*, vol. 20(4), pp. 133–149, 1990.

[170] P. Brucker, *Scheduling Algorithms*, Berlin: Springer-Verlag, 1998.

[171] J.L. Bruno, E.G. Coffmann and R. Sethi, "Scheduling independent tasks to reduce mean finishing time", *Communications of the ACM*, vol. 17(7), pp. 382–387, 1974.

[172] M.B. Dias and A. Stentz, "A market approach to multi-robot coordination", The Robotics Institute, Carnegie Mellon University, Pittsburgh, Pennsylvania, Tech. Rep. CMU-RI-TR-01-26, 2001.

[173] K.L. Moore, Y. Chen and Z. Song, "Diffusion-based path planning in mobile actuator-sensor networks (mas-net): some preliminary results", in *Proceedings of SPIE*, April 2004.

[174] R.S. Aylett and D.P. Barnes, "A multi-robot architecture for planetary rovers", in *Proceedings of the 5th ESA Workshop on Space Robotics, ASTRA'98*, Noordwijk: European Space Agency, 1998.

[175] M. Mataric, M. Nilsson and K.T. Simsarian, "Cooperative multi-robot box-pushing", in *Proceedings of the IEEE/RSJ International Conference on Intelligent Robots and Systems*, vol. 3. Los Alamitos, CA: IEEE Computer Society Press, 1995, pp. 556–561.

[176] D. Rus, B. Donald and J. Jennings, "Moving furniture with teams of autonomous robots", in *Proceedings of the IEEE/RSJ International Conference on Intelligent Robots and Systems*, vol. 1, Pittsburgh, PA: IEEE, 1995, pp. 235–242.

[177] L.E. Parker, "Alliance: An architecture for fault tolerant multi-robot cooperation", *IEEE Transactions on Robotics and Automation*, vol. 14(2), pp. 220–240, 1998.

[178] Y.U. Cao, A.S. Fukunaga and A. Kahng, "Cooperative mobile robotics: antecedents and directions", *Autonomous Robots*, vol. 4(1), pp. 7–27, 1997.

[179] G. Dudek, M. Jenkin and E. Milios, *Robot Teams: From Diversity to Polymorphism*, Natick, MA, EEUU: A.K. Peters, 2002, pp. 3–22.

[180] S. Yuta and S. Premvuti, "Coordinating autonomous and centralized decision making to achieve cooperative behaviors between multiple mobile robots", in *Proceedings of the IEEE/RSJ International Conference on Intelligent Robots and Systems*, Raleigh, North Carolina, 1992, pp. 1566–1574.

[181] T. Balch, "Taxonomies of multi-robot task and reward", Carnegie Mellon University, Pittsburgh, Pennsylvania, Tech. Rep., 1998.

[182] K. Ali, "Multiagent telerobotics: matching systems to tasks", PhD dissertation, Georgia Institute of Technology, 1999.

[183] E. Todt, G. Raush and R. Suarez, "Analysis and classification of multiple robot coordination methods", in *Proceedings IEEE International Conference on Robotics and Automation*, San Francisco, California, 2000.

[184] G. Beni, "The concept of cellular robotic system", in *Proceedings of the IEEE International Symposium on Intelligent Control*, 1988, pp. 57–62.

[185] H. Asama, A. Matsumoto and Y. Ishida, "Design of an autonomous and distributed robot system: Actress", in *Proceedings of the IEEE/RSJ International Conference on Intelligent Robots and Systems*, 1989, pp. 283–290.

[186] T. Fukuda, Y. Kawauchi and H. Asama, "Analysis and evaluation of cellular robotics (cebot) as a distributed intelligent system by communication amount", in *Proceedings IEEE/RSJ International Conference on Intelligent Robots and Systems*, 1990, pp. 827–834.

[187] P. Caloud, W. Choi, J.C. Latombe, C. Le Pape and M. Yin, "Indoor automation with many mobile robots", in *Proceedings of the IEEE/RSJ International Conference on Intelligent Robots and Systems*, 1990, pp. 67–72.

[188] L.E. Parker, "Alliance: an architecture for fault tolerant, cooperative control of heterogenous mobile robots", in *Proceedings of the IEEE/RSJ International Conference on Intelligent Robots and Systems*, 1994, pp. 776–783.

[189] R. Alami, S. Fleury, M. Herrb, F. Ingrand and S. Qutub, "Operating a large fleet of mobile robots using the plan-merging paradigm", in *Proceedings of the IEEE International Conference on Robotics and Automation*, Albuquerque, New Mexico, April 1997, pp. 2312–2317.

[190] R. Alami, S. Fleury, M. Herrb, F. Ingrand and F. Robert, "Multi robot cooperation in the martha project", *IEEE Robotics and Automation Magazine*, vol. 5(1), pp. 36–47, March 1998.

[191] J.C. Latombe, *Robot Motion Planning*, Boston, MA: Kluwer Academic Publishers, 1991.

[192] M. Batalin, G.S. Sukhatme and M. Hatting, "Mobile robot navigation using a sensor network", in *Proc. of the 2004 IEEE Intl. Conference on Robotics & Automation*, April 2004, pp. 636–641.

[193] Q. Li, M. DeRosa and D. Rus, "Distributed algorithms for guiding navigation across a sensor network", 2003 [Online]. Available: URL: citeseer.ist.psu.edu/li03distributed.html

[194] U. Berkeley, "Path Project", URL: http://www.path.berkeley.edu/, April 2005.

[195] D. Reichardt, M. Miglietta, L. Moretti, P. Morsink and W. Schulz, "CarTALK 2000 – Safe and Comfortable Driving Based Upon Inter-Vehicle-Communication", in *IEEE Intelligent Vehicle Symposium*, Versailles, France, June 2002.

[196] N. *et al.*, "FleetNet Project", URL: http://www.ccrle.nec.de/Projects/fleetnet.htm, April 2005.

[197] L. Xu, R. Tonjes, T. Paila, W. Hansmann, M. Frank and M. Albrecht, "DRiVE-ing to the Internet: Dynamic Radio for IP Services in Vehicular Environments", in *The 25th Annual IEEE Conference on Local Computer Networks (LCN'00)*, Tampa, Florida, USA, November 2000, http://www.ist-drive.org/papers/Lcn2000/LCN2000.pdf.

[198] S. Tsugawa, K. Tokuda, S. Kato, T. Matsui and H. Fujii, "An overview on demo 2000 cooperative driving", in *IEEE Intelligent Vehicle Symposium (IV01)*, Tokyo, Japan, May 2001, pp. 327–332.

[199] L. Wischhof, A. Ebner, H. Rohling, M. Lott and R. Halfmann, "Adaptive Broadcast for Travel and Traffic Information Distribution Based on Inter-Vehicle Communication", in *IEEE Intelligent Vehicles Symposium (IV 2003)*, Columbus, Ohio, USA, June 2003, http://www.et2.tu-harburg.de/Mitarbeiter/Wischhof/IV2003_098.pdf.

[200] J. Blum, A. Eskandarian and L. Hoffman, "Challenges of Intervehicle Ad Hoc Networks", *IEEE Transaction on Intelligent Transportation Systems*, vol. 5, no. 4, pp. 347–351, December 2004.

[201] D. Cottingham, "Research Directions on Inter-vehicle Communication", URL: http://www.cl.cam.ac.uk/users/dnc25/references.html, December 2004.

[202] M. Rudack, M. Meincke, K. Jobmann and M. Lott, "On traffic dynamical aspects intervehicle communication (IVC)", in *57th IEEE Semiannual Vehicular Technology Conference (VTC03 Spring)*, Jeju, South Korea, April 2003, http://portal.acm.org/citation.cfm?id=778434.

[203] Q. Xu, K. Hedrick, R. Sengupta and J. VanderWerf, "Effects of Vehicle-vehicle/Roadside-vehicle Communication on Adaptive Cruise Controlled Highway Systems", in *IEEE VTC Fall 2002*, Vancouver, Canada, September 2002, http://path.berkeley.edu/dsrc/pub/vtc2002f.pdf.

[204] D.N. Godbole and J. Lygeros, "Longitudinal Control of the Lead Car of a Platoon", University of California at Berkeley, URL: http://www.path.berkeley.edu/PATH/Publications/PDF/TECHMEMOS/TECHMEMO-93-07.pdf, Tech. Rep. PATH Tech Report 93-7, November 1993.

[205] U. Nations, "A Summary of the Kyoto Protocol", URL: http://unfccc.int/essential_background/feeling_the_heat/items/2879.php.

[206] P. Vidales and F. Stajano, "The Sentient Car: Context-Aware Automotive Telematics", in *First EE Workshop on Location Based Services*, London, UK, September 2002, http://www.lce.eng.cam.ac.uk/pav25/publications/lbs-2002(abstract).pdf.

[207] S. Jain, K. Fall and R. Patra, "Routing in a delay tolerant networking", in *ACM SIGCOMM Technical Conference*, Portland, OR, USA, August 2004, http://www.acm.org/sigs/sigcomm/sigcomm2004/papers/p299-jain111111.pdf.

[208] "Wizzy Project", URL: http://www.wizzy.org.za/.

[209] "Sami Network Connectivity Project (SNC)", URL: http://www.snc.sapmi.net/.

[210] A. Lindgreny, A. Doria and O. Scheln, "Probablistic Routing in Intermittently Connected Networks", in *The First International Workshop on Service Assurance with Partial and Intermittent Resources SAPIR 2004. In Conjunction with ICT 2004*, Fortaleza, Brazil, August 2004.

[211] J. LeBrun, C.N. Chuah, D. Ghosal and H.M. Zhang, "Knowledge-Based Opportunistic Forwarding in Vehicular Wireless Ad Hoc Networks", in *IEEE Vehicular Technology Conference (VTC'05) – Spring*, Stockholm, Sweden, May 2005, http://www.ece.ucdavis.edu/chuah/paper/2005/vtc05-move.pdf.

[212] K. Fall, "A delay-tolerant network architecture for challenged internets", in *ACM SIGCOMM Technical Conference*, Karlsruhe, Germany, August 2003.

[213] S. Das, A. Nandan, G. Pau, M. Sanadidi and M. Gerla, "SPAWN: A Swarming Protocol for Vehicular Ad Hoc Networks", in *Proc. 1st ACM VANET*, Philadelphia, PA, USA, October 2004.

[214] J. Luo and J.-P. Hubaux, "A Survey of Inter-Vehicle Communication", EPFL, Tech. Rep. Tech report IC/2004/04, March 2004, http://icwww.epfl.ch/publications/abstract.php?ID=200424.

[215] J. Zhu and S. Roy, "MAC for dedicated short range communications in Intelligent Transport Systems", *IEEE Communications Magazine*, vol. 41, pp. 60–67, January 2003.

[216] P. Vidales, L. Patanapongpibul, G. Mapp and A. Hopper, "Experiences with Heterogenous Wireless Networks – Unveiling the Challenges", in *Second International Working Conference on Performance Modeling and Evaluation of Heterogenous Networks (HET-NETs)*, West Yorkshire, UK, July 2004.

[217] P. Vidales, J. Baliosian, J. Serrat, G. Mapp, F. Stajano and A. Hopper, "A Practical Approach for 4G Systems: Deployment of Overlay Networks", in *First International Conference on Testbeds and Research Infrastructures for the DEvelopment of NeTworks and COMmunities*, IEEE Computer Society Press, February 2005.

[218] P. Vidales, J. Baliosian, J. Serrat, G. Mapp, F. Stajano and A. Hopper, "Autonomic Systems for Mobility Support in 4G Networks", *Journal on Selected Areas in Communications (J-SAC), Special Issue in Autonomic Communications*, November 2005.

[219] F. Borgonovo, A. Capone, M. Cesana and L. Fratta, "Ad hoc MAC: A new, flexible and reliable MAC architecture for ad-hoc networks", in *IEEE Wireless Communications and Networking Conference (WCNC03)*, New Orleans, Louisiana, USA, March 2003, http://www.elet.polimi.it/upload/cesana/papers/WCNC2003.pdf.

[220] K. Tokuda, M. Akiyama and H. Fujii, "DOLPHIN for Inter-vehicle Communications System", in *IEEE Intelligent Vehicle Symposium (IV00)*, Dearborn, MI, USA, October 2000, pp. 327–332, http://ieeexplore.ieee.org/iel5/7217/19432/00898395.pdf?arnumber=898395.

[221] M. Lott, R. Halfmann, E. Schulz and M. Radimirsch, "Medium access and radio resource management for ad hoc networks based on UTRA TDD", in *ACM SIGMOBILE Symposium on Mobile Ad Hoc Networking and Computing (MobiHoc01)*, Long Beach, California, USA, October 2001, http://portal.acm.org/.

[222] D. Lee, R. Attias, A. Puri, R. Sengupta, S. Tripakis and P. Varaiya, "A Wireless Token Ring Protocol For Intelligent Transportation Systems", in *IEEE 4th International Conference on Intelligent Transportation Systems*, August 2001.

[223] Q. Xu, T. Mak, J. Ko and R. Sengupta, "MAC Protocol Design for Vehicle Safety Communications in Dedicated Short Range Communications Spectrum", in *IEEE ITSC 2004*, Washington, D.C., USA, October 2004, http://path.berkeley.edu/dsrc/pub/ITSC04.pdf.

[224] C. Perkins, *Ad Hoc Networking*, Addison-Wesley, 2001.

[225] B. Karp and H. Kung, "Greedy Perimeter Stateless Routing for Wireless Networks", in *Sixth Annual ACM/IEEE International Conference on Mobile Computing and Networking (MobiCom 2000)*, Boston, MA, August 2000, pp. 243–254.

[226] C. Lochert, H. Hartenstein, J. Tian, H. Fubler, D. Herrmann and M. Mauve, "A Routing Strategy for Vehicular Ad Hoc Networks in City Environments", in *IEEE Intelligent Vehicles Symposium*, Columbus, Ohio, USA, June 2003.

[227] M. Mauve, J. Widmer and H. Hartenstein, "A survey on position-based routing in mobile ad-hoc networks", *IEEE Network Magazine*, vol. 15, no. 6, pp. 30–39, November 2001.

[228] S. Giordano and I. Stojmenovic, *Position-based Routing Algorithms for Ad Hoc Networks: A Taxonomy*, Kluwer, 2004, pp. 103–136.

[229] G. Coulouris, J. Dollimore and T. Kindberg, *Distributed Systems, Concepts and Design*, 3rd ed., Addison-Wesley, 2001.

[230] C. Maihfer, "A Survey on Geocast Routing Protocols", *IEEE Communications Surveys and Tutorials*, vol. 6, no. 2, June 2004, http://www.comsoc.org/livepubs/surveys/public/2004/apr/maihofer.html.

[231] A. Ebner, L. Wischhof and H. Rohling, "Aspects of Decentralized Time Synchronization in Vehicular Ad hoc Networks", in *1st International Workshop on Intelligent Transportation (WIT 2004)*, Hamburg, Germany, March 2004, http://www.et2.tu-harburg.de/Mitarbeiter/Wischhof/Ebner_WIT2004.pdf.

Chapter 4

Vertical System Functions

4.1. Summary

This chapter discusses the roles and effects of *vertical system functions (VFs)*, which provide the required system functionality to address the needs of the CO applications.

4.2. Introduction

In the scope of this book, a CO is defined as a collection of sensors, actuators, controllers or other COs that communicate with each other to achieve, more or less autonomously, a common goal. This recursive definition accounts for the cases where groups of COs can be regarded as a single CO. Such arrangements enable them to combine hardware and software components to support advanced sensor applications. Thus, organizing such components into a framework that can cope with the inherent complexity of the overall CO system will be an important exercise for developers.

The list below of key points in the design space of a CO is based on the requirements set for the wireless sensor platforms developed independently at UCLA and UC Berkeley [37, 48]:

• *Small physical size:* reducing physical size has always been one of the key design issues. For instance, some applications will need more powerful CO units than others.

Chapter written by Marcelo PIAS, George COULOURIS, Pedro José MARRÓN, Daniel MINDER, Nirvana MERATNIA, Maria LIJDING, Paul HAVINGA, Şebnem BAYDERE, Erdal ÇAYIRCI and Chiara PETRIOLI.

Hence, COs are likely to be heterogenous devices in terms of processing, communication and sensing capabilities. Such diversity poses the challenge to find the right balance between the physical device size and the minimal set of required hardware subsystems to be implemented in a CO.

• *Low power consumption:* energy limits processing, lifetime and interconnect capabilities of the basic CO device. The system should make efficient use of the resources striving to minimize the overall power consumption. As a result, this will increase the CO active time without battery recharging, which is an issue for a set of applications such as large-scale forest fire monitoring.

• *Concurrency-intensive operation:* data will be frequently gathered from local sensors or received from other COs. They are then processed through filtering/aggregation, and sent to other COs through the network. These tasks should be carried out simultaneously in order to achieve the target sensing and actuation goals at the highest performance. Therefore, strict resource sharing and task scheduling are key design issues.

• *Diversity in design and usage:* networked sensor devices will tend to be application-domain specific providing only the necessary hardware support. For instance, sensors and actuators built for medical healthcare applications exhibit more complexity when compared to simpler sensors used in environmental monitoring. Therefore, the hardware and software framework of a CO should facilitate trade-offs among component reuse, cost and efficiency.

• *Robust operation:* CO devices will be numerous and deployed over a large environment. The individual devices should be carefully designed to have reliability as one of the key properties. Although device failures can be overcome with distributed fail-over techniques, this approach should be avoided whenever possible because of the communication costs incurred. COs need autonomous management abilities to self-test, self-calibrate and self-repair as recently advocated in [56, 62].

• *Security, privacy and trust:* each CO should have sufficient security mechanisms in place to prevent and counter unauthorized access, denial-of-service attacks and unintentional damage of the locally stored information. These mechanisms need to be considered in the design phase in order to make them pervasive throughout the system.

• *Compatibility:* the cost to develop software components dominates the cost of the overall system. It is important to be able to reuse code developed for other systems.

• *Flexibility:* the CO system will evolve over time both in terms of hardware and software. Support for this growth can be introduced in the system through hardware programmability and reconfiguration by using programmable processors and FPGA-based platforms. Also, software modularity introduces flexibility and should be prioritized during the system design.

The question that arises is whether any available real-time operating system (RTOS) suits this list of requirements. Some researchers believed that traditional

RTOS was unsuitable and therefore designed alternatives such as the micro-threaded operating system TinyOS [48]. Whether this new OS is entirely appropriate for a CO system remains to be investigated. As advanced sensor applications emerge, it is likely that the emphasis will be put on the design of more complex micro-electro-mechanical (MEMS) sensors as already suggested in [129]. To support this sensor development, a CO system will be built from powerful resources that need more implemented functionality and efficient interactions than those currently offered by TinyOS. Readers are referred to Chapter 5 for a comprehensive discussion on CO systems and architectures.

The rest of the chapter is organized as follows: section 4.3 defines a VF in the context of COs. Section 4.4 briefly discusses the characteristics and requirements of the CO applications studied in Chapter 2. Different types of VFs to address these application needs are then discussed including context and location management, data consistency, and communication and security, to name just a few. Section 4.5 gives a summary of the VFs discussed in the chapter and concludes with some final remarks.

4.3. Vertical System Function (VF)

The current operating systems proposed for WSNs and COs cannot offer all the required functionality to the applications. Thus, a VF is defined in this chapter as the *functionality that addresses the needs of applications in specific domains and in some cases a VF also offers minimal essential functionality that is missing from available RTOS.*

Figure 4.1 introduces a simplified architectural view of an application example originally discussed in [48]. The goal is to monitor the temperature and light conditions of an area and periodically transmit their measurements to a central base station. In this example, there are three VFs that applications may "invoke": the communication subsystem (core system function), the light sensors and temperature sensors (application specific functions). Each VF is represented in the diagram by a vertical stack of components. We envisage that *standardized APIs* will create mechanisms for linking applications to vertical functions.

A component, represented by the labeled box in the diagram, is defined as a self-contained unit of code that encapsulates its implementation and interacts with its environment by means of well-defined interfaces. Such an interaction can be achieved through a *software wiring* process that connects the output ports of a component to the input ports of another one. The composition of components can be represented as a system configuration graph, where components are vertices and their interconnects form the graph edges [24, 48, 61].

The essence of component interactions is twofold. The *control* interaction handles the requests for data to lower-level components (top-down). In contrast, the *data* interaction deals with the requested data by creating a bottom-up data flow path between components. This is shown in the figure as vertical lines with arrows.

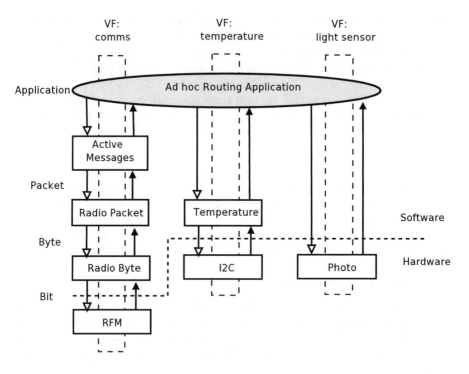

Figure 4.1. *Example of VFs: temperature and light sensing application*

The component wiring process will facilitate the synthesis of individual components into a larger-scale system. Components may be grouped together to form more concise modules [24, 81]. The CO system may be arranged into several *levels of abstraction*, from the most abstract (closest to the application) down to the most concrete (closest to the hardware devices). Abstraction allows better software structuring for clarity and reuse. Although layers are natural consequences of arranging components in the system, they are not a design requirement.

The implementation of a VF will follow a layered approach when the system architecture is designed in such a way. Although layered architectures facilitate key design properties such as flexibility and abstraction levels, they require attention to conformance and can severely impact performance in systems that lack hardware resources, for instance, low-power wireless sensor platforms.

The question to address in this case is how we map a general layered CO system architecture to a resource-constrained hardware platform without sacrificing overall performance. Ideally, we seek to minimize the overhead imposed on the system by the various components and levels of abstraction. It is important to note that critical real-time applications such as control of industrial plants may not tolerate high system response times.

To achieve an energy-efficient design, the traditional strict modularization or layering is not appropriate. In WSNs, for instance, monolithic design of communication software is used to reach the required energy-efficiency needs. However, such a design choice makes system development and management very difficult.

An approach that can be used to improve performance in systems low in resources is to implement the VF using *cross-layer* interactions, where the software components do not necessarily interact with components immediately above or below in the abstraction level [75]. For instance, most abstract components (closest to the application) may bypass other components and interact directly with the most concrete components (closest to the hardware). The cross-layer design explored in [28, 40] have shown promising results in reducing power consumption.

The architectural framework introduced in Figure 4.1 refers to a stand-alone CO. We believe that in practice there will be collections of COs in constant interaction to accomplish a pre-assigned goal. In some application scenarios, VFs will be implemented through a chain of software components that may or may not be within the operational boundaries of a single CO, but rather distributed in the network. For example, a VF that is responsible for collecting temperature readings of rooms in an office building needs distributed coordination among COs located in each room in order to implement the intended functionality.

4.4. Types of vertical system functions

The set of characteristics exhibited in CO applications are more diverse than those found in applications of traditional wireless and wired networks. Critical factors impact the architectural and protocol design of such applications. These factors also introduce some strict constraints.

Chapter 2 examined this set of characteristics and requirements. Below, we briefly review the relevant ones and discuss the most suitable VFs to address them. The reader should refer to Chapter 2 for a comprehensive study on selected COs applications.

Network topology: in a CO application, nodes may communicate directly provided they are geographically close to each other. Such a communication can be established in a *single-hop* network topology. When nodes are located far from each other, they need to rely on third nodes to forward their data packets, therefore requiring a *multi-hop* sensor network. VFs which may offer direct support to this characteristic are *communication* (section 4.4.3) and *distributed storage and data search* (section 4.4.5).

Scalability: the number of COs that may support an application can vary depending on the environment where it is deployed and on its task. This property is important in outdoor applications and it is often a design issue for techniques for *distributed storage and search*. As the system scales up, consistency of the data gathered from multiple sources should be addressed in an efficient manner. *The data consistency* functionality is extensively discussed in section 4.4.2.

Fault tolerance: it is very possible that some COs may fail during the operation of the network for various reasons including battery discharge and harsh environmental operation conditions. The *data consistency* VF tackles some of the issues associated with node failures. In addition, the *communication* VF should provide the data communication resilience required by the applications. Fault tolerance is also closely related to the *security* VF (section 4.4.4) since node failures may be caused by attackers.

Localization: there exist several CO applications for target tracking and physical event detection, including intrusion and forest fire, that require node and/or target localization. GPS may be the natural choice for calculating a node's location. These devices, however, do not work in indoor areas and still carry a high cost for low-power sensor nodes. The *context and location management* VF (section 4.4.1) reviews the most recent research advances in this area.

Time synchronization: applications need to establish a common sense of time among the COs participating in their sensing and actuation goals. Such a functionality can be offered through a *time synchronization* VF (section 4.4.8).

Security: the CO system may be threatened by unauthorized users trying to access the network. Also, there are security risks on the physical layer of the network. For example, a jamming signal may corrupt the radio communication between the entities in the mission-critical networks. In a CO, security is pervasive and must be integrated into every system component to achieve a secure system. Thus, the security functionality is likely to be offered at different levels and not exclusively by the *security, privacy and trust* VF.

Data traffic characteristics: the amount of data traveling inside the network determines the traffic characteristics of an application. In a particular application, the data transferred among nodes may be limited to a few bytes for simple measurements whereas heavy video-audio traffic may be conveyed in another application scenario. At least three VFs can provide the adequate support for different types of traffic. The *communication* and *distributed storage and data search* for high-level dissemination of sensor data offer mechanisms for transferring the data of interest in the network. In addition, the *data aggregation* (section 4.4.6) provides an energy-efficient optimization tool for various types of data traffic.

Networking infrastructure: CO networks can be *infrastructured* or *infrastructure-less (ad hoc)*. In some applications, the data can be collected by some mobile nodes when passing by the source nodes. This is an important characteristic that determines the type of system approach used (with or without supporting infrastructure) in the majority of VFs surveyed in this chapter including the techniques for *context and location management* VF and MAC layer/routing protocols in the *communication* VF.

Mobility: in some applications, all physical components of the system may be static whereas in others, the architecture may contain mobile nodes. Applications which can benefit from autonomous robots for actuation may require special assistance for mobility. Adequate support for medium and high mobility in multi-hop networks is still an open issue that should be addressed in the implementation of future VFs such as *context and location management, communication* and many others.

Node heterogenity: the majority of CO applications include nodes that have distinct hardware and software technical specifications. In a precision agriculture application, for instance, there may exist various types of sensors such as biological and chemical. Energy may be limited in some of the nodes. Thus, the *data search VF mechanism* needs to be energy-efficient. Also, data will originate from different sensor nodes so that adequate schemes for ensuring consistency of heterogenous sensor data must be used. This can be offered to the CO application through the *data consistency* and *aggregation* VFs.

Power awareness: power consumption is one of the performance metrics and limiting factors in almost any CO application. Systems require prolonged network lifetimes. Thus, efficient power consumption strategies must be developed. Power-aware communication protocols are supported in the *communication* VF. Also, as more complex sensors are designed – for instance in healthcare applications – there is a growing need for tighter control on nodes' resources to save energy. The *resource management* VF (section 4.4.7) deals with the power consumption issue.

Real-time: the system delay requirements are very stringent in real-time applications. The broad meaning of delay in this context is comprised of the system data processing and network delay. For instance, in an industrial automation scenario, actuation signals are required in real-time. VFs are capable of offering the required functionality to applications through cross-layer system approaches which can significantly reduce the overall system delay. Thus, resource monitoring and system adaptation achieved with cross-layer component-based interactions are important schemes that should be made available to the applications. The *resource management* VF can offer the necessary functionality to real-time applications.

Reliability: end-to-end reliability guarantees that the transmitted data is properly received by the receiving-end. In some applications, end-to-end reliability may be a dominating performance metric, whereas it may not be important for others. In security and surveillance applications in particular, guaranteed end-to-end delivery is of high importance. The non-functional requirement can be fulfilled through the coordination and interaction of various VFs including *data consistency*, *communication* and *security, privacy and trust*.

We will follow the order below to discuss the relevant VFs in the sections that follow:

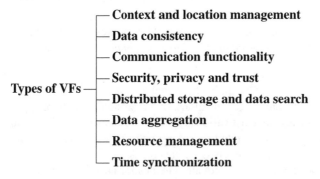

Types of VFs
- **Context and location management**
- **Data consistency**
- **Communication functionality**
- **Security, privacy and trust**
- **Distributed storage and data search**
- **Data aggregation**
- **Resource management**
- **Time synchronization**

4.4.1. *Context and location management*

Distributed CO systems are designed to measure properties of the physical world. They are therefore suitable for gathering the context of an entity, which is the information that can be used to characterize its situation. Individuals, locations or any relevant object can be such entities [13].

Most researchers argue that context information has three major attributes through which it can be accessed: the identity of the entities, their location and the time at which the information has been gathered. Since a reasonable amount of data is collected in large systems, context management systems are needed to handle them. Such systems can separate applications from the process of data processing and context fusion. Additionally, this allows a number of COs to share the gathered context.

Changes in context may also trigger actions to influence the monitored entity. Specialized actuators, for instance, may be programmed to control pipe valves when a fluid pressure reaches a certain threshold. To achieve precise actuation and detailed analysis of collected measurement data, however, the spatial distribution of sensors needs to be known. Thus, determining the *location* of COs is a requirement of a large number of applications, including monitoring of habitat, urban and indoor areas.

This section explores the design space of vertical functions for context and location management.

4.4.1.1. *Context management*

Classical context management systems use infrastructure-based directories to store the information, for example, Aura [39] and Nexus [49]. This has the advantage of the device knowing where the information can be queried from. On the other hand, to avoid bottlenecks, the data has to be structured in some way and distributed among several devices. With usually small COs, the distribution of the context is system-driven. Also, the context often has only limited spatial relevance which is reflected by the typical wireless communication which is also spatially restricted. It is reasonable to store the context at or near the location where it is generated.

To start with the gathering of the sensor data, MiLAN [44] allows for the decoupling of application and data gathering. The user has to provide a specification of which sensors, or a set of sensors, can provide what QoS for which data. Additionally, the user specifies which data with what QoS are needed when the application is in various states. Then, MiLAN makes sure that the needed data is available. The actual source of the data is transparent to the application.

The management of data in a single device is the focus of the MobileMan project [27]. It creates a cross-layer architecture for the network protocol stack in mobile ad hoc networks. MobileMan primarily aims at developing a network protocol stack that is optimized with cross-layer interactions. A more general approach is followed in TinyCubus [75]. Cross-layer data such as context information is stored in a state repository. The cross-layer framework ensures that data needed by an application component is provided by another.

With TinyDB [71], a whole network can be regarded as a database. It mainly focuses on external queries since they are parsed and optimized externally, but a fixed set of pre-parsed and pre-optimized queries could also be used inside the network. Such database approaches thus provide an easy way to get context data when the storage location is unknown.

In geographic hash tables [97], the storage location can be calculated from the index key. Each node has to have a geographic location, and the data is stored in the node geographically nearest to the hash of its key. Thus, context data can be queried directly from the storage node.

The COs themselves belong to the context as well. [135] presents a self-monitoring system for sensor networks. It continuously computes aggregates (sum, average, count) of network properties like loss rates, energy levels, etc., and disseminates them in the network in an energy-efficient manner. All objects can access the system context of the network.

We have shown the different parts of context management, starting with the gathering of sensor data. This data is made available to all components of an application and to other nodes in the network in different ways. Several approaches exist for these purposes, but to the best of our knowledge none covers the complete context management area. Therefore, more research is needed here.

4.4.1.2. Context-aware applications

Applications are called context-aware if they adapt their behavior based on the context. Several forms of this adaptation exist including the selection of information, the change of the presentation or the triggering of some action based on gathered context.

The GUIDE project [25] developed a tourist guide for the city of Lancaster in the UK. Personal and environmental context are used including, for example, the visitor's interests, his or her current location, the time of the day and the opening hours of attractions. The information is presented with respect to the age and the technical background of the visitor.

Sharing of context of a mobile phone user was the focus of the TEA project [110]. The user can set his current context using his mobile phone, for example, to "Free" or "Meeting". This information is presented to the caller who can then decide to call anyway, to leave a message, or to cancel the call.

Gaia OS [100] provides support for Active Spaces which combines physical and virtual contextual information related to the physical space and allows interaction with the physical space. Applications can be developed without strict knowledge of the infrastructure – Gaia is responsible for mapping these applications to a physical space. A framework that separates model, view, and controllers allows for runtime adaptation due to changes in the environment.

In ad hoc environments, context-based adaptation may have different goals, e.g. lower energy consumption or lower delay. Impala [68] contains an application adaptor that adapts the application protocols to different runtime conditions, which include system and application parameters. Adaptation decisions are made using a finite state machine where states represent different protocols/applications. Each directed edge carries a parameter expression for the condition under which the switch occurs. The adaptation goal is, therefore, implicitly given by these conditions.

TinyCubus [75] uses several variables in three dimensions: "system parameters", "application requirements", and "optimization parameters" of the object context to perform the adaptation. For each combination of parameter values, an algorithm which performs best is known to the system. Based on policies and different adaptation strategies, a set of algorithms is selected that provides the functionality required by the application and that fulfills the best desired optimization.

4.4.1.3. *Location management*

Location services for mobile ad hoc networks only offer limited context information, i.e. the position of mobile objects. Prior to storing the location in such a service, the location has to be determined. Small COs usually do not have a GPS device, so different approaches are needed.

We refer the reader to [47, 84] for an extensive discussion on infrastructured and infrastructure-less location mechanisms. This section covers some of the most recent research advances in location management and determination.

A scalable, distributed location service is GLS [67]. Each node has a small set of other nodes as its location servers and updates them periodically with its location. Therefore, it does not have to know their actual identities but only their identifiers. All routing, including routing of location queries, uses a predefined ordering of node identifiers and a predefined geographic hierarchy.

Two basic approaches are commonly used to determine the position of objects. Having the distances to three objects of which the location is known, its own location can be calculated. The other possibility is to measure the angle to two known objects. Since most COs are equipped with omnidirectional antennae, a very accurate measurement of the angles is not feasible. However, [74] shows that the location estimation is significantly more accurate and feasible on Mica2 motes using two directional antennae.

Several methods exist to determine the distances to other nodes, for example time of flight or attenuation. Besides the normal radio, ultrasound can be used [109], which is a more accurate system for distance measurement and does not suffer from problems like relying on radio received signal strength. In order to calculate the node locations, in [109] a global non-linear optimization problem is set up and solved. A fully distributed approximation for the solving algorithm is presented. Some special nodes capable of long distance ranging (e.g. using long range ultrasound) are used as initial location beacons.

The location discovery algorithm in [36] works with distance measurements based on the received signal strength indication (RSSI) values delivered by the RF chip. Especially in indoor environments, RSSI can have an error as large as 50% of the measured distance. Since the error conforms to a Gaussian random variable, it can be calculated with the location. The standard deviation is used as the degree of precision of the location. The precision estimates of all values are considered in all subsequent calculations of location for undetermined nodes, thus accumulating the errors in the results of the new calculation.

While the last approach assumes static nodes, an algorithm for semi-static sensor networks is presented in [33]. It uses the properties of a randomly deployed sensor network, more precisely, the average density of nodes that are uniformly distributed. Assuming a fixed transmission range for all nodes, the distance between any two nodes can be calculated using only the hop count between them. In a further step, the hop count is combined with a distance estimate obtained in the traditional way. By limiting the number of hops a distance message can travel, the precision can be increased again.

Although considerable research has been done in the area of location estimation, positioning is still too inaccurate. Also, for mobile COs no satisfactory approaches exist. Therefore, more research is needed in this direction.

4.4.2. *Data consistency*

The benefits of having several CO nodes mostly come from the fact that many nodes simultaneously monitor the same physical area. Nodes can be put into sleep mode without any loss of precision in the network. This results in the conservation of energy and an increased network lifetime.

The reliability of the system is also improved with several sensor nodes. This scenario, however, raises issues regarding data inconsistencies which may occur due to various reasons, for instance inherent imprecision associated with sensors, inconsistent readings and unreliable data transfer, to name just a few.

This VF provides the functionality to ensure consistency of the sensor data at various system abstraction levels:

• Data consistency may mean that data retrieved from a location in the sensor network should be consistent with data sent to the same location.

• Data consistency may also mean that all sensors sensing the same physical phenomenon should more or less agree on the measured value.

• In a rule-based system, data consistency may mean that all actuators agree on the action that needs to be taken.

Different mechanisms have been used to solve data inconsistency problems in various levels. As will be shown in the next section, while the focus of the first definition of "data consistency" is on a low level, the second concerns the "data consistency" on a higher level and is more commonly known as "consensus" and "data aggregation". The third definition, also on a higher level, relates to the future direction of handling data inconsistency for complex WSN applications.

4.4.2.1. *Consistency handling mechanisms (operation of WSN)*

The following requirements are needed to ensure accurate and consistent operation of the WSN and COs:

Localization

In order to interpret sensor data and collaborate with other nodes, it is crucial for sensor nodes to know approximately the position of the nodes with whom they collaborate. Usually, the manual configuration of nodes' positions is not feasible, and in situations where nodes are mobile this approach is even impossible. The availability of calculating and communication capabilities on nodes makes it possible to use automated location techniques.

Synchronization

WSNs and distributed CO systems must have a mechanism to ensure all nodes have an equal understanding of time and the moment at which events take place. Consequently, the nodes must keep their local clocks approximately synchronized with respect to a reference time, which may be one of the sensor nodes or an external source of time (e.g. GPS). The time synchronization VF is discussed in section 4.4.8.

Reliable data transfer

Applications require guaranteed delivery of information and/or customizable degrees of reliability for data transfer. As sensor nodes are ubiquitously deployed, they can overcome lack of reliability through cooperation. Nevertheless, achieving dependability through collaboration among error-prone entities is a challenging task. On the one hand, collaboration mechanisms should be ingenious enough to provide best-effort robust communication of important data, even in harsh conditions. On the other hand, the overhead introduced by cooperation along with the additional energy consumption should be kept to a reasonable level. Due to the fact that standard approaches cannot be applied to WSNs, reliability remains an open research issue.

Routing

Routing is an essential mechanism in networking, WSNs included. Unlike traditional networks, there are no dedicated routers in ad hoc sensor networks. Instead, data forwarding from source(s) to destination(s) is accomplished through local collaboration among neighbors. The various techniques proposed in other works strive to achieve energy-efficiency while maintaining a best-effort level of reliability.

Coverage

The coverage problem in WSNs generally refers to how well an area is monitored. Monitoring an area by several sensors has the advantages of *(i)* being able to turn off some of the sensors, thus saving valuable energy, and *(ii)* enhancing the accuracy of the sensed data by averaging multiple readings, for instance.

4.4.2.2. *Consistency handling mechanisms (data processing)*

Data consistency at the data processing phase can be achieved through the following mechanisms.

State monitoring

Sensor nodes can detect any change in the state of the environment in which they are placed directly from the sensed and measured data. In many situations, the whole network will not have a single global state that needs to be monitored, but the network will monitor the states of local processes, restricted to a confined area. The state might also be unique to each sensor node, or to the object the node is operating on behalf of.

An attempt to formalize state monitoring is presented in [102]. Instead of composing a state machine that is triggered by the occurrences of events, the authors propose to describe the creations between states and the events that trigger state-changes through a set of rules and predicates over events and their parameters. In this way, arbitrarily complex state-change conditions can be defined. In this system, states have a binary nature as they are either occurring or not occurring. A change of state can be used to trigger additional events or to perform actions. Strohbach *et al.* [119] used a similar approach, using simple predicate logic to monitor hazardous situations such as chemical drums.

Crucial to the design of a collaborative distributed state monitoring is the use of a high-level description of the state transitions, and other relevant aspects.

Data fusion

If redundancy is used to cover each point or region with multiple sensors, then the accuracy and the consistency of sensed data can be improved by merging or fusing correlated sensor data. Various schemes for data fusion have been proposed which deal with the reduction in transmission rates of the radio module (an expensive sensor node resource).

In this case, however, we are faced with the problem of fusing or combining the data reported by each of the sensors monitoring a specified point or region.

This is a challenge as measurements recorded by the sensors can differ (because of inherent imprecision in the sensors and/or the relative location of a sensor with respect to the monitored region) and the fact that sensors might be faulty. The objective of sensor data fusion is to take the multiple measurements and determine either the correct measurement value or a range in which the correct measurement lies.

The sensor fusion problem is closely related to the Byzantine agreement problem that has been extensively studied in other works. In [107], a hybrid distributed sensor-fusion algorithm is presented. Each sensor needs to compute a range, in which the true value lies as well as the expected value. For this calculation each sensor sends its measurement and its estimated accuracy to every other sensor. This algorithm is executed by every sensor using the measurement ranges received from the remaining sensors monitoring the same region combined with the sensor's own measurement. A typical drawback of this approach is the considerable communication overhead introduced by exchanging so many messages. This is an important issue in the context of energy constrained WSN nodes.

Event detection

Event detection is similar to state monitoring in the sense that the sensor network itself is in charge of monitoring the environment in which its nodes are placed, and it is used to detect the occurrence of certain events.

With the event detection mechanism, exceptional situations can be detected and consequently reported. Each sensor node has the task of detecting a possible event based on the data it obtains through its sensors, and through communication with other nodes. Additionally, the occurrence of an event, along with its position, magnitude or other properties need to be reported. In many situations, the value of these additional properties can only be determined, or more accurately be determined, when the sensor data of multiple sensor nodes are considered. The detection and reporting of these tasks will need to be performed in an efficient way, using only limited communication between the nodes, whenever data of interest is gathered, instead of reporting every sensed data sample. Additionally, a description of the event and its properties will be needed to perform the task.

Event detection has been tried in several WSN applications already. In [128], a description is given on a networked sensor array for monitoring volcanic eruptions. Seismic and infrasonic data is gathered continuously, and when higher levels of activity are measured, the recently measured data from different sensors is correlated and analyzed to find data on the recent event. The implementation described does not make use of in-network processing of the sensor data, but a feasibility analysis is performed. Other uses of event detection are shown in systems that are concerned with the detection, identification, localization or tracking of objects in sensor fields [101, 123, 42, 31]. In these applications, sensor nodes around the object or event of interest collaborate to find the location of the object or event.

Fault tolerance and consensus

Clouqueur *et al.* in [26] presents two distinct approaches, value-fusion and decision-fusion, for achieving fault tolerance in collaborative target detection algorithms. When performing a target detection task, multiple sensors in a region detect the presence of an object using sound, motion, or heat associated with the object of interest. Therefore, combining their views and obtaining a consistent conclusion through the fusion process is highly desirable. The value-fusion method consists of two phases: *(i)* exchanging the measured values and *(ii)* arriving to a consensus by computing the average of values and comparing it to a threshold. In the presence of faulty sensors, in order to preserve precision and accuracy, the extreme values are dropped from the set that is going to be averaged. In contrast, in the decision-fusion method, each device first makes an independent decision as to whether or not a target is present and then the devices exchange their decisions to arrive at a fault tolerant consensus decision. As in value-fusion, the fused data is obtained by averaging data received from all the sensors. For the situation of faulty nodes, exact agreement is used to preserve precision. The comparative results show that value-fusion is clearly preferable if the sensor network is highly reliable and fault-free. However, when faulty nodes are present, the performance of value-fusion degrades faster than the performance of decision-fusion and decision-fusion becomes superior to value-fusion. Achieving fault tolerance through consensus is a broad problem. Generally, each node has the task of trying to obtain a confident statement of the state or situation the environment is in. This information constitutes the state information present at the node. The state information needs to be refreshed periodically to account for changes in the topology due to movement of wireless sensors or due to nodes encountering crash failures.

The major challenge within the WSN context is to devise protocols that minimize the effort for refreshing and exchanging state information over the network. The protocol described in [64] generates consensus under the assumption that the number of faulty nodes is less than the number of correct nodes. The protocol is designed to enable self-correction in the network by isolating the faulty nodes and putting them in sleep mode, thereby increasing the concentration of nodes having correct information and improving their ability to generate consensus. The consensus process relies on forming a quorum, i.e. subgroups of sensor nodes with a certain minimum size. For instance, if there are k nodes having the same area of interest, the size of the quorum should be at least $(k + 1)/2$. A quorum is created after exchanging information among nodes as follows: *(i)* the node that needs to generate consensus starts the process by sending its information, *(ii)* other nodes answer with positive or negative acknowledgements, according to their information and *(iii)* the consensus is generated if the initial node receives more than $(k + 1)/2$ positive acknowledgements. As we have mentioned before, under WSN constraints, protocols should reduce as much as possible the number of messages exchanged.

One way to achieve this is using aggregation [3] within the consensus process, either by concatenating values, summarizing them, or by the counting of a set. Other operations as well as aggregation of data other than scalar numbers could also be possible. Key factors in using aggregation is that the operation reduces the amount of data from a whole series of numbers (or different type of data) to a single number, or a single value, or small set of values, and that the aggregation can be performed hierarchically, creating an aggregate of a larger data set from the aggregates of smaller sets. Aggregate operations do need the support of some sensor network management structure or service to ensure proper operation. Aggregation can also make the protocol more susceptible to channel errors: for low to medium bit error rates concatenation has good performance, whereas for a high bit error rate aggregation is not recommended.

In conclusion, the consensus problem in WSN has additional constraints compared to traditional distributed systems. However, it is a very important process for increasing the confidence of the overall system, both by fusing sensed data and by providing fault tolerance.

4.4.2.3. *Consistency handling mechanisms (application programming)*

The more intelligent collaborative WSN applications become, the greater the need for algorithms capable of supporting collective reasoning and actions in an efficient way. The execution space for these collaborative algorithms is represented by groups of COs that combine their efforts to be able to agree on an action to be taken.

Such collaborative algorithms need to benefit from a novel high-level description mechanism (language) oriented towards the collective model of solving tasks. A promising mechanism, extensively explored in the database research community, is to have a rule-based language [119] to implement collaborative WSN applications. In this case, a set of rules that are stated as machine-understandable statements describes legal or allowable states or situations as well as the alerts that must be given when otherwise.

Although such a scenario has the advantage of providing a high adaptive framework for collaborative WSN applications, which in turn can enhance system performance, it may cause severe problems if the rules and policies have not been carefully specified. A contributing factor to data inconsistency in rule-based collaborative applications is *versioning*. To prevent this kind of inconsistency from happening, the fact that the COs use the same (latest) version of the rules should be enforced. On the other hand, since rules and policies should be executed collaboratively, the best performance of the system highly depends on consistency between such rules for all COs involved. Otherwise, rules may cause conflict and prevent the system from functioning.

Therefore, strategies are needed to check for rule consistencies of COs.

4.4.3. *Communication functionality*

The communication vertical function refers to the capability of any pair (or group) of devices to exchange information. Different kinds of communications can be performed: one-to-all, one-to-many, many-to-one, many-to-many. If we consider the case of WSNs with a single sink, one-to-all or one-to-many communications are needed for the sake of interest and query dissemination, while many-to-one communication is exploited to gather sensed data at the sink.

The first problem that needs to be addressed is CO addressing. Communication in a COs environment is expected to be *data-centric* and attribute-based. This means that along with addressing a specific CO the communication infrastructure should be able to deliver data to and from groups of COs which share a set of attributes specifying the destination/source address of the information. For example, a user could issue a query about the average temperature of an area in an office building. This query should be delivered to the objects with temperature sensors. Similarly, once measures on the temperature have been taken, the COs in the specific area will send packets to the sink(s) reporting the measured values.

Other important VF parameters which should be included in a query are the time constraints and accuracy with which a given query needs to be answered. These are application-dependent and even query-dependent parameters: (a) for a query to be successfully resolved, the sensors must deliver new data, (b) the query must be answered fast enough, and (c) the precision with which the query is answered must meet the query requirements. This translates into a new concept of "quality of service" requirements.

Once a communication request between two or more COs is initiated, protocols have to be adopted to deliver the transmitted information from the sender to the final destination. Such protocols will typically involve physical layer protocols, data link protocols (FEC, ARQ protocols), MAC protocols, "topology control" schemes (e.g. addressing the self-organization of nodes into a hierarchical network topology or into dynamically changing communications infrastructures according to given awake-asleep schedules) and routing protocols.

The second issue to address is the fact that a one-fits-all protocol stack may not be suitable for this scenario. It is possible to identify the "vertical functions" that should be provided and possible implementations of such functions for specific sets of possible applications. The communication protocols will benefit from and sometimes require the information provided by different vertical functions such as time synchronization and location awareness. Not only time and location information is included in the delivered data, but some protocols such as geographic-based routing can exploit location-awareness to reduce routing overhead and the nodes' storage demand.

Mobility of the objects is the third issue to be tackled. Although in many CO scenarios the devices themselves are unlikely to be mobile, they can be located on mobile users or mobile stations so that their location changes in time in a predictable or unpredictable way. On one hand this may have a beneficial effect (e.g. load balancing the energy consumption among the different nodes), but on the other hand it requires mobility management or mobility-aware protocols to be added to the protocol stack. The mobility of some of the devices have been explored by some architectures such as the data mules [112], in which a group of mobile nodes move in the deployment area collecting data from the sensor nodes and delivering the collected data to the sinks.

Although TinyOS [48] has been serving as the basis for experimentation of existing and new communication protocols, a common and consolidated framework for comparisons needs to be developed further.

4.4.4. *Security, privacy and trust*

COs are usually placed in locations that are accessible to everyone, including attackers. For example, sensor networks are expected to consist of a few hundred nodes that may cover a large area. It is impossible to protect each of them from physical or logical attacks. Thus, every single node is a possible point of attack.

In a CO, security is pervasive. [87] states that security must be integrated into every component to achieve a secure system. Components designed without security can become a point of attack, as [54] shows. However, specific vertical functions to enforce security are available for applications.

We start by describing how the hardware can be secured. Then, several encryption approaches for COs are presented. We conclude with a look at routing protocols.

4.4.4.1. *Resource protection*

Having physical access to the devices without countermeasures, an attacker could read the node's memory including cryptographic keys and reprogram the node with malicious code. Therefore, security starts with the hardware of sensor networks.

Tamper-resistant devices would make the integration of security in sensor networks much easier – we could rely on the strength of security protocols or cryptographic functions which have been around for many years. Unfortunately, total tamper-resistance is very hard to achieve [9]. There are a lot of known techniques to read data from devices which actually are meant to be tamper-resistant. For instance, [60] describes methods for extracting protected software and data from smartcard processors. They show invasive techniques such as microprobing and non-invasive techniques such as software attacks, eavesdropping and fault generation. They even go further and also show that additional countermeasures such as additional metallization layers, which form a sensor grid above the actual circuit, do not protect the circuit fully. Of course, some of these methods require a well-funded adversary, but reviewing for instance military applications, the existence of such an adversary is highly probable. Additionally, achieving such a tamper-resistance level is very costly, which in most cases is not in line with the requirement that a single device should be low cost.

Thus, the conclusion is that in sensor networks tamper-resistant devices can be used only in very limited and specialized areas. In the majority of cases, the design of the security protocols has to take into account that some devices get compromised by the adversary. Thus, security protocols for sensor networks have to be designed in a way that they tolerate malfunctioning/attacking nodes while the whole sensor network remains functional. Karlof and Wagner [54] discuss a *graceful degradation* of the network in contrast to a totally compromised network.

Besides the above security issues, another danger for sensor networks, which is hard to prevent is the denial of service attack (DoS) on the physical layer. In such an attack, the attacker broadcasts a high-energy signal to prevent any node from communicating. The attacker can also use the features of the MAC protocol by continuously requesting channel access to eliminate all normal communication. Cryptographically secure spread-spectrum communication would be a defence against such jamming, but unfortunately, such RF-devices are not commercially available yet [87]. In the case of the jammed region not covering the whole network, the border nodes could create a map of this region and reroute traffic around the jammed region [130]. Also, battery exhaustion is a DoS attack. This can be counteracted, for instance, with rate limiting [130], which causes the network to ignore requests beyond a threshold value. Thus, although there are some approaches to cope with some kinds of DoS attacks, there is still further research needed in this area.

4.4.4.2. *Encryption*

Encryption is the basic technique for securing and authenticating transmitted data. Using asymmetric cryptography on highly resource constrained devices is often not possible due to delay, energy and memory constraints [20, 18]. It is also not expected that the devices of such sensor networks are going to have more resources in the future. There is an interesting observation by Karlof and Wagner that sensor network devices will more likely ride Moore's law *downward* [54]. They make the point that instead of doubling computational power every 18 months, it may be more likely that the devices become even smaller and cheaper solutions are sought.

Using symmetric cryptographical methods, the easiest solution is a shared key for the whole network as in secure pebblenets [11]. The disadvantages are obvious: if a single node is compromised, all the network traffic can be decrypted. Additionally, no device-to-device authenticity can be achieved.

If a unique key is only shared between any two nodes of the network, a single compromised node only has a limited impact. The attacker can only read and modify data that was sent to it, i.e. the compromised node directly. All other network traffic remains secure. As a drawback of this approach, a node needs to store keys for all other nodes, which may not be in line with scalability and memory constraints. Moreover, these unique keys have to be established in some way.

SPINS [88] uses a central base station to establish new session keys. This introduces a single point of failure that has to be protected exceptionally. Additionally,

COs exhibit a strong ad hoc character. Therefore, neither an infrastructure nor a base station can be assumed. A decentralized key management system is, therefore, more practical.

Several approaches are based upon random pairwise key pre-distribution [35, 23]. That is, before deployment – for instance, by the sensor manufacturer – every device is supplied with a random set of keys from a key pool. After deployment, the devices try to establish connections by finding a commonly shared key or by creating a new key through a secure path including other devices. Due to random pre-distribution, real authenticated communication between arbitrary devices is not always possible. It can only be assured with some probability that two arbitrary devices are able to communicate in a secure way. Moreover, an attacker could also reconstruct the complete key pool by compromising enough nodes.

If the key pre-distribution does not rely on a random set but on some algorithmic chosen set, the secure communication between arbitrary devices can be guaranteed. In [125], such an approach is presented. The authors suggest physical contact[1] for establishing the initial key set: the manufacturer could pre-distribute this initial set of keys according to their algorithm. In their approach, when two devices do not share a key yet, they can establish a new key by sending key-shares along node-disjoint paths via secured point-to-point communication to each other. When all key-shares are available at both nodes (which want to establish a new key) they just have to perform, for instance, a bitwise XOR operation on all key-shares: the result of this operation is the newly unique key, shared only by those two devices. It should be clear that the attacker would also need access to all key-shares in order to gain knowledge of the new key.

Since in-network processing is used to save power, end-to-end encryption can only be used sparingly. Along an aggregation tree, only point-to-point encryption is feasible, but an attacker in the tree has full access to the data. If the aggregation is locally bounded, the results of the aggregation can be sent directly and encrypted to the target node since no intermediate node is expected to change the data.

4.4.4.3. *Secrecy*

Encryption is not sufficient for ensuring secrecy of data. Traffic analysis on the ciphertext can reveal sensitive information about the data. When, for example, a motion detector sends a message, we do not need to know the data, but can assume that it detected a motion. On higher layers, additional dummy messages have to be generated to hide the important messages. This seems to be diametrical in resource

---

1. (Physical contact for establishing an initial security association, i.e. a shared secret/key is also suggested by Stajano and Anderson [118]).

constrained COs, but secrecy may be a more important goal than energy saving in some cases.

A dynamic virtual infrastructure for WSNs is presented in [126]. The system contains a coordinate system, a cluster structure and a routing structure. An energy-efficient protocol is proposed to maintain the anonymity of the network virtual infrastructure by randomizing communications so that it cannot be observed by an external attacker.

4.4.4.4. *Privacy*

The primary goal of sensor networks is to observe real-life phenomena. As long as nature is the target, there is no privacy issue, but if humans are observed, privacy is threatened. However, COs used in medical applications, for example, have to monitor the status of a patient. Encryption has to ensure that data cannot be overheard and is only available to legitimate applications. If nodes are compromised, other mechanisms like intrusion detection have to exclude these nodes from receiving data.

Data of other sensor networks might be available to everyone. Such networks should only deliver data in a coarse-grained detail level that is not dangerous to privacy. For example, the system should only return average values of larger areas but not of single locations. The result has to be calculated in a distributed manner so that no single node has access to the complete result.

Sensor networks are small enough and will become even smaller in the future. Therefore, they are very suitable for spying on people. However, besides jamming transmitters and bug detectors, pure technological solutions are not able to solve the problem. [87] suggests a mix of societal norms, new laws and technological responses. However, they will require more research in the future.

4.4.4.5. *Data integrity*

An attacker in the network can generate a false sensor or, more severe, aggregation data. The more aggregated the data is, the more valuable it is. Therefore, there is a need to especially protect the aggregated data. [93] proposed SIA, a framework for secure information aggregation in large sensor networks: efficient protocols for the computation of the median, the average, the minimum and maximum of a value, and the estimation of the network size. Using random sampling mechanisms and interactive proofs, the user is able to verify that the answer of an aggregator is a good approximation of the true value.

4.4.4.6. *Trust*

With several COs contributing to a common goal, it is necessary to assess the reliability of the information provided by an individual CO. With respect to services and transactions, trust has been researched for several years. Research on data-centric and fully distributed architectures has just started.

A distributed voting system is proposed in [23]. When a node detects a misbehavior of another node, it can cast a vote against this node. Above a certain number of votes, all other nodes refuse to communicate with this node. To avoid a malicious node casting votes against many legitimate nodes, each node is limited to a number of votes.

In [38], a more general reputation-based framework for sensor networks is presented. Each node monitors the behavior of other nodes and builds up their reputation thereupon. This reputation is used to evaluate the trustworthiness of other nodes. Thus, nodes with a bad reputation can be excluded from the community.

COs exhibit strong cooperation when performing an action. Thus, a single node becomes less important to the overall result. Therefore, strong cooperation will be one of the feasible approaches to secure the entire system through extensive data validation.

4.4.4.7. *Protocols*

Encryption mechanisms are not enough to defend against attackers. Careful protocol design is needed as well. [54] describes several attacks against known routing protocols for sensor networks: spoofed, altered or replayed routing information, selective forwarding, sinkhole attacks, sybil attacks, wormholes, HELLO flood attacks and acknowledgement spoofing. For several routing protocols, the relevant attacks are highlighted. Countermeasures could not be given for all protocols since security was not a design issue.

In [30], a routing protocol is presented that is resilient to attempts to obstruct data delivery as it sends every packet along multiple, independent paths.

Ariadne [50] uses efficient symmetric key primitives to prevent compromised nodes from tampering with uncompromised routes consisting of uncompromised nodes. It also deals with a large number of types of DoS attacks. Since it is designed for ad hoc networks, it may be suitable only for some COs.

4.4.5. *Distributed storage and data search*

In COs and WSNs, efficient storage and querying of data are both critical and challenging issues. Especially in WSNs, large amount of sensed data are collected by a high number of tiny nodes. Scalability, power and fault tolerance constraints make distributed storage, search and aggregation of the sensed data essential. It is possible to perceive a WSN as a distributed database and run queries which can be given in

SQL format. These queries can also imply rules about how to aggregate the sensed data while being conveyed from sensor nodes to the query owner.

Since the lifetime of a WSN is generally dependent on irreplaceable power sources in tiny sensor nodes, power efficiency is one of the critical design factors for WSNs [6]. Data aggregation techniques [17, 135] that reduce the number of data packets conveyed through the network are therefore important and are also required for effective fusion of data collected by a vast number of sensor nodes [6, 43]. Data aggregation in sensor networks combines the sensed data coming from the nodes based on the parameters passed in queries. It can be classified according to one of the following approaches:

• **Temporal or spatial aggregation:** data can be aggregated based on time or location. For example, temperature readings taken every hour or temperature readings from various regions in a sensor field can be averaged. Also, a hybrid approach, which combines time and location based aggregation, can be used.

• **Snapshot or periodical aggregation:** data aggregation can be done on the receipt of a query (snapshot mode). Alternatively, temporarily aggregated data can be reported periodically.

• **Centralized or distributed aggregation:** a central node can gather and then aggregate data, or data can be aggregated while being conveyed through a sensor network. A hybrid approach is also possible where clusters are formed, and a node in each cluster aggregates the data from the cluster.

• **Early or late aggregation:** data can be aggregated at the earliest opportunity, or the aggregation of data may not be allowed before a certain number of hops hinder the collaboration between neighboring nodes.

Data queries can be made not only for aggregated data but also for non-aggregated data. A query in a sensor network may be perceived as the task or interest dissemination process. Sensor nodes can be queried by using continuous or snapshot queries. Continuous queries can be periodical where the sensed data are reported at certain time intervals or event driven where certain events stimulate nodes to report the sensed data.

The following characteristics of WSNs should be considered while designing a data storage, querying and aggregation scheme for WSNs:

• Sensor nodes are limited in both memory and computational resources. They cannot buffer a large number of data packets.

• Sensor nodes generally disseminate short data packets to report an ambient condition, e.g. temperature, pressure, humidity, proximity report, etc.

• The observation areas of sensor nodes often overlap. Therefore, many sensor nodes may report correlated data of the same event. However, in many cases, the replicated data are needed because the sensor network concept is based on the cooperative effort of low fidelity sensor nodes [6]. For example, nodes may only report proximity, and the size and the speed of the detected object can be derived from the locations of the nodes reporting them, and from the timing of the reports. The collaboration among the nodes should not be hampered by the data aggregation scheme.

• Since there may be thousands of nodes in a sensor field, associating data packets from numerous sensors to the corresponding event and correlating the data about the same event reported at different times may be a very complicated task for a single sink node or a central system to handle.

• Due to large number of nodes and other constraints such as power limitations, sensor nodes are generally not globally addressed [6]. Therefore, address-centric protocols (end-to-end routing) are mostly inefficient. Instead of address-centric protocols, data-centric or location-aware addressing protocols, where intermediate nodes can route data according to its content [63] or the location of the nodes [21], should be used.

• Querying the whole network node by node is impractical. Thus, attribute-based naming and data-centric routing [114] are essential for WSNs.

Queries made to search data available in a WSN should be resolved in the most power-efficient way. This can be achieved by reducing either the number of nodes involved in resolving a query or the number of messages generated to convey the results. There is a considerable research interest to develop efficient data querying schemes for WSNs.

In the section that follows, we examine data dissemination techniques which are closely related to data querying, and then query processing and resolution techniques.

4.4.5.1. *Data dissemination*

Data dissemination protocols are designed to efficiently transmit and receive queries and sensed data in WSNs. We briefly discuss five of the best known data dissemination protocols for WSNs. There are many others that can be categorized as data-centric, hierarchical and location-based. Since the focus is on distributed storage and search we do not list the other protocols.

The routing protocols for WSNs are generally designed for networks that have fixed homogenous sensor nodes and are based on the assumption that all nodes try to convey data to a central node, often called the sink. However, in CO networks, there will be heterogenous nodes that can be mobile, and the sensed data will be needed by many nodes, i.e. multiple sinks. Therefore, we can say that new routing protocols will be needed for COs.

Classic flooding

In classic flooding, a node that has data to disseminate broadcasts the data to all of its neighbors.

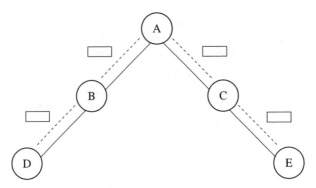

Figure 4.2. *Classic flooding*

Whenever a node receives new data, it makes a copy of the data and sends the data to all of its neighbors, except the node from which it just received the data. In Figure 4.2, an example is depicted. A sends the message to its neighbors B and C. Then, B and C copy the message and send the message to their neighbors D and E respectively. The algorithm finishes when all the nodes in the network have received a copy of the message.

Gossiping

Gossiping [6] uses randomization to conserve energy as an alternative to the classic flooding approach. Instead of forwarding data to all its neighbors, a gossiping node only forwards data to one randomly selected neighbor.

SPIN

SPIN [43] is based on the *advertisement* of data available in sensor nodes. When a node has data to send, it broadcasts an advertisement (ADV) packet. The nodes interested in this data reply back with a request (REQ) packet, then the node disseminates the data to the interested nodes by using data (DATA) packets. When a node receives data, it also broadcasts an ADV, and relay DATA packets to the nodes that sent REQ packets. Hence, the data is delivered to every node that may have an interest. This process is shown in Figure 4.3.

Directed diffusion

In SPIN, the routing process is stimulated by sensor nodes. Another approach, namely directed diffusion [53], is sink oriented. A sink is the name given to the central node responsible for gathering data from all the other nodes in directed diffusion,

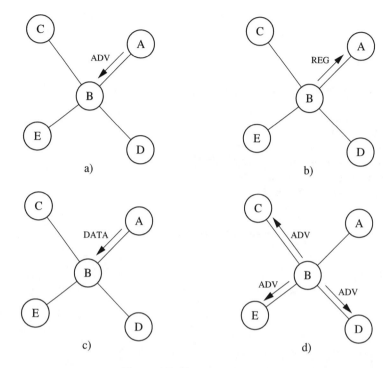

Figure 4.3. *SPIN data exchange*

where the sink floods a task to stimulate data dissemination throughout the sensor network. While the task is being flooded, sensor nodes record the nodes which send the task to them as their gradient, and hence the alternative paths from sensor nodes to the sink are established. When there is data to send to the sink, this is forwarded to the gradients. One of the paths established is reinforced by the sink. After that point, the packets are not forwarded to all of the gradients but to the gradient in the reinforced path. A sample interest description is shown in Figure 4.4, and data dissemination in directed diffusion is illustrated in Figure 4.5.

LEACH

LEACH [45] is a clustering-based protocol that uses randomized rotation of local cluster heads to evenly distribute the load among the sensors in the network. In LEACH, the nodes organize themselves into local clusters, with one node acting as a local cluster head. LEACH includes randomized rotations of the high-energy cluster head position such that it rotates among the various sensors in order not to drain the battery of a single sensor. In addition, LEACH performs local data fusion to compress the amount of data being sent from the clusters to the base station. Sensors

```
Type
Interval
Duration
Rect
 : four-legged animal
 : 20ms
 : 10 seconds
 : [-100,100,200,400]
 // Detect animal location
 // Send back events every 20ms
 // For the next 10 seconds
 // From sensors within range
```

Figure 4.4. *A sample interest description*

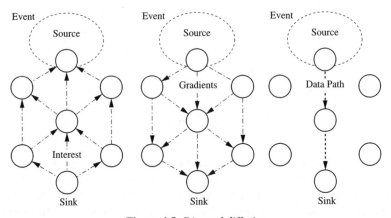

Figure 4.5. *Directed diffusion*

elect themselves to be local cluster heads at any given time with a certain probability. These cluster head nodes broadcast their status to the other sensors in the network. Each sensor node determines to which cluster it wants to attach itself by choosing the cluster head that requires the minimum communication energy. Once all the nodes are organized into clusters, each cluster head creates a schedule for the nodes in its cluster. This allows the radio components of each non-cluster head node to be turned off at all times except during each node's transmit time, thus the energy dissipated is minimized. Once the cluster head has all the data from the nodes in its cluster, the cluster head node aggregates the data and then transmits the compressed data to the base station. However, being a cluster head drains the battery of that node. In order to spread this energy usage over multiple nodes, the cluster head nodes are not fixed. The decision to become a cluster head depends on the amount of energy left at the node. In this way, nodes with more energy will perform the energy-intensive functions of the network. Each node makes its decision about whether to be a cluster head independent from the other nodes in the network and thus no extra negotiation is required to determine the cluster heads. The cluster formation algorithm is depicted in Figure 4.6.

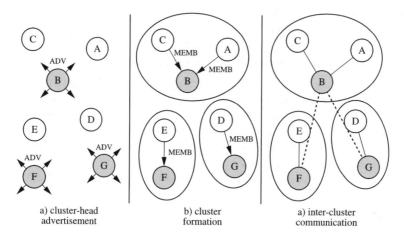

a) cluster-head
advertisement

b) cluster
formation

a) inter-cluster
communication

Figure 4.6. *LEACH cluster formation*

4.4.5.2. *Query processing and resolution*

After a query arrives to a sensor node, it is first processed by the sensor node. If the node can resolve the query, the result of the query is disseminated. This approach is one of the simplest ways of resolving and processing a query. Sensor nodes usually take advantage of collaborative processing to resolve queries so that a smaller number of messages are transmitted in the network. Queries can be flooding-based, where a query is flooded to every node in the network. Alternatively, they can be expanded ring search (ERS) based, where a node does not relay a query that it can resolve. In this subsection, we briefly explain data storage and querying techniques.

TinyDB

TinyDB [69] is a query processing system for extracting information from a network of TinyOS sensors. TinyDB provides a simple, SQL-like interface to specify the data, along with additional parameters, such as the rate at which data should be refreshed, similar to traditional databases. Given a query specifying data interests, TinyDB collects the data from nodes in the environment, filters and aggregates them. TinyDB does this via power-efficient in-network processing algorithms. Some key features of TinyDB areas are as follows:

• Metadata management: TinyDB provides a metadata catalog to describe the kinds of sensor readings that are available in the sensor network.

• High-level queries: TinyDB uses a declarative query language that lets the data be described without having to state how to get it. This makes it easier to write applications.

• Network topology: TinyDB manages the underlying radio network by tracking neighbors, maintaining routing tables, and ensuring that every node in the network can efficiently and (relatively) reliably deliver its data to the user.

• Multiple queries: TinyDB allows multiple queries to be run on the same set of nodes at the same time. Queries can have different sample rates and access different sensor types, and TinyDB efficiently shares work between queries when possible.

• Incremental deployment via query sharing: TinyDB nodes share queries with each other: when a node hears a network message for a query that it is not yet running, it automatically asks the sender of that data for a copy of the query, and begins running it. The TinyDB system contains two applications: one application runs on the sensor platforms and another application runs on the PC side. A user requests his query using the Java application on the PC. This query is disseminated to sensor nodes and the application on the sensor platforms retrieves and returns the requested information.

```
SELECT Temp
FROM Sensors
WHERE temp>threshold
TRIGGER ACTION SetSnd(512)
EPOCH DURATION
```

Figure 4.7. *Triggering SQL query*

TinyDB includes a facility for simple triggers or queries that execute some command when a result is produced. A sample query is given in Figure 4.7 which calls the SetSnd function to give an alarm when the temperature is over some threshold value, and this value is checked every 512 seconds. TinyDB includes the ability to run queries that log into the flash memory in the sensor nodes. TinyDB provides commands for creating tables that reside in flash, running queries that insert into these tables, running queries that retrieve from these tables and deleting these tables. Only one query can log to a buffer at a time and new queries will overwrite data that was previously logged to a table. Currently, a query that selects from a flash table and a query that writes to the same table cannot be run. The logging of the query should be stopped using the TinyDB client utility prior to collecting data from flash tables.

When running queries longer than four seconds by default, TinyDB enables power management and time synchronization. This means that each sensor is on for exactly the same four seconds of every sample period. Results from every sensor node for a particular query should arrive at the base station within four seconds of each other. This time synchronization and power management enables long running deployments of the sensors. TinyDB aggregates results on the way to the sink node. TAG [72] is an aggregation service offered by TinyDB. It operates as follows: users pose aggregation queries from a powered, storage-rich base station. Operators that implement the query are distributed into the network by piggy-backing on the existing ad hoc networking protocol. Sensors route data back towards the user through a routing tree rooted at the base station. As data flows up this tree, it is aggregated according to an aggregation function and value-based partitioning specified in the query. In order for users to pose declarative queries, an SQL-like programming language was designed.

Aggregates are classified in four categories according to their state requirements, tolerance of loss, duplicate sensitivity and monotonicity:

- Duplicate insensitive/sensitive aggregates are unaffected by duplicate readings.

- Duplicate sensitive aggregates will change when a duplicate reading is reported.

- Exemplary/summary aggregates return one or more representative values from the set of all values.

- Summary aggregates compute some property over all values.

- Monotonic aggregates aggregate the property so that when a function f is applied to two partial state records for all resulting values, it will be greater or lower than each of the evaluation of pairs of values. This provides increasing or decreasing values for aggregate results.

- Distributive/algebraic/holistic/unique/context-sensitive aggregates: depending on the function a pair of values has to be carried. For example, AVERAGE function requires the number of elements used to compute the result and the result to further continue in processing. Distributive aggregates do not require other data to calculate a result; therefore, the size of the partial records is the same as the size of the final record. COUNT, MAX and MIN are examples of distributive aggregates. Algebraic aggregates require intermediary state information to continue the operation. The AVERAGE function is an example of an algebraic aggregate. Holistic aggregates require whole values to be kept together prior to computing the result. MEAN is one of these operators. Unique aggregates are similar to holistic aggregates except that the amount of state that must be propagated is proportional to the number of distinct values in the partition.

- In context-sensitive aggregates, the partial state records are proportional in size to some property of the data values in the partition. Many approximate aggregates are content-sensitive. Fixed-width histograms and wavelets are examples of these operators.

Queries in TAG contain named attributes. When a TAG sensor receives a query, it converts named fields into local catalog identifiers. Nodes lacking attributes specified in the query simply tag the missing entry as NULL. This increases the scalability as not all the nodes are required to have a global knowledge of all attributes. Attributes can be sensor values, remaining energy or network neighborhood information. TAG computes aggregates in the network whenever possible to decrease the number of message transmissions, delay and power consumption. Given the goal of decreasing the number of transmitted messages, during the collection phase each parent waits for some time period prior to transmitting its own message in order to aggregate the child nodes' responses. How long each node waits for other nodes' responses is (EPOCH DURATION)/d, where d is the maximum depth of the tree. In order to group received data, group id is tagged to each sensor's partial state record, so that response data is aggregated for the nodes with the same group id. When a node receives an aggregate from a child, it checks the group id. If the child is in the same epoch as the node, it combines the two values. If it is in another epoch, it stores the value of the child's group along with its own value for forwarding in the next epoch.

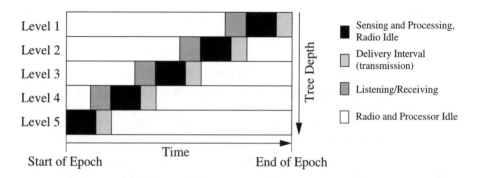

Figure 4.8. *Partitioning of time into EPOCHS*

By explicitly dividing time into epochs as in Figure 4.8, while waiting for other nodes' responses to arrive, the CPU can be set to idle during this time. However, the depth of the tree has to be known for this principle to be made successful. However, waking up the processor will require synchronization to be employed. Finally, real users of sensor networks are most likely not sophisticated software developers. Therefore, TinyDB has been supported by toolkits for the easy access of data including TASK and others. The complexity of sensor network application development must be reduced and deployment must be made easy to ensure the success of sensor network technology in the real world.

Cougar

Cougar [132] is a query layer for sensor networks. The query layer accepts queries in a declarative language that are then optimized to generate efficient query execution plans with in-network processing, which can significantly reduce resource requirements. Cougar is motivated by three design goals. First, declarative queries are especially suitable for sensor networks. Clients issue queries without knowing how the results are generated, processed and returned to the client. Second, it is very important to preserve limited resources such as energy and bandwidth. Since sensor nodes have the ability to perform local computation, part of the computation can be moved from the client and pushed into the sensor network, aggregating records or eliminating irrelevant records. Third, different applications usually have different requirements, from accuracy, energy consumption to delay. For example, a sensor network deployed in a battlefield or rescue region may only have a short lifetime but a higher degree of dynamics. On the other hand, for a long term scientific research project that monitors an environment, power-efficient execution of long-running queries might be the main concern. More expensive query processing techniques may shorten processing time and improve result accuracy, but might use a lot of power. The query layer can generate query plans with different trade-offs for different users. The component of the system that is located at each node is called a query proxy. Architecturally, the query proxy lies between the network layer and the application layer, and the query proxy provides higher level services through queries. Gateway nodes are connected to components outside of the sensor network using long-range communication and all communication with users of the sensor network goes through the gateway node.

```
SELECT FROM WHERE GROUP BY HAVING DURATION EVERY attributes,
aggregates SensorData S Predicate
attributes
predicate
time interval
time span e
```

Figure 4.9. *Query template*

Declarative queries are the preferred way of interacting with a sensor network. The queries having the form in Figure 4.9 are considered. It is very similar to the SQL language but it has limitations when compared to SQL. One difference between the query template and SQL is that the query template has additional support for long-running, periodic queries. The DURATION clause specifies the lifetime of a query and the EVERY clause determines the rate of query answers. A simple aggregate query is an aggregate query without GROUP BY and HAVING clauses. In order to compute these aggregates, further processing such as in-network aggregation has to be done. In order to process data in-network, several sensor nodes transmit the packet to a central node named the leader-node, which calculates the aggregates of incoming messages. There are three approaches that can be taken to collect sensor data at a leader-node: messages can be directly delivered to the leader-node using an ad hoc routing protocol, messages can be merged into the same packet to limit the amount of packets transmitted, and partial aggregation on the way to the central node can be done by intermediary nodes. Synchronization is needed if the messages are to be merged and duplicate sensitive operators such as SUM and AVERAGE require data to be transmitted once. Synchronization is used to determine for each node in each round of the query, how many sensor readings to wait for and when to perform packet merging or partial aggregation. Since the query processing facility has been designed as a layer, Cougar assumes that several ad hoc routing protocols with modifications can be used for the delivery of the messages. An AODV protocol has been used with extensions for simulations. According to Cougar, routes are set up in an initialization phase and each message carries the hop count of the message. Each node records the ID of the received message as a parent node and a reverse path to the leader is set up. Two methods are used to maintain the tree. Local repair is used when a broken link is detected. The depth of the tree with a sequence number is used between nodes that are spatially close to find a new parent in the case of a communication failure. Another method is to reconstruct the tree whenever the number of messages expected reach below some user defined threshold.

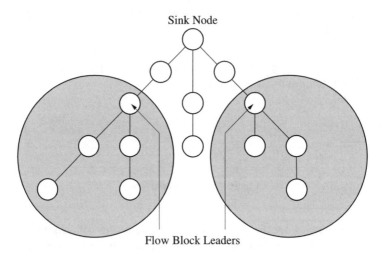

Figure 4.10. *A Cougar query plan*

In order to resolve queries like "What is the minimum average temperature during the next seven days?" two levels of aggregation have to be done. First, the average temperature has to be computed and then the minimum operator has to be applied. In order to resolve these kinds of queries, query plans are used. A query plan is needed to compute complex aggregate queries that a user poses. It also decides how much computation is pushed into the network and it specifies the role and responsibility of each sensor node, how to execute the query and how to coordinate the relevant sensors. It is constructed by flow blocks, where each flow block consists of a coordinated collection of data from a set of sensor nodes at the leader of the flow block, as depicted in Figure 4.10. The task of a flow block is to collect data from relevant sensor nodes and to perform some computation at the destination or sensor internal nodes. A flow block is specified by different parameters such as the set of source sensor nodes, a leader selection policy, the routing structure and the computation that the block should perform. Each flow block is called a cluster and maintained by heartbeat messages transmitted by the leader of the flow. Several optimizations can be applied to query plan construction such as creating flow blocks that are sharable between different queries and the use of the join operator which enforces two conditions coming from different data flows to be true before returning a value. The join operator represents a wide range of possible data reductions. Depending on the selectivity of the join, it is possible to reduce the resulting data size.

ACQUIRE: active query forwarding

The active query forwarding in sensor networks (ACQUIRE) [104] aims to reduce the number of nodes involved in queries. In ACQUIRE, each node that forwards a query tries to resolve it. If the node resolves the query, it does not forward it further but sends the result back. Nodes collaborate with their n hop neighbors, where n is referred to as the look ahead parameter. If a node cannot resolve a query after collaborating with n hop neighbors, it forwards it to another neighbor. When n is equal to 1, ACQUIRE performs as flooding in the worst case. Query resolution in ACQUIRE is depicted in Figure 4.11.

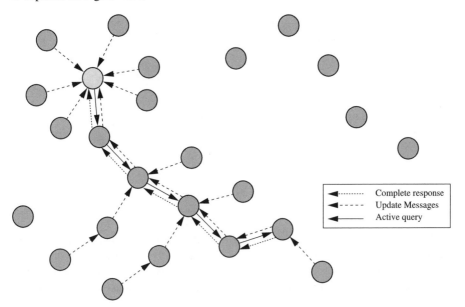

Figure 4.11. *Query resolution in ACQUIRE*

MARQ: Mobility Assisted Resolution of Queries

MARQ [46] makes use of the mobile nodes to collect data from the sensor network. In MARQ, every node has contact with some of the other nodes. When contacts move around, they interact with other nodes and collect data. Nodes collaborate with their contacts to resolve the queries.

SQTL: Sensor Query and Tasking Language

Sensor query and tasking language (SQTL) [114] is proposed as an application layer protocol that provides a scripting language. SQTL supports three types of events, which are defined by the keywords *receive, every* and *expire*. The receive keyword defines events generated by a node when the node receives a message; the *every* keyword defines events occurred periodically; and the *expire* keyword defines the events occurred when a timer expires. If a node receives a message that is intended for it and contains a script, the node executes the script.

SQS: SeMA Querying Protocol for Micro-Sensors

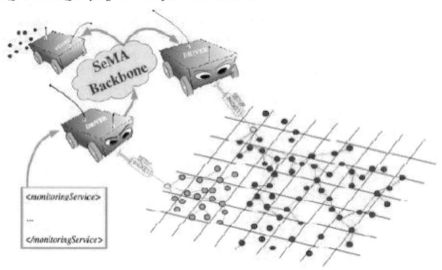

Figure 4.12. *SeMA architecture*

SeMA is an enhanced client-server architecture in which clients are supported to adapt themselves to the new network resources they discover on-the-fly. The nodes get service information either by catching the periodically advertised services or by generating a service request. Applications establish light-weight sessions with the resources using the address returned to them. Binding is done when the service is needed. SeMA is a cross-layer protocol, capable of operating on generic wireless data link layer protocols, such as IEEE 802.11. SeMA addresses issues such as service announcement, binding, session management and data routing. The monitoring of sensor networks is an application of the SeMA architecture. The terrain of the sensor deployment area is assumed to be suitable for navigating by means of mobile units, which form the information retrieval backbone of the overall monitoring application [12]. These mobile units are either wireless equipment-carrying living, or autonomous, robots as seen in multi-robot exploration studies. The architectural view of the system is shown in Figure 4.12.

All available resources on the SeMA network are considered as services. They play a major role on a SeMA network, since mechanisms of the protocol are built with the aim of providing means to access those services. Specification of a service should include necessary information for SeMA clients to determine whether the service fulfills the client's need. Thus, services are modeled with a generic name, and follow attribute-value pairs. This definition is sufficient for a host to discover a required service without any bindings beforehand. Details of XML processing of services and ad hoc routing in the backbone can be found in reference [12]. In a sensor network application, clients of the ad hoc network backbone become the middleman, which is defined in XML in Figure 4.13. They participate in forming query driven sensor trees adaptively and backbone features are used to announce queries and return collected results.

```
<service name = monitoring application>
<keyword attribute = validUntil>20031128193044EEST</keyword>
<keyword attribute = queryPredicate>sensor:read[value>100 and
 order(value)<5]</keyword>
<keyword attribute = resultFunction>fn:concat</keyword>
</service>
```

Figure 4.13. *XML definition of a monitoring service*

Queries for data of interest are transmitted through the backbone in service announcement packets. Service definitions contain an XQuery predicate, a result function and query timeout period (QTP), as well as geographic region boundaries of the area of interest. An XQuery predicate is used to extract the readings that interest the monitoring application. Then, if specified in the service, these readings may be processed via the given result function. The actual query result to be submitted is the returned data from this XQuery function. XQuery specifies more than 200 functions (including functions in SQL) that include numerical processing, data aggregation (sum, avg, min, max), string operations (string-join, starts-with, ends-with), pattern matching (matches, replace, tokenize) and more. By using XQuery in value fetching and processing, the sensor querying process conforms to XML standard, from the monitoring application down to the sensor nodes. Temporarily posed sensor drivers convert service parameters and the XQuery predicate to a bitwise coded format. The coded query is sent to sensors in the payload field of the set-up packet. In this model, sensor nodes are assumed to be very simple equipment with limited processing and battery power. The set-up packet initiates a tree topology network in the area of interest among the sensors that have accepted the query.

DADMA: Data Aggregation and Dilution by Modulus Addressing

In DADMA, a sensor network is considered as a distributed relational database composed of a single view that joins virtual local tables named Virtual Local Sensor Node Tables (VLSNT) located at sensor nodes. Figure 4.14 shows the distributed database perception of DADMA. Records in VLSNT are measurements taken upon a query arrival and consist of two fields: task and amplitude. Since a sensor node may have more than one sensor attached to it, the task field indicates the sensor, e.g. temperature sensor, humidity sensor, etc. that takes the measurement. Since sensor nodes have limited memory capacities, they do not store the results of measurements. Therefore, there can be a single reading for each sensor attached to a node, and the task field is the key field in the VLSNT created upon a query arrival. Our perception of WSNs makes relational algebra practical to retrieve the sensed data without much memory requirement, which is different from the scheme explained in [70], where the sensed data for each task are maintained at a different column in a table.

Sensor Network Database View (SNDV) can be created temporarily either at the sink, i.e. the node that collects the data from the sensor network, or at an external proxy server. An SNDV record has three fields: task, location and amplitude. While data is being retrieved from a sensor node, the location of the sensor node is also added to the sensed data. Since multiple sensor nodes may have the same type of sensors, i.e. multiple sensors can carry out the same sensing task, task and location become the key fields in an SNDV. In applications where nodes are not location-aware, it is also possible to replace the location field with the local identifications of the reporting nodes. The location field can also be used to identify a group of nodes according to the aggregate and dilute functions explained below. It should be noted that SNDV is a temporary view where the results of a query are collected.

For many WSN applications, the sensed data need to be associated with the location data. For example, in target tracking and intrusion detection WSNs, sensed data are almost meaningless without relating them to a location. Therefore, location-awareness of sensor nodes is a requirement imposed by many WSN applications. There are a number of practical location finding techniques for WSNs as reported in [8].

Since each query results in a new SNDV, to keep the aggregated/diluted history of a WSN it may be needed to maintain a permanent External Sensor Network Database Table (ESNDT) in a remote proxy server. In ESNDT, the records obtained from

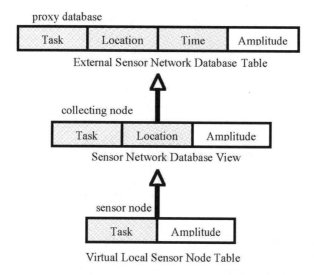

proxy database

External Sensor Network Database Table

collecting node

Sensor Network Database View

sensor node

Virtual Local Sensor Node Table

Figure 4.14. *A sensor network perceived as a distributed database*

queries, i.e. the records in SNDVs, are stored after being joined with a time label. For example, a demon can generate queries at specific time intervals or at the occurrence of a specific event, and insert the records of SNDVs resulting from these queries into the ESNDT. To distinguish the equal amplitudes sensed by the same node about a specific task during different periods, the task, location and time fields make up the key fields in an ESNDT.

In DADMA, a statement that has the structure given in Figure 4.15 starts a query. This structure is largely a part of the SQL standard [10] except for the last field starting with "based on" which will be explained later in this section. Using an SQL style statements for a generic query interface has some advantages as described in [70]. Programmers and system administrators can use this practical and standard interface for all kinds of WSN applications. Hardware design for WSNs can also be optimized to run this language.

In the select keyword of the SQL statement, common aggregation functions such as avg, min, max can be used to indicate how to aggregate the amplitude field. The fields to be projected from an ESNDT are also listed after this keyword. The "from" keyword indicates the nodes to be involved in the query. Any means that even a single node may be enough to resolve the query, and any node in the sensor network can do it. When the "every" keyword is used, all of the nodes in the sensor network are supposed to be involved in the query. When a task or a set of tasks is given in the query, only the sensors in the specified types carry out the measurement. Aggregate

```
Select    [task, time, location, [distinct | all], amplitude,
             [[avg | min | max | count | sum ] (amplitude)]
From     [any, every, task, aggregate-m, dilute-m]
Where [power available [<|>] PA]
             Location [in | not in] RECT |
   tmin<time<tmax|
   amplitude [<|==|>] a]
Group by task
Based on [time limit = lt | packet limiy = lp |
             resolution =r |    region = xy]
```

Figure 4.15. *The structure of an SQL statement for DADMA*

and dilute keywords are also introduced to spatially group the nodes. The "where" keyword is for defining selection conditions according to available power and/or time and/or amplitude and/or location. The "group by" field is used to specify the set of tasks for which the aggregation of the sensed data will be carried out. The "based on" keyword is followed by the parameters required for the aggregation and dilution algorithm run by the sensor nodes. In [22], the distributed algorithm that process the queries in this format is explained.

4.4.6. *Data aggregation*

Since WSNs are comprised of large numbers of energy-constrained, inexpensive devices deployed in possibly inaccessible areas, battery re-charge or replacement is not a viable option. For this reason, energy-efficient protocols have to be developed to minimize device energy consumption, hence maximizing the amount of time during which the network can be fully operational (denoted as network lifetime in the following). In such resource constrained devices, the dominant component in terms of energy consumption is the wireless transceiver. This imposes the development of an energy-efficient protocol stack.

In general terms, an energy-efficient communication protocol is a protocol designed to minimize the energy-consumption associated with its operations without significantly degrading other relevant metrics performance (such as delay, throughput, etc.). Energy-conservation often involves trade-offs (energy vs. accuracy, energy vs. delay, etc.) that have to be considered when in the protocol design.

The basic building blocks to energy-efficient solutions for WSNs are the following. The wireless interface consumes a fixed amount of energy for receiving (due to the transceiver circuitry), while the cost for transmitting is comprised of two components: the first is fixed and is due to the transceiver circuitry, and the second component depends on the emission power. Depending on the transmission range of the considered technology, the cost due to the fixed component may be either negligible or higher than the cost due to the emission power. Short transmission ranges (between 20-30 m) have comparable costs when in an idle state, or when receiving and transmitting. In these scenarios (which reflect the typical operational scenarios of a CO network), the network lifetime is prolonged by keeping the nodes in the "asleep" state in which the interface is not operational, nodes cannot transmit or receive packets, but the energy consumption is much lower (around two orders of magnitude smaller than the cost in the idle state). Other possible states include the idle state, in which the transceiver is operational but no communication exchange is taking place, and the transmit (receive) state in which the transceiver is operational and is currently transmitting (receiving) a packet. The exploitation of the scheduling between awake states (idle, transmit, receive) and the asleep state is one of the basic tools used by energy-efficient protocols for decreasing energy consumption. Other techniques include minimizing the amount of transmitted information, avoiding the waste of resources, transmitting only when it is likely that the data will be received correctly, minimizing protocol overhead, and designing awake/asleep schedules that do not significantly increase data latency. In many sensor network applications, the low transmission range translates in the need of keeping the transceiver asleep for as long as possible. Thus, decreasing the amount of information transmitted and the corresponding overhead are still important issues for managing energy-efficiency. By reducing the load, it is possible to decrease the amount of time devices have to be in the awake state.

Some solutions have been developed for minimizing the load of the WSNs, which are based on the elimination of the redundancy in the transmitted data. In a WSN, many nearby nodes detect the same or similar events, so that the information they transmit is highly correlated. Some of the relay nodes can thus buffer a few packets and process them, eliminating redundancy. This process is often referred to as data aggregation or data fusion. Data fusion reduces the amount of transmitted data (thus possibly the energy consumption) at the price of increased delay. Other trade-offs which have to be considered in the design of a good data aggregation strategy are energy vs. robustness and energy vs. accuracy. Depending on how data fusion is performed at the aggregation points, the precision with which an event is reported may be compromised. It is possible to simply concatenate in the same packet the data of several packets (saving in terms of header bytes). However, the majority of the advantages that can be obtained in decreasing the load come with techniques which combine the data. The drawback here is that this approach results in a partial decrease in precision/accuracy. A second problem is that the inherent redundancy of the transmitted information makes the data delivery process more robust in the presence of transmission errors. If redundancy is avoided, then aggregated packets should be strongly protected to avoid their corruption or wrongful delivery. An effective solution for data fusion has to consider all these different trade-offs, so that data fusion is maximized while maintaining data latency, accuracy and robustness degradation within limits tolerable for the specific application.

Data fusion has two aspects: a scheme to select the roles of the different devices that act as aggregators (how many aggregation points there are, and where), and the data fusion strategy used at these points (how to determine which data to aggregate, how long is it worth waiting for or how many packets before aggregating and forwarding, etc.). Data fusion should also be combined with other protocols (data link, routing, etc.) for further optimization.

In the following, we review the most recent works in the area of data aggregation.

4.4.6.1. *Types of aggregation*

Packet-level aggregation

This scheme works by dropping duplicate packets and by merging several packets to reduce the overhead due to header transmission. This technique can always be applied to only affect data latency. No loss occurs in terms of data precision. No compression of highly correlated data is performed.

Total aggregation

In this case, all data received within a given time interval can be fully aggregated. This kind of aggregation can be realistic for given types of queries. For instance, when the average temperature in the area monitored by the networks is needed at the network collection point (the sink), an aggregation tree could be built rooted at the sink in which each internal node can be an aggregation point. Each point has simply to collect from each of its children the average value of the temperature, and the number of samples it was obtained from. All these values are then combined to form a single packet that conveys the average temperature in the area monitored by the point subtree. The size of the packet which will be sent out, containing the new average and number of samples, is the same as that of the packets received at the point.

Geographic aggregation

This scheme of aggregation, proposed in [2], assumes that data related to certain events can be aggregated if their sources are geographically close and the events happen within a certain time interval. This implies that data precision is maintained. (The precision here refers to both the geographical place of the events and the times of their occurrence.) Such aggregation tries to model scenarios in which the sources can classify events and send a code classifying an event, together with the time of occurrence and the location to the sink(s). With this time of aggregation, aggregation points away from the sources of the events may not be able to aggregate data. To obviate these problems, [2] proposes enhancing the method by co-locating aggregation points with cluster heads, i.e. specific network nodes chosen in such a way that aggregation points are well-spread throughout the network.

In [115], the authors point out the major advantages of data fusion. Many of the queries asked by users require aggregating and combining the data generated by the sources (so why not perform such processing aggregation in the network instead of at the final destination, saving energy and network resources?). In-network aggregation also allows for the comparison of different measurements of a given event before transmitting the information to the sink(s). A comparison among different samples at the aggregation points allows for the identification and filtering of false or inaccurate measurements performed by faulty or malicious sensor nodes.

There is a huge difference in terms of network load when answering an average or a median query: the first can be answered by delivering aggregated information and the second needs to bring to the sink(s) information on the distribution of the sensed values. The solution proposed in [115] addresses this problem, proposing a represen-tation of value distributions which allows us to answer the listed types of queries with a bounded accuracy while significantly compressing the amount of information which have to be disseminated in the network. This is achieved by defining and transmit-ting compressed information on the measured value distribution, denoted as a quantile digest. Quantile digests can be combined by aggregation points without compromis-ing the resulting accuracy, and can be used at the final destination to answer quantile, range and consensus queries. The amount of transmitted information is significantly reduced in the case in which all the sensed data have to be reported.

[32] describes a security problem associated with data fusion. The current data fusion process puts a great deal of trust in aggregation points. Not only is a consensus-based mechanism needed to filter out information transmitted by malicious nodes, but also a scheme has to be designed for the sink(s) to have proof of the validity of the reported aggregated information, since aggregation points themselves may be faulty or malicious. The way this problem is solved in the paper is by adopting a witness-based data fusion node assurance scheme. It is assumed that there are more nodes which gather the same data and result in the same aggregated packets, even if only one aggregation point reports such results to the sink(s). Such nodes know of each other. Each node computes a summary of the results and encrypts it, sending it to the selected aggregation point which will have to transmit all such summaries together with the aggregated data to the sink(s). The final destinations will then be able to compare the summaries of the witnesses and of the aggregated data and use a voting scheme to decide if the reported data is reliable.

4.4.6.2. *Selection of the best aggregation points*

In [99], a way to select aggregation points is presented. The goal is to maximize the aggregation gain (the reduction in terms of network load achieved via data fusion) while maintaining the end-to-end delay below an application-dependent threshold T. A static sensor network with a single sink is considered. The authors provide a math-ematical formulation to select one single aggregation point per path in order to max-imize performance. Heuristics are then derived from this formulation and integrated with WSN routing (tree-based or directed diffusion based routing). A performance evaluation shows the advantage that can be achieved by means of such a way of selecting aggregation points with regard to more empirical approaches. The paper also points out how delay of data fusion should be rewarded by a significant amount of aggregated data, so lower or higher delay spent at aggregation points should depend on the amount of data that can be fused at that aggregation point.

In [117], the authors assume a set of aggregation points and address the problem of setting the timeouts at the different aggregation points encountered along the source-destination paths. The assumptions made are the periodic transmission of data from the source to the sink and the adoption of a tree-based converge-casting. Three different approaches are compared: 1) periodic simple, in which all the packets received within time T are aggregated, 2) periodic per hop, in which an aggregation point waits to receive data from all its children in the dissemination tree before it sends the aggregated packet out (recall the traffic is periodic), and 3) the proposed approach in periodic per hop adjusted. In this latter approach, the authors envisage the use of cascading timeouts. Not only do timeouts depend on the aggregation point position in the dissemination tree but a node timeout will happen just before its parent. In this way, a cascading effect (i.e. data originating at the leaves will be clocked out first, reaching nodes in the next tree level in time to be aggregated at that level together with the data originated at that level and so on) reduces the overall delay to reach the sink. Comparative simulations among the different schemes show that, as expected, a cascading timeout-based scheme reduces delay without impacting the resulting received data accuracy. Similar ideas have been reported in [133] for the case in which the WSN traffic is event-based. The authors discuss the potential inefficiencies associated with data fusion in the case where timers are not synchronized: though the packets sent by sources are generated at similar times as they detect the occurrence of an event, unsynchronized settings of the aggregation point timers may result in an aggregation point at level i not being able to gather information from the nodes at level $i + 1$ before performing aggregation. To avoid this problem, each aggregation point starting its timer informs its neighbors, also triggering their timer start, and the timer's duration depends on the aggregation point level in the tree. The solution proposed exploits synchronization and different timer settings at the different levels to allow a node at level i to gather information from the previous levels before its timer clocks out.

In [19], it is proposed that clustering to select cluster heads as aggregation points is exploited. In addition, the paper proposes assigning an interval of possible values to each metric of interest and to select critical values for each measured metric. A pattern code is then computed based on the measured values and sent to the cluster head. The cluster head uses this aggregated implicit information of the data gathered by the sensor node to deduce whether some of the data to transmit are redundant and to avoid redundant transmissions before they occur. Redundancy is thus eliminated by a pattern code exchange between the sensor nodes and their cluster head and by the cluster head deciding based on such codes, which sensor should send the data. This also avoids security problems associated with cluster heads receiving data and having to decrypt them to perform the data fusion operation. In the proposed scheme, denoted EPSDA (energy-efficient and secure pattern-based data aggregation), data encryption is performed end-to-end by means of symmetric cryptography. A way to select aggregation points in a hierarchical WSN organization is also proposed in [7], based on an ILP formulation which jointly addresses aggregation points selection and routing. In [89], the authors define a metric to evaluate the level of compression achieved via data fusion. Variants of protocols for data dissemination proposed in other works (LEACH and PEGASIS) are then proposed in order to maximize the possibility to perform aggregation. It is shown how the extended solutions decrease network load and energy consumption without significantly affecting accuracy over the basic LEACH and PEGASIS schemes.

In [113], the authors claim that instead of grouping devices into clusters based on a distance-based criterion, the hierarchical tree-based communication infrastructure used to deliver the data back to the sink should be designed to group the same subtree nodes which belong to the same group. Group affiliation is attribute-based and reflects the possibility to answer similar queries and to lead to possibly redundant, merged data (in addition to the affiliation to the same group, devices are organized into a tree also based on a distance criterion so that highly correlated data are likely to have such redundancy eliminated close to the sources). This groupware-based tree results in a significant increase in terms of capability to effectively aggregate data. In addition, the TiNA (Temporal coherency aware in Network Aggregation) scheme is adopted. New data are reported only when they differ significantly from the last reported data (how significantly is an application-dependent factor which is communicated via interest dissemination). Overall the two schemes combined lead to significant energy saving.

In [41], an adaptive scheme to determine how many packets to wait for (and how much time to wait for) before performing aggregation is proposed. First, the authors propose a solution in which, instead of waiting for a fixed amount of time or a fixed amount of packets, an aggregation point simply tries to exploit the availability of the wireless channel. If the channel is available, the aggregation point transmits a packet aggregating the queued packets. This reflects, for example, the following way of reasoning. If data aggregation and awake/asleep schedules are decoupled, there will be times in which the aggregation point has its radio interface and in which the channel is free. Transmitting or delaying transmission does not significantly change the energy consumption (as the costs associated with transmitting or being idle when the transceiver is operational are similar in these networks), but can severely affect the end-to-end delay so it may be convenient to transmit anyway. This scheme is named the on-demand scheme. What is not considered in this scheme is that a single device decision to transmit may impact the possibility of another device to do the same in a CSMA/CA-based network. What would be interesting to have is a dynamic scheme which adapts its decision to make the MAC layer operate at a desirable operational point, limiting on one side the amount of transmitted data (aggregating several packets in the queue in each packet sent out) but also maintaining a low delay period. This is the idea behind the proposed dynamic feedback scheme which adopts a dynamic controller to achieve this goal. Simulation results show that in schemes which rely on fixed thresholds, the proposed solutions achieve the best compromise between the decrease of network load and delay. The dynamic feedback scheme is particularly valuable as it is able to operate efficiently under varying network load conditions.

4.4.7. *Resource management*

Generally speaking, COs are battery powered devices that usually lack resources, such as a CPU, memory and power, to name a few.

COs may form a self-organizing network, in which COs arbitrarily join and leave the network or even move during an operation. Since system components are distributed, any action often involves multiple COs at a time. Due to the distributed, dynamic and uncertain nature of system components (since both COs and communication are associated with uncertainty), the design of such embedded wireless collaborative systems proves to be difficult and requires a scheme to facilitate the system design and application development.

Resource management aims at providing a way to manage the resources of a system by enabling high-level system primitives to hide unnecessary low-level details. It enables the application to be independent of the actual underlying distributed system. It should also address the dynamic nature of available resources, such as variable network bandwidth. Because the system is built of fault-prone components, failures should be handled as normal and not as exceptions.

4.4.7.1. *Design challenges*

Traditionally, *system software* aims at providing a unified way of using underlying resources. System software operates the system and manages available resources in a way that provides higher-level services that are usually required. System software consists of three major components:

• *Operating system (OS)*: the OS aims at providing a set of services that are generally needed by application developers, such as starting and stopping processes or allocating memory and other resources. These services make the system much easier to manage, since the actual low-level details are kept hidden from the application developer. The operating system provides a unified interface to use resources and also manages its allocation among incoming requests.

• *Protocol stack*: communication software is decomposed into a set of standardized layers, referred to as a *protocol stack*. Different layers provide different abstraction levels of generally needed networking services. Layers use some of the underlying services to provide some higher-level services. It aims at hiding the heterogenous low-level details of the network. The standardization of these layers played a significant role in the success of Internet deployment.

• *Middleware*: this is a reusable system software that bridges the gap between the end-to-end functional requirements and the lower-level OSs and protocol stacks. Thus, it is a piece of software running in each member of the network, and its goal is to provide high-level services that are independent of the heterogenous underlying systems.

The layered design has proved to be a very successful approach to designing complex systems. Defining high-level interfaces to easily use underlying complex functionality makes application development easier, but unfortunately makes the tuning of *non-functional properties* extremely difficult. These properties usually remain spread across several system layers, thus it is impossible to make modifications to them in a single place. Energy-efficiency or fault tolerance are such non-functional properties. Optimization issues, such as energy-efficiency, usually require special details that may be hidden by the interfaces. Higher layers can make better decisions by working together with the OS, and sometimes it is beneficial to influence the behavior of the OS to meet some higher level needs. Optimizations in separated layers may not achieve a global and synchronized goal. A well known example is the unfortunate action of the OS putting the system into stand-by while the user is giving a presentation.

In networking, the layers of the protocol stack are defined in a way to let them be developed independently. However, a number of dependencies and redundancies among different protocol layers make the communication *energy-inefficient*. In [28, 40], in which the cross-layer nature of low-power ad hoc wireless network design is characterized, it is shown that cross-layer design is more energy-efficient.

Therefore, for energy-efficient designs, the traditional strict modularization or layering is not appropriate. For example, in sensor networks, the monolithic design of communication software is common to reach the required energy-efficiency needs. However, such a design makes system development and management very difficult. In the next sections, we will present solutions that allow both development and management easy, and at the same time allow for optimization of non-functional properties.

4.4.7.2. *Adaptation in resource management*

In embedded wireless collaborative systems, the availability of the underlying resources is highly dynamic. For example, the networking conditions may change frequently, thus the available bandwidth and the certainty of communication varies. Such a change may influence the chosen networking layers or possibly change the whole behavior of the application. The traditionally applied approach of a resource request and its allocation does not provide the needed flexibility anymore [94]. The system must adapt to changes in the surrounding physical and virtual environment.

Instead of allocation, a *resource negotiation* should take place and the allocation of available resources should be adaptive. *Adaptation* is the ability of a system to change its behavior in response to environmental changes.

Early experiments highlighted that in the case of dynamic underlying resources, the ultimate goal of resource management is to combine adaptation with allocation. It is not beneficial to handle the system as a transparent entity with fixed services; instead, the system should allow the negotiation of resource demands and supplies. To manage such dynamic resources, the system software should meet new requirements. It should drive the negotiation of resource supply and demand together with the applied mechanisms based on the environment.

4.4.7.3. *Adaptation and enabling technologies*

Over the years, various technologies have contributed to the evolution of adaptation and reconfigurable software design [5, 76]. Since the main aim is to enable the software to reconfigure itself dynamically based on some conditions, the core of the majority of approaches is the interception and redirection of interactions among software entities.

Middleware

Middleware is defined as a piece of higher-layer software present in each member of a possibly heterogenous network [29]. Its purpose is to provide high-level services to make the development of networked applications easier by hiding details of the underlying infrastructure, such as the OS and network protocol details. Since it represents another level of indirection and transparency, it is a straightforward way to provide adaptive behavior.

Middleware is commonly divided into four layers [111]. *Host infrastructure middleware* is directly on top of the OS and provides a high-level API to hide lower-level heterogenity. *Distribution middleware* provides high-level programming models, such as remote objects, enabling the developers to use local and remote objects in a similar way. *Common middleware services* define higher-level domain independent components to help manage resources. Finally, *domain-specific middleware* is usually designed specifically for a particular class of applications. Any of these middleware layers may be a suitable place to provide adaptive behavior.

Many of the adaptive middleware platforms work by intercepting and modifying messages to provide a level of indirection to influence behavior of the application. Most of the adaptive middleware approaches are based on some popular object-oriented middleware platform such as CORBA or JavaRMI. A large number of adaptive middleware approaches exist, and their taxonomy can be found in [77].

However middleware-based adaptive frameworks may not be sufficiently flexible; their application is straightforward from the software development point of view.

These solutions may provide a way to extend heterogenous, layered systems with the ability to adapt.

Component-based design

Software decomposition to separate modules has become a fundamental approach to make the development of reusable components independent of each other. Software components are reusable software entities that third parties can develop independently and can easily be composed and deployed as general components in the future [122]. Popular platforms include COM/DCOM [79], Microsoft.NET [78] and Enterprise Java Beans [120].

Well-defined interface specifications enable service clients and providers to be developed independently; however, it turns out that the specification of the functional interface usually does not provide enough flexibility to reuse the component. The details of how the component would behave in certain conditions are usually crucial to design choices. To trust a component, specific guarantees should be available regarding its non-functional properties. These guarantees are called *contracts* [15].

Contracts may be categorized into four levels. Level 1 is the basic or syntactic level that defines the functional interface needed to use the module. Level 2 provides behavioral level contracts, which gives more information about how the functionality reacts in different conditions. Level 3 is the synchronization level, which describes synchronization properties required in parallel environments. Finally, level 4 describes QoS properties, such as response delay.

The composition of modules can be categorized into two large groups. In *static composition*, the developer chooses the appropriate composition at compile-time and it cannot be changed later in run-time. In contrast, *dynamic composition* makes it possible to add, remove or reconfigure components during run-time. *Late binding* is the technology that enables coupling of compatible components at run-time.

The system may also support multiple different versions of a given component with different contracts. In such cases, this may be an enabling technology to adaptively select the versions that are the most appropriate for the given environmental conditions.

Computational reflection

Computational reflection is the ability of a software system to reason about and possibly alter its own behavior [73]. Reflection exposes a system's implementation at a level of abstraction that contains enough details and enables necessary changes. The ability of a system to observe and change itself is a key enabling technology towards self-modifying code.

Computational reflection involves two activities. *Introspection* allows the system to observe its own structure or behavior. This may involve revealing its structure or evaluation of software-sensors to collect statistics. The other main act is *intercession*, which enables the system to act on the observations and make necessary modifications. This may be the replacement of a module, or installation of new monitoring components.

In reflective systems, the actual *base-level* software includes its abstract self-representation as a special object. This is called a *meta-level* representation, which represents the actual objects of the base-level. The two levels should remain causally connected.

Direct exploration of the meta-level would provide many details about the base-level objects, but it also makes the process difficult and error-prone. A *meta-object protocol* (MOP) [57] is a higher-level interface that enables easier introspection and intercession by enabling systematic access to meta-level objects. MOP can be categorized as enabling either structural or behavioral reflection. *Structural reflection* reveals inner structural details, such as class hierarchy, object interconnection and data types. In contrast, *behavioral reflection* focuses on higher-level computational semantics of the application, such as the application of a new mechanism.

Reflection may be provided by a programming language in a native way, such as in Lisp [116], Python [95] or in some Java derivatives such as Reflective Java [131]. Even if the programming language does not directly support reflection, similar functionality can be implemented in a higher level using a middleware platform [16] such as KAVA [127] or OpenCorba [66]. Instead of programming construct representation, reflective middleware deals with the self-representation of middleware services.

Reflective abilities of a software system obviously play a key role to provide adaptation.

Separation of concerns

Over the past few years, some major difficulties in software decomposition have been revealed [86]. Consequently, it was realized that the same functional behavior can be implemented with several different decompositions.

Separation of concerns [1] is a software development technology that aims at separating the development of the application's functional behavior (business logic) from other non-functional requirements. The non-functional concerns are usually cross-cutting concerns spread across several modules, such as QoS, energy consumption or fault tolerance. This approach allows for the separate development and maintenance of concerns, while providing a straightforward opportunity for adaptation.

The most widely used approach is *Aspect Oriented Programming* (AOP) [59]. It includes abstraction techniques and language constructs to manage different concerns of the system separately. The codes implementing different concerns, called aspects, can be developed separately. *Pointcuts* are placed into the functional code to signal the locations where different aspects may be woven. Finally, a specialized compiler called the *aspect weaver* is used to combine aspects with the business logic of the application (Figure 4.16). Examples of such AOP frameworks are the AspectJ compiler [58] and KAVA [127].

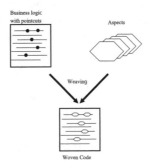

Figure 4.16. *Aspect weaving in AOP*

Composition filters [14] is an approach that dynamically intercepts messages that are sent and received by components. Filters can be applied to all input and output messages, or can select particular messages. Such filters are developed independently and can be compiled into the source code or can be preserved as run-time manipulation objects.

Similarly, aspect weaving at compile-time results in a tangled executable, which is not reconfigurable later. However, some recent approaches delay the weaving process until link- or run-time [124], thus providing a way to tune different aspects of the system during execution.

4.4.7.4. Adaptation frameworks

This section introduces some of the most important cross-layer adaptation frameworks and projects. These frameworks assume that mobile devices are more resource rich than wireless sensor nodes. However, these frameworks present interesting adaptivity concepts that could be at least partially implemented in smaller devices to achieve a desirable degree of resource management.

Odyssey

Odyssey [85] was one of the early attempts to demonstrate *application-aware adaptation*. It demonstrated how application-awareness would result in enhancements in system performance.

Odyssey supports the change of application quality based on low-level system conditions. When the system cannot guarantee the required supply, the application should regulate its demand. The project demonstrated the effectiveness of the approach by scaling down video quality in case the bandwidth decreased.

The key point in Odyssey was to extend the OS resource allocation API with resource call-backs. Applications register a tolerance window in Odyssey and receive a call-back when the required resource levels cannot be kept within the tolerance window. At that point, the application can adapt and the resources can be registered again. Only minor changes are needed to port applications to Odyssey, and it is demonstrated that in some cases it is also possible to avoid any changes.

In [82], Narayanan extends Odyssey to achieve a cross-layer adaptation solution in order to support mobile interactive applications. These kind of applications require low response times and a long battery life, while the available resources are varying. Their work introduces the term *fidelity* as a metric of accuracy. *Data fidelity* is the extent to which the degraded version of a data object matches its reference version. On the other hand, *computational fidelity* is a run-time parameter of an algorithm that can change the quality of its output. The framework uses *multi-fidelity computation* to achieve adaptive behavior.

A major difficulty addressed in their framework is to bridge the gap between the three independent parameter dimensions:

- user satisfaction metrics (utility);
- application parameters (fidelity choice);
- resource supply and demand of the system (resources).

The presented solution uses *resource supply predictors* to predict the available amount of system resources, such as CPU availability. *Performance predictors* give performance metrics, such as delay or battery lifetime, as a function of resource supply and demand, and *resource demand predictors* compute application resource needs as a function of fidelity and non-tuneable properties.

Some of these predictions are application-dependent. This is solved by providing an application-dependent binary module called a *hint module*. Moreover, the system has to be aware what fidelity values are possible for the given application and what are the actually used non-tunable parameters. These are declared in the *application configuration file*.

The architecture of the run-time system can be seen in Figure 4.17. Note that only the hint module is application-dependent. The central component that drives the adaptation is the *solver*. Its task is to search the state space of tunable parameters and find a set of values that maximizes the utility. When the application starts, it submits the

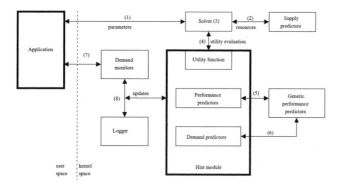

Figure 4.17. *System architecture for multi-fidelity API*

values of non-tunable parameters. A set of generic supply predictors predict the applications' resource supply for the near future. The solver searches the space of tuneable parameters for the best possible combination.

The evaluation of candidates is done by computing their utility, which is application-dependent. The solver sends the tuneable and non-tunable parameters as well as the resource supply prediction to the utility function. The utility function invokes performance predictors to compute delay and battery drain for the particular choice. Performance predictors call application-specific resource demand predictors to compute performance.

Resource demand monitors compute the resource demand of the operations performed. The logger records these values together with the applied parameters to update demand-predictors based on online learning techniques.

The effectiveness of the framework is demonstrated by several experiments ranging from augmented reality to speech recognition.

RAPIDware

The aim of the RAPIDware project [96] is to design an adaptive, component-based middleware for dynamic mission critical systems. These systems must continue to operate correctly even during exceptional situations. Such systems require run-time adaptation, including the ability to modify and replace components, in order to survive hardware component failures, network outages and security attacks. RAPIDware is a significant research effort, which applies the combination of computational reflection and aspect-oriented programming to support software recomposition at run-time.

Sadjadi *et al.* [105] present an adaptable communication component called *MetaSocket*. It is an extension of a regular Java socket to provide adaptable communication services. A MetaSocket is created using Adaptive Java [55], a reflective extension to Java.

To convert an existing Java class into an adaptable component, two main steps are required. Figure 4.18 illustrates the two steps on a multicast socket component, which enables it to adapt to meet different QoS needs. The first step is called *absorption*, which transforms the class to a base-level Adaptive Java component. This base-level class has a set of mutable method invocations, which expose the functionality of the absorbed class. In the example, the absorption steps create a send-only multicast socket, which implements only the *send()* and *close()* adaptable invocations. Links between invocations and the methods are highlighted by lines in the figure. The following step is called *metafication*, which enables the systematic investigation and changing of base-level invocations. In the example, this means inserting or removing filters to the multicast stream, such as different error-correction filters. By this, the socket component is adaptable to meet different quality requirements.

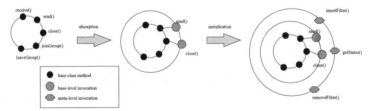

Figure 4.18. *MetaSocket absorption and metafication*

These steps use TRAP/J [106], a Java implementation of TRAP, to weave in the required adaptive behavior. The framework takes an existing Java class together with the list of required adaptable invocations as its input, and makes the required adaptable class as its output. It uses the AspectJ compiler [58] to generate the adapt-ready version of the application.

A MetaSocket enables the middleware or an application to monitor the communication status and to insert, remove and configure adaptation filters at run time. With the help of MetaSockets, a kernel-middleware interaction framework is provided in [108] to support both horizontal and vertical cooperating of components. The Kernel Middleware Exchange (KMX) project addresses the interaction between the middleware and OS layers to support universal adaptation. This approach is especially well-suited to mobile ad hoc networks and overlay networks, where a set of nodes collaborate to accomplish collaborative services.

This framework thus provides an interesting mixture of aspect-oriented programming and reflection to easily turn software components into adaptable ones. Their work also points out how a universal adaptation framework can be feasible by using such components.

GRACE

The Illinois GRACE project [4, 103] aims at providing an adaptation framework for saving energy in mobile multimedia systems. In the considered system, all the system layers and applications are allowed to be adaptive. These adaptive entities cooperate with each other to achieve a system-wide optimal configuration in the presence of variations in the available resources and application demands.

One of the main questions addressed in this research is how to achieve global adaptation with an acceptable overhead, since adaptation with global scope potentially results in a larger overhead. GRACE supports three levels to provide a balance between the scope and the temporal granularity of adaptation:

Global adaptation is invoked only infrequently, in response to large changes in resource supply or demand. It considers all applications and all the system levels. A global coordinator performs the resource allocation by searching through the space of all possible configurations and evaluating the overall utility and resource usage for the particular choice.

Per-application adaptation considers only one application but all system layers at a time. At this time, the accuracy about the resource demands of the actual application is more accurate, thus making finer granularity decisions possible.

Internal adaptation adapts only a single system layer or application at a time. Since it does not need to consider the cross-product of configurations, it can be very efficient. Unfortunately, the system may only remain in a local minimum of optimality.

Another main question this framework addresses is how to make the prediction of future demands and supplies. All the adaptation levels require predicting the resource demands for each application. Global adaptation requires long-term prediction, while per-application adaptation requires prediction only for the next job. For the moment, history-based prediction techniques are commonly used.

The prediction is not straightforward, since the resource usage depends on the actual system and application configuration in use. The approach in GRACE to address this problem is to divide the prediction problem into two parts. The *system-independent* part can be predicted entirely by the application. The other is the *system-dependent* part, which is handled by the specific system layer. For example, the CPU time demand of a job can be divided into the number of instructions and the time per instruction parts. The former is independent of the system, while the latter depends on the actual configuration.

GRACE provides an architecture for the optimization process as well. This framework controls when different levels of adaptation are triggered and also finds the optimal configuration. In the proposed solution, local and global adaptations together allow the system to find a better configuration. The simulation-based evaluation shows the benefits of the solution.

DEOS

The DEOS project addresses the problem of how to best meet the dynamic QoS requirements of end-users in face of dynamic changes in the underlying hardware/ software platform's resources. However, the project mostly focuses on QoS issues. The proposed solution involves both the system and the application.

In the project, *Q-fabric* [90, 91] is presented. Q-fabric is a kernel-level abstraction for cooperative, distributed resource management and adaptation. The solution is based on a kernel-level event notification service, which follows the publish/subscribe paradigm to exchange specific events. Such events may be related to monitoring activity, or can also be the trigger event of an adaptation process. Since the event service is fully anonymous, it is a flexible way to let separate local resource managers cooperate.

A resource manager can be seen in Figure 4.19. Its task is to distribute system resources among a number of competing applications based on a policy. The monitor entity keeps track of changes in resource allocations and keeps statistics, and it also submits monitoring events. The adaptor receives such events and makes decisions about triggering changes either in local or remote resource allocations.

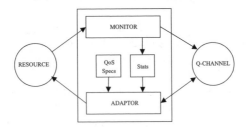

Figure 4.19. *A resource manager*

The approach can be combined with middleware-level mechanisms [92]. Combining a middleware with such a kernel event service allows the flexibility of a global resource management. Such a resource management framework may also be important to coordinate adaptation.

ECOSystem

Zeng *et al.* [134] presents ECOSystem (Energy-Centric Operating System), which is an approach to let the operating system manage energy as a first-class system resource. It manages energy consumption at the OS level and also influences applications in their energy consumption.

Their energy-centric resource management is based on *currentcy* (current currency). Currentcy reflects the ratio of supply and demand and provides a way of efficient resource management in the case of resource shortages.

The project has demonstrated the effectiveness of the approach in a Linux environment. Its basis is the accounting of the energy consumption of processes. The system specifies each task's relative importance and provides an energy-conserving allocation strategy as well.

Neugebauer *et al.* [83] introduces a similar approach; it also applies energy pricing to efficiently manage resources. In the presented model, a price is calculated for resources that are congested. These prices capture the ratio of demand and supply. These resources are charged afterwards, thus price represents a meaningful feedback to applications.

The project uses the Nemesis OS [51]. The Nemesis OS shifts many functionalities (e.g. protocol stack) to the user level, thus avoiding anonymous resource allocations. This makes more accurate resource accounting possible.

Both of the approaches use the OS to account for resource prices and to figure out the relative usage costs. They enable a distributed solution to resource management and adaptation.

4.4.7.5. *Adaptation categorization and its parameters*

One criterion for the classification of adaptation is whether the system is capable of free recomposition of its components or just uses some built-in settings. *Parameter adaptation* is the modification of variables that influence the behavior of the component. This requires built-in support for the given parameter and also the interface to modify it. In contrast, *compositional adaptation* involves the adaptation of new algorithms and the run-time recomposition of the software. The latter enables a very general ability to adapt; however, the former is much easier to implement. In the former, the time of composition may entirely influence the flexibility of the solution. Development time adaptation results in a hardwired ability to change behavior. Composition may occur at compile, link, or even load-time; however, they all result in a *static* composition. In contrast, run-time adaptation results in a completely *dynamic* system.

In order to choose the most suitable adaptation approach, the following parameters should be taken into account:

- *The best configuration*: when the adaptation of the system occurs, the goal is to find the most appropriate configuration for the new environmental conditions. In the majority of cases, selecting the most optimal configuration leads to further complications. When investigating the possible candidates, the outcome of the choices should be predicted by the system. This is not easy, even in a simple case; moreover, a number of adaptable components and all their available cross-compositions may result in a huge configuration space. The system should search through this space and figure out the new configuration. This may result in a significant overhead.

- *Scope of adaptation*: to limit the number of available configurations, it is crucial to investigate the scope of adaptation. A global adaptation involves all the adaptable components of several networked entities. The scope may be limited to *(i)* a node, *(ii)* one of its applications, or *(iii)* even a single module. Unfortunately, limiting the scope may result in an oversight of the best configuration.

- *Time of adaptation*: an important decision to make is when the adaptation should occur. Adaptation may happen continuously; however, in more complex cases, it may result in a large overhead. A way to avoid this is to allow adaptation on a periodic basis; however, this may still result in a large amount of unnecessary overhead. The most sophisticated approach is to do adaptation on a reactive basis. This means that the occurrence of some conditions would trigger the system to find a more optimal configuration.

- *Resource supply and demand*: there may be critical reasons for the need to change the behavior of the system. To gather these conditions, some monitoring functionality is required. Continuously monitoring a large number of quantities may result in a significant overhead, so it is better to monitor only those parameters that are actually necessary. Conditions that may trigger adaptation may result in a change of either resources or demands.

The triggering event may be related to some *hardware* level resource changes. For example, the change from AC to battery supply would result in changes that the application would do. On the other hand, a significant change in the *application* resource needs may also trigger an adaptation, since it would influence the applied allocation strategies. In many cases, changes in *network* conditions should result in an adaptation as well. Network bandwidth is the classic example of a highly dynamic resource that highly influences services that higher layers may use. Another and more general category of operating conditions is the *context*, for example, the detection of the user or location in which the system may operate in.

- *Agility*: an important property of adaptation is the speed at which it can occur. A straightforward approach would be to let the system react as quickly as possible, but other problems may arise. It is possible that the new configuration is valid only for a very short period of time, thus the system would never settle down in a stable state. For example, when resources change very quickly, it is better to select a safe parameter set and not to follow the bandwidth changes. The adaptation should result in a stable configuration.

• *Place of redirection*: the basis of most adaptations is the redirection of module interactions. The question arises where the redirection should occur. The entity that drives the adaptation may be built entirely in the OS, but may also lay outside and handle the entire OS as an adaptable layer. A large number of higher-level solutions are available that push the interception and redirection of modules up to the middleware.

4.4.7.6. *Future direction of adaptivity in WSN*

DCOS (Data Centric Operating System) [34] is an operating system for WSNs. In these networks, the resources on the nodes are extremely constrained. DCOS was designed to fit inside the limited memory of a sensor node while being able to provide real-time guarantees and energy-efficient operation.

Sensor networks are usually driven by events, which can range from environmental phenomena to software events. DCOS applies the *Data-Centric Architecture (DCA)*, where data is the main abstraction of events. The system is built of software components called *Data-Centric Entities (DCEs)*. These entities produce data and they can be triggered by other data. Therefore, the operation of the resulting system depends on what kind of data types are produced and what entities are triggered.

The central element of this architecture is the *Data-Centric Scheduler*. It is the component that keeps track of all the entities inside a node and decides which ones to activate. Moreover, this element is responsible for managing how components should be connected and enables the data flow between them.

DCOS is a dynamic system, which is able to adapt its functionality. Several DCEs may exist with the same behavior, and possibly several sequences of DCEs result in the same functionality. DCOS supports run-time reconfiguration to adapt its functionality to create the most efficient configuration.

Thus, the architecture provides a simple and straightforward path towards indirect software interactions. This ability is the main enabler of adaptive behavior. A next step to have adaptive behavior on a WSN-wide scale is to extend the concept of DCA to the whole network.

4.4.8. *Time synchronization*

Time synchronization is an important vertical function. Applications need to establish a common sense of time among the COs participating in their sensing and actuation goals.

Forest fire monitoring is a scenario that requires not only the information of whether there has been an indication of fire, but also where and when this is happening. The collected sensor data provides the basis for the decision making process, which may trigger actions to be taken in the monitored environment in order to address abnormal circumstances. Thus, the true time of the observed events is crucial to the prompt action of fire fighting.

It is sufficient to associate local clock timestamps with the sensor data when decisions can be made locally. For instance, the decision of regulating the fluid flux of an industrial pipe may be taken locally and actions would follow through local CO actuators, for example, a regulating valve.

However, there are cases where the collected sensor data will be aggregated from various COs at various points inside the network. To perform such a function, the data need to be timestamped at the source.

In this case, the data originates from multiple sources which may or may not agree on a similar time. Thus, a common view of time among the COs is an essential aspect to correlate multiple sources of sensor data in order to produce a consistent and reliable result.

We list two time frameworks that could suit the CO applications:

• *Relative time:* the sense of time is established with respect to an agreed time reference which could be an elected CO among a specific group. This is applicable to scenarios where preserving the order of events is the only required function from a time synchronization service.

• *Absolute time:* in other cases, preserving the order of events is crucial but it is not the only functionality needed. Applications might require event data to be timestamped when they occurred with an absolute time value with respect to a true time standard, such as the universal coordinated time (UTC).

Messages containing the event data could take different routing paths in the network to be transmitted from the CO sources to the sinks. When the application requires strict order preserving, relative time could be used. It is important to note that absolute time is applicable to scenarios where the applications need both order preserving and a sense of true time.

The measurement of structural vibration in a building requires that the data collected from the instrumented cooperating objects be associated with an absolute timestamp. As it is important to correlate the time of day of the events in this case, the clocks of the COs must be synchronized with each other with respect to a true time. This would allow engineers to analyze data regarding the entire building structure and from the neighboring buildings.

Some of the challenges that impact the design of ad hoc time synchronization protocols are discussed in [98, 121]. We present a summary of these issues here:

• Energy-efficiency: sensor network applications often require nodes to be small and cheap. The device size poses physical limitations to the amount of energy that can be stored in a battery using the current technology. In some scenarios, the charging of a depleted battery or the replacement of discharged ones could be extremely difficult and expensive. In some oceanography monitoring applications where the physical access to the deep water is through a boat expedition, the COs deployed should strive to be energy-efficient. Thus, it is important that the time synchronization service strives to achieve energy-efficiency. This raises the question of whether GPS devices often used for time synchronization on the Internet could be adopted in inexpensive sensor networks. The issues to be addressed are cost, size and power consumption of GPS receivers.

• End-to-end delay: in some scenarios, the network may be partitioned because of changes to the network topology (e.g. mobility of nodes or addition/removal of COs). In this case, the delay between any two points in the network may be difficult to estimate because of the large variations in delay. This can introduce problems to the design of protocols that rely on stable delay characteristics of network paths to achieve high and predictable accuracy.

• Infrastructure: in some applications, the network will be created in an ad hoc set-up. Although such foreseen applications would represent a niche market today, self-organizing protocols that can create an overlay for time synchronization should be considered in the design space of protocols. The Network Time Protocol (NTP) [80] hierarchical structure may not be easily adapted to this type of ad hoc network topology.

• Configuration: a few human operators will be responsible for a large number of COs possibly in the order of thousands. The configuration, usually carried out by system administrators who manually specify the time servers for each node, could be a serious problem. The dynamics of this scenario demands a more automated system configuration than the one found in time protocols such as NTP.

An important question to ask is whether the NTP [80], currently in use on the Internet, could be adopted without further modification in low-cost WSNs. This protocol has been used for more than ten years and proved to work reasonably well on the Internet with thousand of nodes synchronized worldwide. The protocol is well-known in the research community (see [29]).

In summary, the NTP relies on an external time reference to synchronize the top layer of time servers called *stratum 1* servers. Servers in the second highest layers are called *stratum 2* and so on. The primary source of time in NTP is often the time pulse received with a 1 Hz frequency by a GPS device locally attached to a server.

As the NTP protocol has been originally developed considering other design issues, some of the challenges discussed above cannot be addressed if the protocol is adopted unmodified in sensor network applications. One of the issues is energy-efficiency, as the external source of time (e.g. GPS) tends to consume more power than other components of a sensor node and does not work in indoor areas which have no line-of-sight to the satellites. This is certainly not an issue in the case with the Internet where the NTP server is a standard PC plugged into a power socket.

However, the issue raised in [98], where the NTP protocol maintains a synchronized system clock by regularly adding small increments to a system counter, is an important one. This behavior precludes the sensor node processor of being switched to a power-saving idle mode. In addition, NTP servers must be prepared to handle synchronization requests at any point in time. This could potentially introduce a problem as the communication module of a CO (e.g. radio) shows the highest cost among the other components when it is transmitting or listening to any signal from the network. Thus, this module could not be put in sleep mode in order to save energy.

GPS devices of small size and low power consumption have been developed and are now commercially available. The Leadtek 9546 low-power GPS receiver [65] was recently integrated into a sensor board of the Crossbow Mote [52]. These devices consume a current of 60 mA at 3.3 V in full operation. Despite the fact that such a consumption profile is higher compared to the micro-controller current draw of 8 mA, it is comparable to the energy consumption profile of the radio module.

However, such sensor nodes are optimized to operate in sleep mode most of the time. If the GPS is configured to occasionally operate to acquire the time signal, its energy consumption may not be an issue. Although adding this particular GPS module to a sensor board increases its cost by 50% today, we believe that the cost will drop significantly because of economy of scale. In contrast, this is also the case for the current cost of sensor nodes, which are similar to the cost of a sensor board equipped with GPS.

We list below some open issues with time synchronization in large-scale distributed COs and WSNs:

• The characteristics of the network topology in ad hoc WSNs may introduce severe delays because of disconnected parts of the network. To compensate for such uncertainties, the time synchronization protocol needs to be *tolerant* to the message delay irrespective of its dominant source (e.g. processing, transmission).

• The clock drift in cheap oscillators could have values ranging from 10 ppm to 100 ppm which represent a drift between 0.036 seconds (10 ppm) and 0.36 seconds (100 ppm) every hour. Therefore, the update frequency used in time synchronization protocols needs to be carefully considered with respect to the characteristics of different types of hardware.

• The scope of time synchronization is quite important. Some applications require the ordering of events occurring within a specific area, which means that only a subset of nodes deployed in that area may be synchronized. However, other applications will need to correlate events collected from different areas in a global scale.

• When some sensor nodes become responsible for providing the time reference to other nodes in the network, the issue of trusting the source of time arises. Also, if a GPS time signal is used in some reference nodes, a related question is how we can guarantee the authenticity of the GPS time.

• Romer *et al.* [98] suggest that calibration of sensors is a much more general and complex problem than time synchronization. The interesting question is whether the current proposed time synchronization techniques can be used for general sensor calibration problems. Also, it may be the case that time synchronization (at least for outdoor sensor deployment) can be achieved by instrumenting low-cost, low-power GPS devices to sensor nodes as previously discussed.

4.5. Summary and conclusions

This section presents summaries of the VFs reviewed in this chapter. The summary of each VF follows a similar format. An overview of the VF is given, which is then followed by a discussion of identified trends and relevant open issues.

Chapter 2 discussed a list of characteristics and requirements of selected COs applications. Table 4.1 relates each of these requirements to the VFs. See Chapter 2 for a comprehensive discussion on selected COs applications and their requirements.

4.5.1. *Context and location management*

Distributed COs systems are designed to measure properties of the physical world. They are therefore suitable for gathering the context of an entity, which is the information that can be used to characterize its situation. Individuals, locations or any relevant objects can be such entities [13]. Since a reasonable amount of data is collected in large systems, context management systems are needed to handle them. Such systems can separate applications from the process of sensor processing and context fusion. To achieve precise actuation and detailed analysis of collected measurement data, however, the location of sensors and actuators need to be known.

Changes in context may trigger actions to influence the monitored entity. Specialized actuators, for instance, may be programmed to control pipe valves when a fluid pressure reaches a certain threshold.

Trends

Traditional context management systems use infrastructure-based directories to store the information, for example Aura [39] and Nexus [49]. In the last few years, there has been a clear trend towards ad hoc (infrastructure-less) systems. Such systems

Characteristic/requirement (Chapter 2)	Vertical System Function	Section
Network topology – topologies to support either single or multi-hop communication.	Communication; distributed storage and search	4.4.3; 4.4.5
Scalability – necessary system support to growing number of COs.	Distributed storage and search; data consistency	4.4.5; 4.4.2
Fault tolerance – provides the mechanisms for supporting the resilience of the system in failures.	Data consistency; communication; security	4.4.2; 4.4.3; 4.4.4
Localization – determining a node's location is the basic required functionality for various applications.	Context and location management	4.4.1
Time synchronization – COs need to establish a common sense of time.	Time synchronization	4.4.8
Security – this is pervasive and must be integrated into every system component to achieve a secure system.	Security, privacy and trust	4.4.4
Data traffic characteristics – the system should provide support for various types of application traffic.	Communication; distributed storage and data search; data aggregation	4.4.3; 4.4.5; 4.4.6
Networking infrastructure – CO networks can be infrastructured or infrastructureless (ad hoc).	Context and location management; communication	4.4.1; 4.4.3
Mobility – the physical components of the system in some applications may be static whereas in others, the architecture may contain mobile nodes.	Context and location management; communication	4.4.1; 4.4.3
Node heterogenity – most CO applications include nodes that have distinct hardware and software factors.	Data lookup mechanisms; data consistency; data aggregation	4.4.5; 4.4.2; 4.4.6
Power awareness – power consumption is one of the performance metrics and limiting factors in almost every CO application.	Communication; resource management	4.4.3; 4.4.7
Real-time – the system delay requirements are very stringent in real-time applications. The broad meaning of delay in this context comprises the system data processing and network delay.	Resource management	4.4.7
Reliability – guarantees that the data is properly received by the applications.	Data consistency; communication; security, privacy and trust	4.4.2; 4.4.3; 4.4.4

Table 4.1. *Application requirements and VFs*

better reflect the limited spatial relevance of context since this is stored at or near the location where it is generated. Current systems focus on the gathering of sensor data using QoS specifications (MiLan [44]), the management of context for cross-layer data in the network stack (MobileMan [27]) or more generally for the whole node (TinyCubus [75]), the storage of information in the network of COs at calculated geographic locations (Geographic Hash Table [97]), the querying of information using a SQL-like language based on the "network as a database" abstraction (TinyDB [71]), and the self-monitoring of COs ([135]) since they belong to the context themselves. Most of the systems consider hybrid COs as well. MiLan also presents a further direction to the research: the simplification of the use of COs. The user specifies what it wants and not how.

Applications that use context to adapt their behavior are called context-aware. Such applications include visitor information systems (e.g. GUIDE), support in workspaces such as context dependent configurations of mobile phones (TEA) or smart environments (e.g. Gaia), where facilities in the environment interact with user devices, also, applications based on ad hoc networks in development. Due to this fact, adaptation of applications to the context is crucial. This can either be done application-driven, that is, the application decides which actions should be taken, or system-driven, where the system manages the adaptation transparently. The latter class has drawn attention over the last few years. A few adaptive middlewares or frameworks have been proposed including Impala [68], which is based on a finite state machine, or TinyCubus [75], which tries to select the best set of algorithms based on several parameters, policies and different adaptation strategies.

Location services for mobile ad hoc networks only offer limited context information, in this case, the position of mobile objects. A scalable, distributed location service is, for example, GLS [67]. Prior to storing the location in such a service, the location has to be determined. Small COs usually do not have a GPS device, so different methods have been developed in the last few years. Two basic approaches are commonly used: having the distance of three objects of which the location is known, its own location can be calculated. The other possibility is to measure the angle to two known objects. Since most COs are equipped with omnidirectional antennae, a very accurate measurement of the angles is not feasible. However, [74] shows that is it feasible and more accurate to use two directional antennae on small Mica2 nodes. Several methods exist to determine the distances to other nodes: [36] uses RSSI values delivered by the RF chip in combination with an error model, [33] presents an algorithm for semi-static sensor networks that works with hop counts, and [109] uses more accurate ultrasound and maps the location estimation to a global non-linear optimization problem which is approximated fully distributed.

Some open issues

Current context management solutions for ad hoc networks focus only on one part of context management. None cover the complete context management area of the gathering of data, in-node data management, in-network data distribution, and in-network data querying.

Positioning is still too inaccurate; better algorithms are required here. Also, for mobile COs, no good approaches exist.

4.5.2. *Data consistency and adaptivity in WSNs*

The benefits of having several nodes in a WSN mostly come from the fact that many nodes simultaneously monitor the same physical area. Nodes can be put into sleep mode without any loss of precision in the network. This results in an increase in energy conservation and, therefore, the network lifetime.

The reliability of the system is also improved with several sensor nodes. This scenario, however, raises issues of data inconsistency which may occur due to various reasons: for instance, inherent imprecision associated with sensors, inconsistent readings and unreliable data transfer, to name just a few.

This VF provides the functionality to ensure consistency of the sensor data at various system abstraction levels:

• Data consistency may mean that data retrieved from a location in the sensor network should be consistent with data sent to the same location.

• Data consistency may also mean that all sensors sensing the same physical phenomenon should more or less agree on the measured value.

• In a rule-based system, data consistency may mean that all actuators agree on the action that needs to be taken.

The following parameters are needed to ensure accurate and consistent *operation* of WSNs and distributed COs:

• **localization:** to interpret sensor data and collaborate with other nodes;

• **synchronization:** to ensure that all nodes have an equal understanding of time and the moment at which events take place;

• **reliable data transfer:** to guarantee delivery of information and/or customizable degrees of reliability for data transmission;

• **routing:** to accomplish forwarding data from source(s) to destination(s) through collaboration among neighbors in an energy efficient manner while maintaining a best effort level of reliability;

• **coverage:** to cover a large area while being able to turn off some of the sensors, thus, saving valuable energy, as well as enhancing the accuracy of the sensed data.

Data consistency at the *data processing level* can be achieved through the following mechanisms:

- **state monitoring:** to detect any change in the state of the environment in which sensors are placed, directly from the sensed data;

- **data fusion:** to take the multiple measurements, some of which may be faulty, and determine either the correct measurement value or a range in which the correct measurement lies;

- **event detection:** for the WSN to be in charge of monitoring the environment in which its nodes are placed;

- **fault tolerance and consensus:** to preserve precision and to agree on measured data and required actions in the presence of faulty collaborative sensors.

Trends

For state monitoring, describing the conditions between states and the events that trigger state-change through a set of rules and predicates over events and their parameters has proved to be popular. The authors in [102] proposed this approach for the first time, which was later also used and modified by Strohbach *et al.* [119]. Research in this direction is on the rise. Also, hybrid distributed algorithms similar to the one proposed by Sahni *et al.* [107], which is executed by every sensor using the measurement ranges received from the remaining sensors monitoring the same region as data, accompanied with the sensor's own measurement, have received considerable attention.

Environmental monitoring, detection, identification, localization or tracking of objects in sensor fields have created a new domain for event detection. Research reported in [101, 123, 42, 31, 128] are a few examples.

Introducing, defining and implementing the concept of collaboration in various levels of abstraction is the newest trend in the area of data consistency, which seems to have great potential to solve various issues of the operation of WSN, data processing and application programming.

Some open issues

There are still open issues in the area of data consistency for WSNs and COs. Here we elaborate on some of the most important ones.

A reasonable number of location techniques have been proposed in other works. An integration framework (e.g. standardized APIs) of such mechanisms would be useful for the applications. This stems from the fact that the application requirements are dynamic and often change. The ability to adapt to these frequent changes is thus essential. Since various localization techniques have been defined, the integration framework may choose a proper localization technique to be utilized based on the requested accuracy at the time and the environmental status.

Reliable and energy-efficient data transfer which is application-dependent for improved performance is required. No doubt collaboration between sensor nodes will increase the reliability of the transferred data in the WSN. However, the overhead introduced by cooperation along with the additional energy consumption should be at a reasonable level. Indeed, in order to avoid being suboptimal, protocols should be developed according to particular application requirements.

Data fusion issues that require further investigation are the associated overhead of the fusion process and the heterogenity of data due to different sensor sources.

Also, a high-level descriptive language oriented towards the collective model of solving tasks is necessary in order to implement distributed collaborative WSN applications. This language should support DCAs and provide real-time guarantees for energy-efficient operations.

4.5.3. *Communication functionality*

The communication vertical function refers to the capability of any pair (or group) of devices to exchange information. Different types of communication can be performed: one-to-all, one-to-many, many-to-one, many-to-many. If we consider the case of WSNs with a single sink, one-to-all or one-to-many communications are needed for query dissemination, while many-to-one communication is explored to gather sensed data at the sink.

Trends

Communication in COs environment is expected to be data-centric and attribute-based. This means that as well as addressing a specific CO, the communication infrastructure should be able to deliver data to and from groups of COs which share a set of attributes specifying the destination/source address of the information. For example, a user could issue a query on the average temperature of an area in an office building. This query should then be delivered to the objects with temperature sensors. Similarly, once measures on the temperature have been taken, the COs in the specific area will send packets to the sink(s) reporting the measured values.

The trend of data-centric and attribute-based data dissemination may lead to the selection of different communication overlays depending on the values to be reported. A given overlay interconnects COs of the same "group" sharing a common set of attributes (attribute-based routing). Attribute-based routing could leverage the burden of delivering queries to objects which cannot answer the query. Also, attribute-aware communication infrastructures may optimize sensor data fusion by selecting routes which maximize the chances of aggregating a given type of data, overall decreasing the network load and energy-consumption [113].

Some open issues

Important VF parameters which should be included in a query are the time constraints and accuracy with which a given query needs to be answered. These are application-dependent and even query-dependent parameters: (a) for a query to be successfully resolved, the sensors must deliver fresh data, (b) the query must be answered fast enough, and (c) the precision with which the query is answered must meet the query requirements. This can be regarded as a new concept of QoS requirements that remains to be explored.

The second issue to address is the fact that a one-fits-all protocol stack may not be suitable to this scenario. It is possible to identify the "vertical functions" that should be provided and possible implementations of such functions for specific sets of possible applications. The communications protocols will benefit from and sometimes require the information provided by different vertical functions such as time synchronization and location awareness. Not only are time and location information included in the delivered data, but some protocols such as geographic-based routing can exploit location-awareness to reduce routing overhead and the nodes storage demand. The interactions between complementary vertical functions from the application viewpoint should be further investigated.

Mobility of objects is the third issue to be tackled. Although in many CO scenarios, the devices themselves are unlikely to be mobile, they can be, however, located on mobile users or mobile stations so that their location changes in time in a predictable or unpredictable way. On the one hand, this may have a beneficial effect (e.g. load balancing the energy consumption among the different nodes), but on the other hand, it requires mobility management or mobility-aware protocols to be added to the protocol stack. The mobility of some of the devices have been explored by some architectures such as the data mules [112], in which a group of mobile nodes move in the deployment area collecting data from the sensor nodes and delivering the collected data to the sinks.

4.5.4. *Security, privacy and trust*

COs and WSNs are usually placed in locations that are accessible to everyone – including attackers. For example, sensor networks are expected to consist of a couple of hundred nodes that may cover a large area. It is impossible to protect each of them from physical or logical attacks. Thus, every single node is a possible point of attack.

In a CO, security is pervasive. [87] states that security must be integrated into every component to achieve a secure system. Components designed without security can become a point of attack as [54] shows. However, specific vertical functions which enforce security are available for applications.

Since fully tamper-resistant devices are hard to build [9] and would cost too much money, protocols for sensor networks have to be designed in a way that tolerates malfunctioning/attacking nodes while the whole sensor network remains functional.

Trends

Karlof and Wagner [54] describe several attacks against known routing protocols for sensor networks and present countermeasures for some attacks. This shows that secure routing protocols are a trend themselves since security was only a minor issue before. New routing protocols were developed that are resilient to black-hole attacks ([30]) or that use efficient symmetric key primitives to prevent compromised nodes from tampering with uncompromised routes consisting of uncompromised nodes (Ariadne [50]). [130] also deals with a large number of DoS attacks with an approach that maps jammed regions and reroutes data around them. [93] proposed SIA, a framework for secure information aggregation in large sensor networks, which uses random sampling mechanisms and interactive proofs to verify that the answer of an aggregator is a good approximation of the true value.

Encryption is the basic technique for securing and authenticating transmitted data. Using asymmetric cryptography on highly resource constrained devices is often not possible due to delay, energy and memory constraints [20, 18]. With symmetric cryptographical methods, two communicating COs need a common key. Secure pebblenets [11] use a shared key for the whole network but since the compromise of one single node leads to a security failure of the whole network, pairwise approaches have been developed in the last years. SPINS [88] uses a central base station to establish new pairwise session keys, and [35, 23] is based on random pairwise key pre-distribution. [125] does not rely on infrastructure or pre-distribution: physical contact between nodes is used for establishing the initial key, and new keys are then established by sending key-shares along node-disjoint paths via secured point-to-point communication to each other.

Traffic analysis on the ciphertext can reveal sensitive information about the data, even if it is encrypted. To provide better secrecy, dummy messages can be generated. This seems to be diametrical in resource constrained COs, but secrecy may be a more important goal than energy-saving in some cases. [126] presents an energy-efficient framework that maintains the anonymity of a virtual infrastructure including routing by randomizing communications.

Using COs, especially sensor networks, humans can be observed which leads to privacy problems. Encryption tackles overhearing and data coarsening ensures that no conclusions can be drawn from the data to a single person. The usage of sensor networks to spy on individuals cannot be met with technology alone, but with a mix of societal norms, new laws and technological responses [87].

With several COs contributing to a common goal, it is necessary to assess the reliability of the information provided by an individual CO. With respect to services and transactions, trust has been researched for several years. For data-centric and fully distributed architectures, research has just started. A distributed voting system is proposed in [23], where votes can be cast against misbehaving nodes until all other nodes refuse to communicate with this node. [38] is a more general reputation-based framework with each node monitoring the behavior of other nodes and building up their reputation thereupon. Nodes with a bad reputation can be excluded from the community.

Some open issues

Existing protocols have to be made resilient to attacks, and more new protocols have to be designed as well. Intelligent intrusion detection systems for COs are also needed.

Encryption concepts for aggregation are needed since along an aggregation tree, only point-to-point encryption is feasible, but an attacker in the tree has full access to the data. No general solutions to this problem exist so far.

Some of the key distribution protocols consume a lot of power, therefore energy-efficient algorithms without assumptions about the available infrastructures are required.

Energy-efficient solutions are needed for secrecy. Systems are needed where the user can specify the amount of energy it should spend for dummy messages and the maximum delay it wants to accept to enable mix cascades in COs.

Another important question is how the illegitimate use of sensor networks, for example, to spy on individuals, can be prevented.

Concerning trust, energy is the main problem, since the exchange of votes and reputation data consumes a reasonable amount of power.

4.5.5. *Distributed storage and data search*

In COs and WSNs efficient storage and querying of data are both critical and challenging issues. Especially in WSNs, large amounts of sensed data are collected by a high number of tiny nodes. Scalability, power and fault tolerance constraints make distributed storage, search and aggregation of these sensed data essential.

It is possible to envisage a WSN as a distributed database and run queries which can be given in SQL format. These queries can also imply some rules about how to aggregate the sensed data while being conveyed from the sensor nodes to the query owner. Data aggregation [17, 135] techniques that reduce the number of data packets conveyed through the network are therefore important and also required for effective fusion of data collected by a vast number of sensor nodes [6, 43]. A query in a sensor network may be perceived as the task or interest dissemination process. Sensor nodes process the task or interest and return data to the interested owner.

The following characteristics of WSNs should be considered while designing a data storage, querying and aggregation scheme for WSNs:

• Sensor nodes are limited in both memory and computational resources. They cannot buffer a large number of data packets.

• Sensor nodes generally disseminate short data packets to report an ambient condition, e.g. temperature, pressure, humidity, proximity report, etc.

• The observation areas of sensor nodes often overlap. Therefore, many sensor nodes may report correlated data of the same event. However, in many cases, the replicated data are needed because the sensor network concept is based on the cooperative effort of low fidelity sensor nodes [6]. For example, nodes may report only proximity, then the size and the speed of the detected object can be derived from the locations of the nodes reporting them and timings of the reports. The collaboration among the nodes should not be hampered by the data aggregation scheme.

• Since there may be thousands of nodes in a sensor field, associating data packets from numerous sensors with the corresponding events and correlating the data of the same reported event at different times may be a very complicated task for a single sink node or a central system.

• Due to a large number of nodes and other constraints such as power limitations, sensor nodes are generally not globally addressed [6]. Therefore, the use of address-centric protocols are mostly inefficient. Instead, data-centric or location-aware addressing protocols, where intermediate nodes can route data according to its content [63] or the location of the nodes [21], should be used.

• Querying the whole network node by node is impractical. Thus, attribute-based naming and data-centric routing [114] are essential for WSNs.

Trends

Queries made to search data available in a WSN should be resolved in the most power-efficient way. This can be achieved by reducing either the number of nodes involved in resolving a query or the number of messages generated to convey the results.

There is considerable research interest in developing efficient data querying schemes for WSNs. Data querying systems in general have two major components: interest and data dissemination, and query processing and resolution. Query resolution usually involves data aggregation for energy-efficient processing. We first examine interest and data dissemination techniques, which are closely related to data querying, and then query processing and resolution techniques.

Interest and data dissemination

Protocols for data dissemination protocols are designed to efficiently transmit and receive queries and sensed data in WSNs. In this subsection, we briefly explain five of the best known protocols:

• *Classic flooding:* in classic flooding, a node that has data to disseminate broadcasts the data to all of its neighbors.

• *Gossiping [6]:* this technique uses randomization to conserve energy as an alternative to the classic flooding approach. Instead of forwarding data to all its neighbors, a gossiping node only forwards data to one randomly selected neighbor.

• *SPIN [43]:* this protocol is based on the advertisement of data available in sensor nodes. When a node has data to send, it broadcasts an ADV packet. The nodes interested in this data reply back with a REQ packet. Then, the node disseminates the data to the interested nodes by using DATA packets. When a node receives data, it also broadcasts an ADV and relays DATA packets to the nodes that send REQ packets. Hence, the data is delivered to every node that may have an interest.

• *Directed diffusion [53]:* in SPIN, the routing process is stimulated by sensor nodes. Another approach, namely directed diffusion, is sink-oriented. A sink is the name given to the central node responsible for gathering data from all the other nodes in directed diffusion, where the sink floods a task to stimulate data dissemination throughout the sensor network. While the task is being flooded, sensor nodes record the nodes which send the task to them as their gradient, and hence the alternative paths from sensor nodes to the sink are established. When there is data to send to the sink, this is forwarded to the gradients. One of the paths established is reinforced by the sink. After that point, the packets are not forwarded to all of the gradients but to the gradient on the reinforced path.

• *LEACH [45]:* this is a clustering-based protocol that employs randomized rotation of local cluster heads to evenly distribute the load among the sensors in the network. In LEACH, the nodes organize themselves into local clusters, with one node acting as a local cluster head. LEACH includes randomized rotations of the high-energy cluster head position such that it rotates among the various sensors in order not to drain the battery of a single sensor. In addition, LEACH performs local data fusion to compress the amount of data being sent from the clusters to the base station.

Query processing and resolution

When a query arrives at a sensor node, it is first processed by the node. If the node can resolve the query, the result of the query is disseminated. This approach is one of the simplest ways of resolving and processing a query. Sensor nodes usually take advantage of collaborative processing to resolve queries so that smaller number of messages are transmitted in the network. Queries can be flooding-based where a query is flooded to every node in the network. Alternatively, they can be ERS based where a node does not relay a query that it can resolve. Currently available query processing systems are summarized below:

• *TinyDB [69]:* a query processing system for extracting information from a network of TinyOS sensors. TinyDB provides a simple SQL-like interface to specify the data, along with additional parameters such as the rate at which data should be refreshed much as in traditional databases. Given a query specifying data interests, TinyDB collects data from nodes in the environment, filters and aggregates them. TinyDB does this via power-efficient in-network processing algorithms. Some key features of TinyDB are as follows: TinyDB provides metadata management, provides a declarative query language, supports multiple query resolution on the same set of nodes and supports different levels of in-network aggregation. It also includes a facility for simple triggers or queries that execute some command when a result is produced.

• *Cougar [132]:* a query layer for sensor networks which accepts queries in a declarative language that are then optimized to generate efficient query execution plans with in-network processing which can significantly reduce resource requirements.

• *Active query forwarding scheme (ACQUIRE) [104]:* this aims at reducing the number of nodes involved in queries. In this scheme, each node that forwards a query tries to resolve it. If the node resolves the query, it does not forward it further but sends the result back. Nodes collaborate with their n hop neighbors, where n is referred to as the look ahead parameter. If a node cannot resolve a query after collaborating with n hop neighbors, it forwards it to another neighbor. When n equals 1, ACQUIRE carries out flooding in the worst case.

Some open issues

The routing protocols for WSNs are generally designed for networks that have fixed homogenous sensor nodes and are based on the assumption that all nodes try to convey data to a central node, often named the sink. However, in COs networks there will be heterogenous nodes that can be mobile, and the sensed data will be needed by many nodes, i.e. multiple sinks. Distributed data storage and search (DSS) solutions should have support for heterogenity. Additionally, open research issues on DSS can be summarized as follows: replication, concurrency and consistency control of distributed data, mobility support, general purpose solutions for declarative query languages, interoperability with open standards and query driven architectures with QoS support.

4.5.6. *Resource management*

COs may form a self-organizing network, in which COs arbitrarily join and leave or even move during this operation. Since system components are distributed, any action often involves multiple COs at a time. Due to the distributed, dynamic and uncertain (since both COs and communication are associated with uncertainty) nature of system components, the design of such embedded wireless collaborative systems proves to be difficult and requires a scheme to facilitate the system design and application development.

Resource management aims at providing a way to manage the resources of a system by enabling high-level system primitives to hide unnecessary low-level details. It enables the application to be independent of the actual underlying distributed system. It should also address the dynamic nature of available resources, such as variable network bandwidth. Because the system is built of error-prone components, failures should be handled as normal and not as exceptions.

Trends

Over the years, various enabling technologies have contributed to the evolution of adaptation and reconfigurable software, each having its own advantages and disadvantages. The most important ones are middleware, component-based design, computational reflection and separation of concerns. Recent efforts have been directed towards replacing the traditional strict modularization or layering design with a more energy-efficient one, i.e. cross-layered design. However, most of the important cross-layer adaptation frameworks and projects that have been proposed assume mobile devices that are richer than wireless sensor nodes in terms of resources. Therefore, adaptation frameworks designed specifically for WSNs are emerging.

OSs for WSNs with DCA and specifically designed with limited memory are considered to form one of the main trends in the area of adaptation in WSNs. Due to the increasing interest in having adaptive behavior on a WSN-wide scale, recently attention has been paid to extending the concept of DCA to the whole network.

Some open issues

One of the most important open issues regarding resource management is designing an architecture tailored to support different adaptation models. Since application requirements, environmental conditions and available resources and services may frequently change, depending on goals of the application at the time, sensor nodes must be able to adapt accordingly. However, since these goals may change as well, different adaptation models are required from which the most appropriate one can be automatically selected. As an example, the application may, at one time, aim for frequent and high accuracy sensor readings and at another time, aim for less data accuracy and more energy-efficient readings.

A direct result of the above-mentioned issue is the need for multi-dimensional optimization techniques for resource management.

4.5.7. *Time synchronization*

Applications need to establish a common sense of time among the COs participating in their sensing and actuation goals. Such a functionality can be offered through a time synchronization vertical function.

Forest fire monitoring is a scenario that requires not only the information of whether there has been an indication of fire, but also where and when this is happening. The collected sensor data provides the basis for the decision making process, which may trigger actions to be taken in the monitored environment in order to address abnormal circumstances. Thus, the true time of the observed events is crucial to the prompt action of fire fighting.

It is sufficient to associate local clock timestamps with the sensor data when decisions can be made locally. For instance, the decision of regulating the fluid flux of an industrial pipe may be taken locally and actions would follow through local CO actuators, for example, a regulating valve.

However, there are cases where the collected sensor data will be aggregated from various COs at various points inside the network. To perform such a function, the data need to be timestamped at the source. In this case, the data originates from multiple sources which may or may not agree on a similar time. Thus, a common view on time among the COs is an essential aspect to correlate multiple sources sensor data in order to produce a consistent and reliable result.

Trends

An important question to ask is whether the NTP [80], currently in use on the Internet, could be adopted without further modification in low-cost WSNs. This protocol relies on an external time reference to synchronize the top layer of time servers called *stratum 1* servers.

As the NTP protocol was originally developed considering other design issues, some of the challenges that arise in CO and WSN scenarios cannot be addressed if the protocol is adopted unmodified in sensor network applications. One of the issues is energy-efficiency, as the external source of time (e.g. GPS) tends to consume more power than other components of a sensor node, and does not work in indoor areas which have no line-of-sight to the satellites.

However, the issue raised in [98], where the NTP protocol maintains a synchronized system clock by regularly adding small increments to a system counter, is an important one. This behavior precludes the sensor node processor of being switched to a power-saving idle mode. In addition, NTP servers must be prepared to handle synchronization requests at any point in time. This could potentially introduce a problem as the communication module of a CO (e.g. radio) shows the highest cost among the other components when it is transmitting or listening to any signal from the network. Thus, this module could not be put in sleep mode in order to save energy.

GPS devices of small size and low power consumption have been developed and are now commercially available. The Leadtek 9546 low-power GPS receiver [65] was recently integrated into a sensor board of the Crossbow Mote [52]. These devices consume a current of 60 mA at 3.3 V in full operation. Despite the fact that such a consumption profile is higher compared to the micro-controller current draw of 8 mA, it is comparable to the energy consumption profile of the radio module.

However, such sensor nodes are optimized to operate in sleep mode most of the time. If the GPS is configured to occasionally operate to acquire the time signal, its energy consumption may not be an issue. Although adding this particular GPS module to a sensor board increases its cost by 50%, we believe that the cost will drop significantly because of the economy of scale. This is also the case for the current cost of sensor nodes, which is similar to the cost of a sensor board equipped with GPS.

The current trend is towards the design of self-configuring protocols on the assumption that NTP cannot be directly used in COs and WSNs applications. The latest research results address the time synchronization problem with approaches that do not rely on GPS time signals. As researchers realize that low-power and low-cost GPS devices are becoming commercially available, the research trend may shift to hybrid protocol designs where GPS is used as the primary source of time. Such information is then disseminated throughout the network to non-GPS nodes. How such time servers ought to be organized in the network (e.g. hierarchical such as NTP) would be an important research question.

Some open issues

The characteristics of the network topology in ad hoc WSNs may introduce severe delays because of disconnected parts of the network. To compensate for such uncertainties, the time synchronization protocol needs to be *tolerant* to the message delay irrespective of its dominant source (e.g. processing, transmission).

The second issue is related to trust. When some sensor nodes become responsible to provide the time reference to other nodes in the network, the issue of trusting the source of time arises. Also, if a GPS time signal is used in some reference nodes, a related question is how we can guarantee the authenticity of the GPS time.

Finally, Romer *et al.* [98] suggest that calibration of sensors is a much more general and complex problem than time synchronization. The interesting question is whether the current proposed time synchronization techniques can be used for general sensor calibration problems. Also, it may be the case that time synchronization (at least for outdoor sensor deployment) can be achieved by instrumenting low-cost, low-power GPS devices to sensor nodes as previously discussed.

4.6. Bibliography

[1] Software fundamentals: collected papers by David L. Parnas, *SIGSOFT Softw. Eng. Notes*, 26(4), 2001.

[2] A. Marcucci, M. Nati, C. Petrioli and A. Vitaletti, The Impact of Data Aggregation and Hierarchical Organizazion on Wireless Sensor Networks, Technical report, Rome University, 2004.

[3] A. Köpke, H. Karl and A. Wolisz, Consensus using aggregation – a wireless sensor network specific solution, Technical Report TKN-04-004, Technical University of Berlin, April 2004.

[4] Sarita V. Adve, Albert F. Harris, Christopher J. Hughes, Douglas L. Jones, Robin H. Kravets, Klara Nahrstedt, Daniel Grobe Sachs, Ruchira Sasanka, Jayanth Srinivasan and Wanghong Yuan, The Illinois grace project: Global resource adaptation through cooperation, in *Proceedings of the Workshop on Self-Healing, Adaptive, and self-MANaged Systems (SHAMAN)*, June 2002.

[5] Mehmet Aksit and Zièd Choukair, Dynamic, adaptive and reconfigurable systems overview and prospective vision, in *Proceedings of the 23rd International Conference on Distributed Computing Systems Workshops (ICDCSW'03)*, May 2003.

[6] I.F. Akyildiz, W. Su, Y. Sankarasubramaniam and E. Cayirci, A Survey on Sensor Networks, *IEEE Communications Magazine*, pp. 102–114, August 2002.

[7] J.N. Al-Karaki and A.E. Kamal, On the correlated data gathering problem in wireless sensor networks, in *Proceedings. of the Ninth International Symposium on Computers and Communications (ISCC 2004)*, vol. 1, pp. 226–231, June 2004.

[8] A. Nasipuri and K. Li, A directionality based location discovery scheme for wireless sensor networks, in *1st ACM International Workshop on Wireless Sensor Networks and Applications*, 2002.

[9] Ross Anderson and Markus Kuhn, Tamper resistance – a cautionary note, in *Proceedings of the Second Usenix Workshop on Electronic Commerce*, pp. 1–11, November 1996.

[10] ANSI, SQL Standard, Technical Report X3, 135-1992.

[11] Stefano Basagni, Kris Herrin, Danilo Bruschi and Emilia Rosti, Secure pebblenets, in *MobiHoc '01: Proceedings of the 2nd ACM International Symposium on Mobile Ad Hoc Networking & Computing*, pp. 156–163, New York, NY, USA, 2001. ACM Press.

[12] S. Baydere and M.A. Ergin, An architecture for service access in mobile ad hoc networks, in *Proceedings of IASTED WOC*, Banff, Canada, July 2002.

[13] Christian Becker, *System Support for Context-Aware Computing*, Fakultät Informatik, Elektrotechnik und Informationstechnik, Stuttgart University, 2004.

[14] Lodewijk Bergmans and Mehmet Aksit, Composing crosscutting concerns using composition filters, *Commun. ACM*, vol. 44, no. 10, pp. 51–57, 2001.

[15] Antoine Beugnard, Jean-Marc Jézéquel, Noël Plouzeau and Damien Watkins, Making components contract aware, *Computer*, vol. 32, no. 7, pp. 38–45, 1999.

[16] Gordon S. Blair, G. Coulson, P. Robin and M. Papathomas, An architecture for next gen-
      eration middleware, in *Proceedings of the IFIP International Conference on Distributed
      Systems Platforms and Open Distributed Processing*, London, 1998. Springer-Verlag.

[17] A. Boulis, S. Ganerival and M.B. Srivastava, Aggregation in Sensor Networks: An Energy
      Accuracy Trade-off, in *Proceedings of IEEE SNPA '03*, pp. 128–138, May 2003.

[18] Michael Brown and Donny Cheung, PGP in constrained wireless devices, in *Proceedings
      of the 9th USENIX Security Symposium*, 2000.

[19] H. Cam, S. Ozdemir, P. Nair and D. Muthuavinashiappan, ESPDA: Energy-efficient and
      Secure Pattern-based Data Aggregation for wireless sensor networks, in *Proceedings of
      IEEE Sensors 2003*, vol. 2, pp. 732–736, October 2003.

[20] D. Carman, P. Kruus and B. Matt, Constraints and approaches for distributed sensor net-
      work security, Technical Report #00-010, NAI Labs, September 2000.

[21] E. Cayirci, Addressing in wireless sensor networks, in *COST-NSF Workshop on
      Exchanges and Trends in Networking (Nextworking '03)*, Crete, 2003.

[22] E. Cayirci and T. Coplu, Data Aggregation and Dilution by Using Modulus Addressing in
      Wireless Sensor Networks, in *Proceedings of the 2004 Intelligent Sensors, Sensor Networks
      and Information Processing Conference*, pp. 373–379, December 2004.

[23] Haowen Chan, Adrian Perrig and Dawn Song, Random key predistribution schemes for
      sensor networks, in *SP '03: Proceedings of the 2003 IEEE Symposium on Security and
      Privacy*, p. 197, Washington, DC, USA, 2003. IEEE Computer Society.

[24] E. Cheong, J. Liebman, J. Liu and F. Zhao, TinyGALS: A Programming Model for Event-
      Driven Embedded Systems, in *18th Annual ACM Symposium on Applied Computing (SAC
      '03)*, Melbourne (FL), USA, March 2003.

[25] Keith Cheverst, Nigel Davies, Keith Mitchell, Adrian Friday and Christos Efstratiou,
      Developing a context-aware electronic tourist guide: some issues and experiences, in
      *CHI '00: Proceedings of the SIGCHI Conference on Human Factors in Computing Sys-
      tems*, pp. 17–24, New York, NY, USA, 2000. ACM Press.

[26] T. Clouqueur, P. Ramanathan, K.K. Saluja and K.-C. Wang, Value-fusion versus decision-
      fusion for fault tolerance in collaborative target detection in sensor networks, in *Proceedings
      of the 4th Annual Conference on Information Fusion*, 2001.

[27] Marco Conti, Gaia Maselli, Giovanni Turi and Silvia Giodano, Cross-layering in mobile
      ad hoc network design, *IEEE Computer*, vol. 37, no. 2, pp. 48–51, 2004.

[28] Marco Conti, Gaia Maselli, Giovanni Turi and Silvia Giordano, Cross-layering in mobile
      ad hoc network design, *IEEE Computer*, pp. 48–51, February 2004.

[29] George Coulouris, Jean Dollimore and Tim Kindberg, *Distributed Systems, Concepts and
      Design*, Addison-Wesley, 3rd edition, 2001.

[30] Jing Deng, Richard Han and Shivakant Mishra, A Performance Evaluation of Intrusion-
      Tolerant Routing in Wireless Sensor Networks, in *Proceedings of the 2nd IEEE Interna-
      tional Workshop on Information Processing in Sensor Networks (IPSN 2003)*, pp. 349–364,
      April 2003.

[31] D.M. Doolin, S.D. Glaser and N. Sitar, Software architecture for GPS-enabled wildfire sensorboard, in *TinyOS Technology Exchange*, February 2004.

[32] W. Du, J. Deng, Y.S. Han and P.K. Varshney, A witness-based approach for data fusion assurance in wireless sensor networks, in *Proceedings of the IEEE Global Telecommunications Conference*, vol. 3, pp. 1435–1439, December 2003.

[33] Stefan Dulman and Paul Havinga, Statistically enhanced localization schemes for randomly deployed wireless sensor networks, in *Proceedings of the DEST International Workshop on Signal Processing for Sensor Networks*, Melbourne, Australia, 2004.

[34] Stefan Dulman, Tjerk Hofmeijer and Paul Havinga, Wireless sensor networks dynamic runtime configuration, in *Proceedings of ProRISC 2004*, 2004.

[35] Laurent Eschenauer and Virgil D. Gligor, A key-management scheme for distributed sensor networks, in *CCS '02: Proceedings of the 9th ACM conference on Computer and communications security*, pp. 41–47, New York, NY, USA, 2002. ACM Press.

[36] Leon Evers, Stefan Dulman and Paul Havinga, A Distributed Precision Based Localization Algorithm for Ad hoc Networks, in *Proceedings of the Second International Conference on Pervasive Computing 2004*, pp. 269–286, Linz/Vienna, Austria, April 2004.

[37] J. Feng, F. Koushanfar and M. Potkonjak, *Sensor Network Architecture*, Chapter 12. CRC Press, July 2004.

[38] Saurabh Ganeriwal and Mani B. Srivastava, Reputation-based framework for high integrity sensor networks, in *SASN '04: Proceedings of the 2nd ACM workshop on Security of ad hoc and sensor networks*, pp. 66–77, New York, NY, USA, 2004. ACM Press.

[39] D. Garlan, D. Siewiorek, A. Smailagic and P. Steenkiste, Project Aura: Towards Distraction-Free Pervasive Computing, *IEEE Pervasive Computing,* special issue on "Integrated Pervasive Computing Environments", vol. 1, no. 2, pp. 22–31, 2002.

[40] Andrea J. Goldsmith and Stephen B. Wicker, Design challenges for energy-constrained ad hoc wireless networks, *IEEE Wireless Communications*, vol. 9, pp. 8–27, August 2002.

[41] Tian He, Brian M. Blum, John A. Stankovic and Tarek Abdelzaher, AIDA: Adaptive application-independent data aggregation in wireless sensor networks, *ACM Transactions on Embedded Computing Systems (TECS)*, vol. 3, no. 2, May 2004.

[42] Tian He, Sudha Krishnamurthy, John A. Stankovic, Tarek Abdelzaher, Liqian Luo, Radu Stoleru, Ting Yan, Lin Gu, Jonathan Hui and Bruce Krogh, Energy-efficient surveillance system using wireless sensor networks, in *MobiSYS '04: Proceedings of the 2nd International Conference on Mobile Systems, Applications, and Services*, pp. 270–283. ACM Press, 2004.

[43] W.R. Heinzelan, J. Kulik and H. Balakrishnan, Adaptive Protocols for Information Dissemination in Wireless Sensor Networks, in *ACM MobiCom '99*, pp. 174–185, Seattle, Washington (USA), August 1999.

[44] Wendi B. Heinzelman, Amy L. Murphy, Hervaldo S. Carvalho and Mark A. Perillo, Middleware to support sensor network applications, *IEEE Network*, vol. 18, no. 1, pp. 6–14, January/February 2004.

[45] W.R. Heinzelman, A. Chandrakasan and H. Balakrishnan, "Energy-efficient communication protocol for wireless microsensor networks", in *International Conference on System Sciences*, January 2000.

[46] A. Helmy, Mobility-assisted resolution of queries in large-scale mobile sensor networks, *Computer Networks (Elsevier)*, vol. 43, no. 4, pp. 437–458, November 2003.

[47] J. Hightower and G. Borriello, Location Systems for Ubiquitous Computing, *Computer*, vol. 34, no. 8, pp. 57–66, August 2001. This article is also featured in "IT Roadmap to a Geospatial Future", a 2003 report from the Computer Science and Telecommunications Board of the National Research Council.

[48] J. Hill, R. Szewczyk, A. Woo, S. Hollar, D. Culler and K. Pister, System Architecture Directions for Networked Sensors, in *Ninth International Conference on Architectural Support for Programming Languages and Operating Systems (ASPLOS '00)*, Cambridge (MA), USA, November 2000.

[49] Fritz Hohl, Uwe Kubach, Alexander Leonhardi, Kart Rothermel and Markus Schwehm, Next century challenges: Nexus – an open global infrastructure for spatial-aware applications, in *MobiCom '99: Proceedings of the 5th annual ACM/IEEE international conference on Mobile computing and networking*, pp. 249–255, New York, NY, USA, 1999. ACM Press.

[50] Yih-Chun Hu, A. Perrig and D.B. Johnson, Ariadne: A Secure On-demand Routing Protocol for Ad Hoc Networks, in *Proceedings of the 8th annual international conference on Mobile computing and networking (Mobicom)*, Atlanta, Georgia, USA, September 2002.

[51] Eoin Hyden, *Operating System Support for Quality of Service*, PhD thesis, Wolfson College, University of Cambridge, February 1994.

[52] Crossbow Tech Inc., *MTS 420 Datasheet*, http://www.xbow.com.

[53] C. Intanagonwiwat, R. Govindan and D. Estrin, Directed Diffusion: A Scalable and Robust Communication Paradigm for Sensor Networks, in *ACM MobiCom '00*, pp. 56–67, Boston, Massachusetts (USA), August 2000.

[54] C. Karlof and D. Wagner. Secure routing in wireless sensor networks: Attacks and coutermeasures, in *Proceedings of the 1st IEEE International Workshop on Sensor Network Protocols and Applications*, Anchorage, AK, May 2003.

[55] Eric P. Kasten, Philip K. McKinley, S.M. Sadjadi and Kurt Stirewalt, Separating introspection and intercession to support metamorphic distributed systems, in *ICDCSW '02: Proceedings of the 22nd International Conference on Distributed Computing Systems*, pp. 465–472, Washington, DC, USA, 2002. IEEE Computer Society.

[56] J.O. Kephart and D.M. Chess, The Vision of Autonomic Computing, *IEEE Computer Magazine*, vol. 36, no. 1, pp. 41–50, January 2003.

[57] Gregor Kiczales, Jim des Rivières and Daniel G. Bobrow, *The art of metaobject protocol*, MIT Press, Cambridge, MA, USA, 1991.

[58] Gregor Kiczales, Erik Hilsdale, Jim Hugunin, Mik Kersten, Jeffrey Palm and William G. Griswold, An overview of AspectJ, *Lecture Notes in Computer Science*, vol. 2072, pp. 327–355, 2001.

[59] Gregor Kiczales, John Lamping, Anurag Menhdhekar, Chris Maeda, Cristina Lopes, Jean-Marc Loingtier and John Irwin, Aspect-oriented programming, in Mehmet Akşit and Satoshi Matsuoka, editors, *Proceedings European Conference on Object-Oriented Programming*, vol. 1241, pp. 220–242. Springer-Verlag, Berlin, Heidelberg, and New York, 1997.

[60] O.Kömmerling and M. Kuhn, Design Principles for Tamper-Resistant Smartcard Processors, in *Proceedings of the USENIX Workshop on Smartcard Technology (Smartcard '99)*, pp. 9–20, Chicago, Illinois, USA, May 1999.

[61] E. Kohler, R. Morris, B. Chen, J. Jannotti and M.F. Kaashoek, The Click Modular Router, *ACM Transactions on Computer Systems*, vol. 18, no. 3, pp. 263–297, August 2000.

[62] F. Koushanfar, M. Potkonjak and A. Sangiovanni-Vincentelli, Fault Tolerance in Wireless Ad hoc Sensor Networks, *IEEE Sensors*, vol. 2, pp. 1491–1496, June 2002.

[63] B. Krishnamachari, D. Estrin and S. Wicker, Modeling Data-Centric Routing in Wireless Sensor Networks, in *IEEE INFOCOM'02*, New York, USA, June 2002.

[64] M. Kumar, A consensus protocol for wireless sensor networks, Master's thesis, Wayne State University, 2003.

[65] Leadtek, *GPS 9546 Module Technical Specification*, www.leadtek.com, August 2003.

[66] Thomas Ledoux, Opencorba: A reflective open broker, in *Reflection '99: Proceedings of the Second International Conference on Meta-Level Architectures and Reflection*, pp. 197–214, London, UK, 1999. Springer-Verlag.

[67] Jinyang Li, John Jannotti, Douglas S.J. De Couto, David R. Karger and Robert Morris, A scalable location service for geographic ad hoc routing, in *Proceedings of the 6th ACM International Conference on Mobile Computing and Networking (MobiCom '00)*, pp. 120–130, Boston, Massachusetts, August 2000.

[68] Ting Liu and Margaret Martonosi, Impala: A middleware system for managing autonomic, parallel sensor systems, in *Proc. of the 9th ACM SIGPLAN Symp. on Principles and Practice of Parallel Programming*, pp. 107–118, 2003.

[69] S. Madden, J. Hellerstein and W. Hong, TinyDB: In-Network Query Processing in TinyOS, Technical Report TinyOS Document, version 4, September 2003.

[70] S. Madden, R.S. Szewczyk, M.J. Franklin and D. Culler, Supporting Aggregate Queries over Ad hoc Wireless Sensor Networks, in *4th IEEE Workshop on Mobile Computing Systems and Applications*, New York (USA), June 2002.

[71] Samuel Madden, Michael J. Franklin, Joseph M. Hellerstein and Wei Hong, The design of an acquisitional query processor for sensor networks, in *Proceedings of the 2003 ACM SIGMOD International Conference on Management of Data*, pp. 491–502. ACM Press, 2003.

[72] S.R. Madden, M.J. Franklin, J.M. Hellerstein and W. Hong, TAG: A Tiny Aggregation Service for Ad hoc Sensor Networks, in *OSDI Conference*, Boston, MA, USA, December 2002.

[73] Pattie Maes, Concepts and experiments in computational reflection, in *OOPSLA '87: Conference Proceedings on Object-oriented Programming Systems, Languages and Applications*, pp. 147–155. ACM Press, 1987.

[74] N. Malhotra, M. Krasniewski, C. Yang, S. Bagchi and W. Chappell, Location Estimation in Ad hoc Networks with Directional Antenna, in *Proceedings of the 25th International Conference on Distributed Computing Systems (ICDCS)*, Columbus, Ohio, June 2005.

[75] Pedro José Marrón, Andreas Lachenmann, Daniel Minder, Jörg Hähner, Robert Sauter and Kurt Rothermel, Tinycubus: A flexible and adaptive framework for sensor networks, in Erdal Çayırcı Şebnem Baydere and Paul Havinga, editors, *Proc. of the 2nd European Workshop on Wireless Sensor Networks*, pp. 278–289, Istanbul, Turkey, January 2005.

[76] Philip K. McKinley, S. Masoud Sadjadi, Eric P. Kasten and Betty H.C. Cheng, Composing adaptive software, *IEEE Computer*, pp. 56–64, July 2004.

[77] Philip K. McKinley, S. Masoud Sadjadi, Eric P. Kasten and Betty H.C. Cheng, A taxonomy of compositional adaptation, Technical Report MSU-CSE-04-17, Department of Computer Science, Michigan State University, East Lansing, Michigan, May 2004.

[78] Microsoft.net.

[79] Com: Delivering on the promises of component technology, 2000.

[80] D.L. Mills, Internet Time Synchronization: the Network Time Protocol, *IEEE Trans. Communications*, vol. 39, no. 10, pp. 1482–1493, October 1991.

[81] H. Naguib and G. Coulouris, Towards Automatically Configurable Multimedia Applications, in *ACM Multimedia Middleware workshop at MM '01*, Ottawa, Canada, October 2001.

[82] Dushyanth Narayanan, *Operating System Support for Mobile Interactive Applications*, PhD thesis, School of Computer Science, Computer Science Department, Carniege Melon University, August 2002.

[83] Rolf Neugebauer and Derek McAuley, Energy is just another resource: Energy accounting and energy pricing in the Nemesis OS, in *Proceedings of the 8th Workshop on Hot Topics in Operating Systems*, Schloss Elmau, Germany, May 2001.

[84] D. Niculescu and B. Nath, Position and Orientation in Ad hoc Networks, *Journal of Ad Hoc Networks*, vol. 2, no. 2, pp. 133–151, April 2004.

[85] Brian Noble, M. Satyanarayanan, Dushyanth Narayanan, James Tilton, Jason Flinn and Kevin Walker, Agile application-aware adaptation for mobility, in *Proceedings of the 16th ACM Symposium on Operating System Principles*, St. Malo, October 1997.

[86] D.L. Parnas, On the criteria to be used in decomposing systems into modules, *Commun. ACM*, vol. 15, no. 12, pp. 1053–1058, 1972.

[87] Adrian Perrig, John Stankovic and David Wagner, Security in wireless sensor networks, *Commun. ACM*, vol. 47, no. 6, pp. 53–57, 2004.

[88] Adrian Perrig, Robert Szewczyk, J.D. Tygar, Victor Wen and David E. Culler, Spins: security protocols for sensor networks, *Wirel. Netw.*, vol. 8, no. 5, pp. 521–534, 2002.

[89] T. Pham, Eun Jik Kim and M. Moh, On data aggregation quality and energy efficiency of wireless sensor network protocols – extended summary, in *Proceedings of the First International Conference on Broadband Networks (BroadNets 2004)*, pp. 730–732, October 2004.

[90] Christian Poellabauer, Hasan Abbasi and Karsten Schwan, Cooperative run-time management of adaptive applications and distributed resources, in *MULTIMEDIA '02: Proceedings of the Tenth ACM International Conference on Multimedia*, pp. 402–411. ACM Press, 2002.

[91] Christian Poellabauer and Karsten Schwan, Kernel support for the event-based cooperation of distributed resource managers, in *RTAS '02: Proceedings of the Eighth IEEE Real-Time and Embedded Technology and Applications Symposium (RTAS'02)*, p. 3. IEEE Computer Society, 2002.

[92] Christian Poellabauer, Karsten Schwan, Sandip Agarwala, Ada Gavrilovska, Greg Eisenhauer, Santosh Pande, Calton Pu and Matthew Wolf, Service morphing: Integrated system- and application-level service adaptation in autonomic systems, in *Proceedings of the 5th Annual International Workshop on Active Middleware Services (AMS 2003)*, June 2003.

[93] B. Przydatek, D. Song and A. Perrig, SIA: Secure Information Aggregation in Sensor Networks, in *Proceedings of the 1st ACM International Conference on Embedded Networked Sensor Systems*, pp. 255–265, May 2003.

[94] Anu Purhonen and Esa Tuulari, *Ambient intelligence and the development of embedded system software*, pp. 51–67. Kluwer Academic Publishers, Norwell, MA, USA, 2003.

[95] Python website, URL: http://www.python.org/.

[96] RAPIDware Project website, URL: http://www.cse.msu.edu/rapidware/.

[97] Sylvia Ratnasamy, Brad Karp, Li Yin, Fang Yu, Deborah Estrin, Ramesh Govindan and Scott Shenker, Ght: a geographic hash table for data-centric storage, in *Proceedings of the 1st ACM International Workshop on Wireless Sensor Networks and Applications*, pp. 78–87. ACM Press, 2002.

[98] K. Rmer, P. Blum and L. Meier, *Time Synchronization and Calibration in Wireless Sensor Networks*, Chapter 7. Wiley and Sons, September 2005.

[99] Uts Roedig, Andre Barroso and Cormac Sreenan, Determination of aggregation points in wireless sensor networks, in *Proceedings of the 30th EUROMICRO Conference (EUROMICRO '04)*, 2004.

[100] Manuel Roman and Roy H. Campbell, Gaia: enabling active spaces, in *EW 9: Proceedings of the 9th Workshop on ACM SIGOPS European Workshop*, pp. 229–234, New York, NY, USA, 2000. ACM Press.

[101] K. Römer, Tracking real-world phenomena with smart dust, 2004.

[102] Kay Römer and Friedemann Mattern, Event-based systems for detecting real-world states with sensor networks: A critical analysis, in *DEST Workshop on Signal Processing in Sensor Networks at ISSNIP*, pp. 389–395, Melbourne, Australia, December 2004.

[103] Daniel Grobe Sachs, Wanghong Yuan, Christopher J. Hughes, Albert Harris, Sarita V. Adve, Douglas L. Jones, Robin H. Kravets and Klara Nahrstedt, Grace: A hierarchical adaptation framework for saving energy, Technical report UIUCDCS-R-2004-2409, Computer Science, University of Illinois, Februrary 2004.

[104] N. Sadagopan, B. Krishnamachari and A. Helmy, Active Query Forwarding in Sensor Networks (ACQUIRE), *Elsevier Journal of Ad Hoc Networks*, January 2005.

[105] S.M. Sadjadi, P.K. McKinley and E.P. Kasten, Architecture and operation of an adaptable communication substrate, in *Proceedings of the The Ninth IEEE Workshop on Future Trends of Distributed Computing Systems (FTDCS '03)*, pp. 46–56, May 2003.

[106] S.M. Sadjadi, P.K. McKinley, R.E.K. Stirewalt and B.H.C. Cheng, Trap: Transparent reflective aspect programming, Technical Report MSU-CSE-03-31, Department of Computer Science, Michigan State University, East Lansing, Michigan, November 2003.

[107] S. Sahni and X. Xu, Algorithms for wireless sensor networks, *International Journal on Distributed Sensor Networks*, pp. 35–56, 2004.

[108] F.A. Samimi, P.K. McKinley, S.M. Sadjadi and P. Ge, Kernel-middleware interaction to support adaptation in pervasive computing environments, In *Proceedings of the 2nd Workshop on Middleware for Pervasive and Ad Hoc Computing*, pp. 140–145. ACM Press, 2004.

[109] Andreas Savvides, Heemin Park and Mani B. Srivastava, The bits and flops of the n-hop multilateration primitive for node localization problems, in *WSNA '02: Proceedings of the 1st ACM International Workshop on Wireless Sensor Networks and Applications*, pp. 112–121, New York, NY, USA, 2002. ACM Press.

[110] Albrecht Schmidt, Antti Takaluoma and Jani Mäntyjärvi, Context-Aware Telephony Over WAP, *Personal Ubiquitous Comput.*, vol. 4, no. 4, pp. 225–229, 2000.

[111] Douglas C. Schmidt, Middleware for real-time and embedded systems, *Commun. ACM*, vol. 45, no. 6, pp. 43–48, 2002.

[112] R.C. Shah, S. Roy, S. Jain and W. Brunette, Data MULEs: Modeling a Three-tier Architecture for Sparse Sensor Networks, in *IEEE Workshop on Sensor Network Protocols and Applications (SNPA)*, May 2003.

[113] A. Sharaf, Jonathan Beaver, Alexandros Labrinidis and K. Chrysanthis, Balancing energy efficiency and quality of aggregate data in sensor networks, *The International Journal on Very Large Data Bases (VLDB)*, vol. 13, December 2004.

[114] C. Shen, C. Srisathapornphat and C. Jaikaeo, Sensor Information Networking Architecture and Applications, *IEEE Personal Communications Magazine*, pp. 52–59, August 2001.

[115] Nisheeth Shrivastava, Chiranjeeb Buragohain, Divyakant Agrawal and Subhash Suri, Aggregation: Medians and beyond: new aggregation techniques for sensor networks, in *Proceedings of the 2nd International Conference on Embedded Networked Sensor Systems*, November 2004.

[116] Brian Cantwell Smith, Reflection and semantics in LISP, in *POPL '84: Proceedings of the 11th ACM SIGACT-SIGPLAN Symposium on Principles of Programming Languages*, pp. 23–35, New York, NY, USA, 1984. ACM Press.

[117] I. Solis and K. Obraczka, The impact of timing in data aggregation for sensor networks, in *IEEE International Conference on Communications 2004*, vol. 6, pp. 3640–3645, June 2004.

[118] Frank Stajano and Ross Anderson, The resurrecting duckling: Security issues for ad hoc wireless networks, in *7th International Workshop on Security Protocols*, LNCS, pp. 172–194, 1999.

[119] Martin Strohbach, Hans-Werner Gellersen, Gerd Kortuem and Christian Kray, Cooperative artefacts: Assessing real world situations with embedded technology, in *Adjunct Proceedings of UbiComp 2003*, September 2003.

[120] Enterprise javabeans technology, 2000.

[121] B. Sundararaman, U. Buy and A.D. Kshemkalyani, Clock Synchronization in Wireless Sensor Networks: A Survey, *Ad hoc Networks*, vol. 3, no. 3, pp. 281–323, May 2005.

[122] Clement Szyperski, *Component Software: Beyond Object-Oriented Programming*, Addison-Wesley, January 1998.

[123] The Ohio State University NEST Team, A line in the sand, August 2003.

[124] E. Truyen, B. Jrgensen, W. Joosen and P. Verbaeten, Aspects for run-time component integration, 2000.

[125] Arno Wacker, Timo Heiber, Holger Cermann and Pedro Marron, A fault tolerant Key-Distribution Scheme for Securing Wireless Ad hoc Networks, in *Proceedings of the Second International Conference on Pervasive Computing 2004*, pp. 194–212, Linz/Vienna, Austria, April 2004.

[126] A. Wadaa, S. Olariu, L. Wilson, M. Eltoweissy and K. Jones, On Providing Anonymity in Wireless Sensor Networks, in *Proceedings of the Tenth International Conference on Parallel and Distributed Systems (ICPADS '04)*, pp. 411–418, Newport Beach, CA, USA, July 2004.

[127] Ian Welch and Robert J. Stroud, Kava – a reflective java based on bytecode rewriting, in *Proceedings of the 1st OOPSLA Workshop on Reflection and Software Engineering*, pp. 155–167. Springer-Verlag, 2000.

[128] Geoffrey Werner-Allen, Jeff Johnson, Mario Ruiz, Jonathan Lees and Matt Welsh, Monitoring volcanic eruptions with a wireless sensor network, in *Proceedings of 2nd European Workshop on Wireless Sensor Networks*, Istanbul, Turkey, February 2005.

[129] J. Wong, S. Megerian and M. Potkonjak, Design Techniques for Sensor Appliances: Foundations and Light Compass Case Study, in *40th IEEE/ACM Design Automation Conference*, pp. 66–71, June 2003.

[130] Anthony D. Wood and John A. Stankovic, Denial of service in sensor networks, *Computer*, vol. 35, no. 10, pp. 54–62, 2002.

[131] Zhixue Wu, Reflective Java and a reflective component-based transaction architecture, in *Proceedings of Workshop on Reflective Programming in C++ and Java*, 1998.

[132] Y. Yao and J.E. Gehrke, Query Processing in Sensor Networks, in *First Biannual Conference on Innovative Data Systems Research (CIDR 2003)*, Asilomar, California, USA, January 2003.

[133] Wei Yuan, S.V. Krishnamurthy and S.K. Tripathi, Synchronization of multiple levels of data fusion in wireless sensor networks, in *IEEE Global Telecommunications Conference (GLOBECOM '03)*, vol. 1, pp. 221–225, December 2003.

[134] Heng Zeng, Carla S. Ellis and Alvin R. Lebeck, Experiences in managing energy with ecosystem, *IEEE Pervasive Computing*, January-March 2005.

[135] Jerry Zhao, Ramesh Govindan and Deborah Estrin, Computing aggregates for monitoring wireless sensor networks, in *Proceedings of First IEEE International Workshop on Sensor Net work Protocols and Applications*, pp. 139–149, May 2003.

Chapter 5

System Architectures and Programming Models

5.1. Summary

This chapter provides a discussion about current state of the art of programming models and system architecture for cooperating objects (COs) and highlights their importance for a successful development of these technologies. Section 5.2 provides a brief introduction to the topics and explains the motivation for designing suitable programming abstractions for COs. In section 5.3, the most relevant existing programming abstractions are covered and classified. The main reason for the development of these abstractions is to allow a programmer to design applications in terms of global goals and to specify interactions between high-level entities (such as *agents* or *roles*), instead of explicitly managing the cooperation between individual sensors, devices or services. For example, the database abstraction makes it possible to consider a whole sensor network as a logical database, and performs network-wide queries over the set of sensors. The various paradigms are discussed and a set of criterions making it possible to easily review their strengths and weaknesses are presented.

Section 5.4 presents the existing system architectures for COs at two different levels: first, system architectures of individual nodes, which includes the structure of the operating system running at node level, and its facilities; and second, system architectures supporting the cooperation of different nodes, such as communication models.

Chapter written by S. SANTINI, K. ROEMER, P. COUDERC, P. MARRÓN, D. MINDER, T. VOIGT and A. VITALETTI.

Finally, in section 5.5, the chapter points out some of the limitations of current approaches, and proposes some research perspectives. In particular, programming paradigms should provide more support to ease programming, heterogenity, as well as scalability issues. Regarding system architectures, real-time aspects, which are not currently well addressed, will become increasingly important for COs. Dynamic maintenance (such as code deployment and run-time update support) is another important issue to address in future systems. Lastly, effort is required to better integrate the various paradigms and systems into a unified framework.

5.2. Introduction

Key to the successful and widespread deployment of COs and sensor network technologies is the provision of appropriate programming abstractions and the establishment of efficient system architectures able to deal with the complexity of such systems. Programming abstractions shield the programmer from the "nasty system details" and allow the developer to think in terms of the concrete application problem rather than in terms of the system. This is also true for traditional distributed systems, where numerous software frameworks and middleware architectures are crucial in performing an integrated computing task. Such frameworks and middleware are based on programming models such as distributed objects or events. These conventional and successful programming abstractions for distributed systems can, however, not be simply applied to COs and sensor networks, due to some substantial differences existing among the latter and the former systems.

We will refer to a *programming model* as "a set of abstractions and paradigms designed to support the use of computing, communication and sensing resources in an application" and to a *system architecture* as "the structure and organization of a computing system, as a set of functional modules and their interactions".

The notion of *COs* refers to devices ranging from sensors and smart tags to personal computing devices such as cell phones, PDAs or even digital cameras. Within the Embedded WiSeNts project, a *CO* is formally defined as "a collection of:

- sensors,
- controllers (information processors),
- actuators, or
- COs

that communicate with each other and are able to achieve, more or less autonomically, a common goal".

For the purpose of achieving a given global goal, COs can organize themselves in a particular set-up and, like in traditional distributed systems, coordinate and cooperate in order to perform a form of distributed computing. However, traditional distributed computing applications show many important differences from those considered in the context of COs and sensor networks: COs are in fact closely related to real-life objects and/or to the physical space, of which neither are considered in traditional distributed computing. This close integration with the real world raises several issues for the design of feasible architectures and programming abstractions for COs. Since real-life objects/entities may be mobile, resource availability and context of operation often change. The system must be able to accommodate dynamic (re-)configuration, as well as being able to expose context changes to the programmer. The latter point is an aspect which differentiates COs from mobile computing: mobile computing approaches often try to provide the impression of a standard (static) computing environment in order to shield the programmer from mobility constraints. In addition, typical application domains of COs are different from those of distributed and mobile computing: in ubiquitous and pervasive computing, applications often have to deal with the notion of *context*, which is directly related to data about the physical environment and does not appear as an issue in distributed computing. Another aspect that differentiates COs and, in particular, sensor networks from traditional distributed systems is that a service or function may often be provided anonymously by one (or several) sensor(s): the identity of the device providing the function is thus often unimportant.

Because of these differences, the design and development of dedicated system architectures and programming models appears as a key issue to enable the success of COs as a mainstream technology. In the next sections, we will provide an analysis of the requirements with which feasible system architectures and programming models must be able to comply. We will also survey the most significant existing approaches and evaluate them, underlining the issues that, in our opinion, still require attention from the research community.

5.3. Programming models

In order to design feasible abstractions and paradigms, a first issue to address is to decide which objects will be used in the computing environment. While entities like a simple sensor or a complex cellular phone share the common properties of a CO, it is unclear whether common programming models would apply to such different entities. In section 5.3.1, we will provide a brief overview of the most significant common attributes of COs and underline the requirements that these attributes pose on programming models. In section 5.3.2 we will then discuss the state of the art, providing examples and references for further readings. Finally, in section 5.3.3, a summary of the existing approaches will be provided and the issues requiring further research will be pointed out.

5.3.1. *Requirements*

COs and sensor network applications pose specific requirements on the design of programming abstractions and paradigms. Fundamental work has already been done to identify general significative characteristics of COs and to derive from them some common requirements with which suitable programming models have to comply [9, 29].

Considering the existing work and our experience in this domain, we conclude that an adequate programming abstraction for COs must essentially take into consideration the following aspects:

1. **Ease of programming and expressiveness.** Future applications scenarios for sensor networks and COs typically involve developers that are not necessarily expert programmers, for example, biologists, supply-chain managers or hospital operators. Moreover, since networks of COs are envisaged to include hundreds or thousands of single devices, an application developer should not be bound to program COs individually. On the contrary, developers should be provided with the ability to specify a *global task* for the network, granting the underlying system layers the task of translating this global goal into concrete actions which the single devices have to carry out.

On the other hand, it is also extremely important to keep as high as possible the level of expressiveness of the system "programming language", by providing the programmer with a set of operators and instructions which must be large and diverse enough to enable him/her to exploit all capabilities and functionalities the network is endowed with.

On the application level, this requires the programmer to be somehow aware of the *global features* the system can offer and to be provided with adequate *programming primitives* to access these features without necessarily knowing either the characteristics of the system or the way a specific task is achieved by assigning concrete roles to the single devices. In an environmental monitoring application, for example, a programmer may probably prefer to issue a query like *"detect all bats in the cave"* rather than *"detect and discriminate all ultrasonic-pulses coming from the area delimited by the given reference points"*. In this case, the phenomena the programmer is interested in should be reinterpreted in measurable quantities and detected by multiple cooperating entities. In this case, the programmer will not need to be aware of this "translation" process because of the higher-level primitives he/she is provided with.

With regard to usability and expressiveness aspects, we can conclude that a good programming abstraction for COs must:

- provide "easy-to-use" and expressive programming primitives;
- allow the developer to program the system as a whole (global task specification).

2. **System diversity.** We defined a CO as "a collection of sensors, controllers (information processors), actuators or COs that communicate with each other and are able to achieve, more or less autonomically, a common goal". Thus, a CO may be as

simple as a single sensor endowed with some computing and communication capabilities, or as complex as a sophisticated mobile unit equipped with a wide set of sensors and actuators, a powerful processor and some kind of long-lasting power supply. An application developer is, however, not interested in knowing the concrete ability of each single device the system consists of, only the results she could gain from using the system as a whole. In this sense, the heterogenity of the system should be hidden as much as possible from the programmer, allowing him/her to define a *global network task* and eventually allowing the underlying system layers or even the single devices to cope with the translation of this global task into *single device tasks*. It is worth noting that the aspects of single system devices may differ in, among others, the number and the type of sensors, computing power, communication range, communication technology, the type and durability of power supply, packaging and physical dimensions.

An adequate programming abstraction must, therefore, be able to cope with heterogenous devices and should, in particular:

• furnish a set of programming primitives for task assignment independently of single devices capabilities.

3. **System dynamic.** When deploying a system composed of wireless devices which are potentially unevenly distributed over a wide geographical area, keeping track of the exact state of the system in terms of topology, lifetime, availability and connectivity appears to be a challenging task. In fact, not only could the initial state of the system be partially unknown but, due to the extreme dynamics the system may experience, its state may strongly vary in both space and time domains. Due to node mobility or environmental factors, for example, the topology of the network may change over time in an undetermistic manner, thus bringing uncertainty to network coverage and connectivity. In a sensor network, single nodes may run out of power, be temporarily or permanently unavailable for unforeseen reasons (environmental factors, conscious or unconscious human interaction), or additional nodes may be redeployed to replace crashed ones. Since an application programmer would and should not track or control the ever-changing state of the system, it is a role of the run-time system to adequately adapt the assignment of network resources.

The abstraction designer is therefore compelled to consider that the system must:

• be able to deal with unknown/unstable topology;
• be able to deal with unknown/unstable network size;
• be robust against temporarily or permanently device crashes;
• cope with unstable connectivity.

At this point, we would like to point out that the system dynamic may also vary depending on the concrete application and/or deployment[1]. In application scenarios

---

1. See also Chapter 2.

that include mobile COs, for example, the uncertainty about the topology of the system is typically much more significant than in typical building monitoring scenarios, where the sensing devices may be fixed on walls or furniture.

4. **Environmental dynamic.** When embedded into the real world, sensing devices may experience extreme dynamics with respect to the phenomena being observed, both due to the unpredictable nature of the phenomena and due to their wide varying intensity and space-time extension. Even a single sensed quantity, for example, may span in a very wide interval and exhibit extremely irregular behavior or, on the contrary, show no relevant changes for long periods of time in wide geographical areas. The detection of real-life phenomena may also be a complex process, which may require the use of many different sensors and an aggregation of the measured values.

An application programmer is, however, presumably not interested in knowing how the occurrence of the phenomena can be detected, but just needs to be able to correctly specify which phenomena he/she is interested in and eventually which actions to perform in consequence of their detection. The occurrence of a specific phenomenon thus represents an *event* for the application to which the system should be able to react by either just reporting the measured variables or performing a more complex action. An abstraction designer must thus:

- provide programming primitives to address a wide variety of real-life *events*;
- provide programming primitives to react to real-life *events*.

5. **Resource constraints.** We have already pointed out that distinct COs may show considerable diversity in terms of the kind and number of usable sensors, available power supply and computing and communication capabilities. Therefore, the measure of which resource constraint influences the design of programming paradigms and abstractions for COs depends on the specific objects being considered. When designing algorithms for sensor systems where nodes are typically powered with small batteries, have poor computing and storing capabilities, strict bounded communication range and a limited set of sensing and/or actuating devices, saving resources becomes a key factor for a feasible and successful design. As we have already pointed out when discussing *system diversity*, a suitable programming abstraction should be able to hide device heterogenity to the programmer, and should thus also be able to:

- cope with power constraints;
- cope with computing constraints;
- cope with hardware constraints.

6. **Scalability.** Since networks of COs may include hundreds or thousands of single devices, a suitable programming abstraction must be able to cope with network sizes that range from a few units to thousands of devices.

Depending on the application context and the geographical area that the network is deployed on, the density of devices may also vary. Scalability must thus be ensured for varying (and possibly unknown) network density. In order to program a network of COs, the developer must thus be provided with a programming model that is able to:

- ensure scalability to varying network size;
- ensure scalability to varying network density.

7. **Deployment and maintenance.** Physical deployment and maintenance of networks of COs may have different challenges that depend on the concrete application scenario. However, since devices may be, in most cases, physically inaccessible, hardware repairs and the number of software updates must be kept as low as possible, due to the resulting transmission (energy) costs and due to the problems related to ensuring update coherence throughout the network.

An application developer will also need to be provided with suitable debugging tools, whose design and realization are issues of growing interest in the research community. Due to the manifold of devices a network of COs may be composed of and due to the extreme dynamics the system may experience, fault isolation appears to be an extremely complex task, and therefore reliable and resource-efficient tools must be provided as a fundamental system feature.

In order to guarantee efficient maintenance, a suitable programming model for COs must thus:

- limit the number of code updates;
- support a resource-efficient application-level debugging.

The complexity and diversity of COs pose strong constraints on the design and development of a suitable programming abstraction. Most of the approaches being investigated by the research community focus on specific application scenarios and are thus typically able to comply with only a subset of the above listed requirements. In order to give the reader an overview on the state of the art, we supply in section 5.3.2 a brief overview of the ongoing work on programming models for COs.

5.3.2. *State of the art*

In the last couple of years, a growing interest in the research area of COs brought to a number of designs for programming abstractions specifically targeted to these systems. We analyzed the research literature and ongoing work and came up to the conclusion that currently most relevant existing programming abstractions for COs sensor networks can be classified in the following categories:

1. database view
2. event detection
3. virtual markets
4. virtual machines

5. mobile code and mobile agents

6. role-based abstraction

7. group-based approach

8. spatial programming

9. shared information space

10. other approaches (service discovery, client-server approach, distributed objects)

In the following sections, we will analyze the main characteristics of the above-listed general approaches and we will give a brief description of one or two representative sample-implementations for each given category.

5.3.2.1. *Database view*

A *database* may be commonly defined as a collection of data elements (facts) stored in a centralized or distributed memory in a systematic way, such that a computer program can automatically retrieve them to answer user-defined questions (queries). A system composed of a manifold of entities, such as simple sensor nodes, complex devices or common everyday objects, each one of them endowed with sensing capabilities, may be regarded as a distributed database. In this system, stored data consist of sensor readings and users can issue SQL-like queries to have the system performing a certain sensing task or delivering required data. In this perspective, the system appears merely as a collection of sensors whose readings need to be adequately and automatically stored. Thus, a database approach abstracts away much of the complexity of a system collecting a manifold of different devices, allowing users to see the system like a common database and querying it in a simple, user-friendly query language.

Unfortunately, traditional data-retrieving and processing techniques from the database community cannot be applied directly to COs, since the traditional assumptions about reliability, availability and requirements of data sources cannot be extended to simple sensors. In fact, when comparing *sensor-based* data sources and *traditional* database sources, relevant differences can be identified. First of all, sensor nodes in a sensor network have (typically) limited processors and battery resources. They do not deliver data at reliable rates and the data may often be corrupted. Secondly, since sensors typically produce data continuously or at pre-defined time intervals, near real-time processing may be required, because storing raw sensor streams may be extremely expensive and because sensor data may represent real-life events the user would like to be aware of and eventually respond to. Thirdly, sensor nodes are typically connected in an ad hoc manner and must share common protocols and algorithms to collect, transmit and process data. The use of a multi-hop transmission strategy may, for example, allow in-network processing and data aggregation strategies as query-answers flow through the network to reach a central sink [51].

Thus, the traditional database approach needs to be readjusted to cope with the new requirements of COs. However, the benefits of this kind of approach overwhelm the drawback of a new design and have brought several researchers to the development of query-like interfaces to sensor networks like *Cougar* ([16, 78]), *IrisNet* ([3, 60]) and *TinyDB* ([51, 52, 53]). Section 5.4.1.4 provides some implementation details about the *Cougar* approach.

TinyDB. For a representative example of a query-like interface to sensor networks, we examine *TinyDB*, which is a query processing system for extracting information from a network of tiny wireless sensors developed at the Intel Research Laboratory Berkeley in conjunction with UC Berkeley. "Given a query specifying your data interests, *TinyDB* collects that data from motes in the environment, filters it, aggregates it together, and routes it out to a PC. *TinyDB* does this via power-efficient in-network processing algorithms" [5].

The *TinyDB* query processor runs on top of the TinyOS [35] operating system (for details see section 5.4.1.2) and each sensor node within the network needs to be endowed with an instance of the processor before deployment. *TinyDB* supports a single "virtual" database table sensors, where each column corresponds to a specific type of sensor (e.g. temperature, light) or other source of input data (e.g. sensor node identifier, remaining battery power). Reading out the sensors at a node can be regarded as appending a new row to the sensor table. The query language is a subset of *SQL* with some extensions. In order to understand the way *TinyDB* works, consider the following query example. Several rooms are equipped with multiple sensor nodes. Each node is equipped with sensors to measure the acoustic volume. The table sensors contain three columns: room (i.e. the room number the sensor is in), floor (i.e. the floor on which the room is located) and volume. We can determine the rooms on the 6th floor where the average volume exceeds the threshold of 10 with the following query:

```
SELECT AVG(volume), room FROM sensors
    WHERE floor = 6
    GROUP BY room
    HAVING AVG(volume) > 10
    EPOCH DURATION 30s
```

The query first selects the rows from the sensors at the 6th floor (WHERE floor = 6). The selected rows are grouped by the room number (GROUP BY room). Then, the average volume of each of the resulting groups is calculated (AVG(volume)). Only groups with an average volume above 10 (HAVING AVG(volume) > 10) are kept. For each of the remaining groups, a pair of average volume and the respective room number (SELECTAVG(volume), room) is returned. The query is re-executed every 30 seconds (EPOCH DURATION 30s), resulting in a stream of query results. *TinyDB* uses a decentralized approach, where each sensor node has its own query processor that preprocesses and aggregates sensor data on its way from the sensor node to the user. Executing a query involves the following steps: firstly, a spanning tree of the network rooted at the user device is constructed and maintained as the network topology changes, using a controlled flooding approach. Secondly, a query is broadcast to all the nodes in the network by sending it along the tree from the root towards the leafs. Thirdly, nodes fulfilling the query criteria select the requested data and send them back to the sink. Data can eventually be aggregated as they flow through the network[2]. More details about the *TinyDB* query processor are provided in section 5.4.1.4 and are available in [5, 51, 52, 53], amongst others.

In the context of COs, the database view is also used in the *PerSEND* system to support proximate collaborations between PDAs. In this model, a federated view of a database is maintained from the data available on each node. The database model is relational and the system proposes an *SQL*-like interface to the applications. The database view is dynamic in the sense that it directly reflects a physical context. This context is represented by the set of near-by objects. As objects move, the context evolves and the data associated with the objects are added or deleted from the database view. This system relies on a decentralized architecture, using only peer-to-peer communications (one-hop) over short distance wireless interfaces.

An important aspect of this database approach is that the system supports the notion of continuous queries. This means that the dynamics of query results can be managed at the system level, instead of the application level. For example, consider a continuous query for the evaluation of the maximum offer in a bidding: the query result continuously reflects the best bid, and the data dependency between the best bid and current offers is managed at the system level.

5.3.2.2. *Event detection*

Events are a natural way to both represent and trigger state changes in the real world and in distributed systems, giving rise to model applications as producers, consumers, filters and aggregators of events. Regardless of the specific scenario, "interesting events" may represent node-internal occurrences (timeouts, message sending or receiving) or specific sensing results. Thus, the application can specify interest in certain state changes of the real world (*basic events*) and upon detecting such an event, a sensing-device sends an *event notification* towards interested applications. The application can also specify certain patterns of events (*compound events*), such that the application is only notified if occurred events match this pattern [41, 65].

The *event detection* paradigm is particularly well suited to provide a programming abstraction for sensor network applications. We discuss *DSWare* [45] as a representative example of this category.

---

2. The *TinyDB* engine may be provided with *TAG* [52], an aggregation service for in-network processing of data.

DSWare. *DSWare* is a software framework that supports the specification and automated detection of compound events. A compound event specification contains, among others, an event identifier, a detection range specifying the geographical area of interest, a detection duration specifying the time frame of interest, a set of sensor nodes interested in this compound event, a time window W, a confidence function f, a minimum confidence c_{min}, and a set of basic events E. The confidence function f maps E to a scalar value. The compound event is detected and delivered to the interested sensor nodes if $f(E) \geq c_{min}$ and all basic events occurred within time window W. Consider the example of detecting an explosion event, which requires the occurrence of a light event (i.e. a light flash), a temperature event (i.e. high ambient temperature) and a sound event (i.e. a banging sound) within a subsecond time window W. The confidence function may be defined as:

```
f = 0.6 * B(temp) + 0.3 * B(light) + 0.3 * B(sound)
```

The function B maps an event ID to 1 if the respective event has been detected within the time window W, and to 0 otherwise. With $c_{min} = 0.9$, the above confidence function would trigger the explosion event if the temperature event is detected along with one or both of the light and sound events. This confidence function expresses the fact that detection of the temperature event gives us higher confidence in an actual explosion happening than the detection of the light and sound events. *DSWare* also includes various real-time features, such as deadlines for reporting events and event validity intervals.

Some architectural issues about the *DSWare* abstraction are discussed in section 5.4.1.4.

The event detection paradigm is also used for higher-level programming model for COs, such as in Linda-like tuple space systems and databases. In tuple spaces, a node interested in an event creates a tuple pattern and reads the tuple space for this pattern (which is equivalent to subscribing to an event class, which corresponds to the pattern). An actuator node produces an event by publishing a tuple. When one or more tuples match a given tuple pattern, the corresponding thread is woken up and can process the event:

```
actuator thread:
out<'event'>

listener thread:
rd<'event'>  // handling the event
```

We must note that in this approach, event handling is synchronous, while asynchronous event handling may be expressed either through multiple threads (one per event) or through a "catch all" pattern, and then determines which type of event occurred.

Database approaches which offer *triggers* also support a form of event handling: the trigger is defined by a logical predicate on a query, with the associated code to process when its predicate becomes true. This mechanism is proposed in systems as old as *Xerox Parctab*, to trigger pre-programmed operations in a database of location-dependent data.

5.3.2.3. *Virtual markets*

The market-based approach offers a very expressive and intuitive way to model and analyze typical distributed control problems, as well as guidelines for the design and implementation of distributed systems. This methodology has also been proposed as a generic programming paradigm for distributed systems and is addressed as *Market-oriented programming* [73]. This approach regards modules in a distributed system as autonomous agents holding particular knowledge, preferences and abilities, and the distributed computation may be implemented as a market price system [58]. This abstraction is particularly suited to model systems where different devices need to cooperate and coordinate in order to reach a common goal in a globally efficient way. Under this point of view, the system is seen as a virtual market where agents (i.e. single devices) act as self-interested entities, which regulate their behavior to achieve maximal profit with minimal costs (resource usage) taking into account globally-known price information (set in order to achieve a globally efficient behavior).

Mainland *et al.* [55] applied the basic ideas of *Market-oriented programming* and defined the *Market-Based Macroprogramming* (*MBM*) paradigm, a promising programming abstraction for sensor networks.

Market-based macroprogramming. In the *MBM* approach, sensor nodes are seen as agents that perform actions to produce goods in return for (virtual) payments. The prices of goods are globally advertised throughout the network and single nodes decide to perform only those actions that maximize their (local) utility function, whose value depends on both the node's internal state and the payment the node will (virtually) get to perform those actions. By dynamically tuning the prices of goods, a user can force nodes to perform desired actions, and network re-tasking is accomplished by adjusting prices rather than injecting new code on sensor nodes. *MBM* also allows multiple users to share the network by offering different payments for node actions, providing a sort of "free market". In order to preserve network resources, the energy budget of a node has to be taken in account when calculating the utility function value.

In the context of the cooperation of personal communication devices, some studies are investigating the use of economic-like regulation systems to enforce a global objective. In the *IST Secure* project, which aims at trust management in uncertain environments (such as the cooperation between newly discovered PDAs), a general *trust model* based on the notion of reputation is proposed. Each possible action involved in the cooperation is associated with a set of possible *outcomes*. Each node maintains an *evidence store* to log the interactions with other nodes and interesting events. Based on the history of interactions and observed outcomes, nodes are able to dynamically build a notion of *reputation*. Essentially, positive outcomes lead to a "good" reputation. If an action is required in an interaction, a benefit/risk analysis is performed and a decision is taken based on the active policy for the node. For example, if the risk is beyond a given threshold or if the benefit is below a given value, the action may be rejected.

Similar mechanisms are investigated in a French study called *Mosaic*, which aims at providing system support for collaborative backup of vulnerable personal devices (PDA, phones, digital camera, etc.). Each device uses other devices to save part of their data, and provides spare space to backup other devices. Fair use of the resources (space and energy) is considered critical for the success of this kind of application, and market-like regulation is a promising approach to ensure this goal. The idea is to associate a currency to the resources and account for their use when a device provides backup to another. Backing up data "gives" credits to a device, while saving data to another device "uses" credits.

5.3.2.4. *Virtual machines*

The concept of virtual machines has been well-known since the early 1960s and is used to indicate a piece of computer software able to shield applications from the details of an underlying hardware or software platform. A virtual machine offers to applications a suite of *virtual instructions* and maps them to the real instruction set actually provided by the underlying real machine. In this way, the virtual machine abstraction can mask differences in the hardware and software lying below the virtual machine itself, thus facilitating code and data mobility.

In a system of complex COs, the interchange of data and other information may be very cumbersome since single devices may show a broad range of different internal architectures and protocols. This problem may be overcome by providing each different device with an adequate version of the virtual machine, which will hide the single device's peculiarities and provide a "unique" virtual hardware and software setting to applications. In this way, applications can easily be written to run on the virtual machine itself instead of having to create separate application versions for each different platform. The virtual machine approach has already been deeply investigated for different applications ([21, 24, 25, 47]), but only a few approaches have been proposed and designed explicitly for COs. A tiny virtual machine specifically designed for sensor networks is *Maté*, developed at the Intel Research Laboratory at Berkeley, in conjunction with UC Berkeley. For further implementation details the reader may refer to section 5.4.1.3.

Maté. *Maté* is a bytecode interpreter that concisely describes a wide range of sensor network applications through a small set of common high-level primitives. The design of the *Maté* virtual machine is focused on producing a very concise instruction set, in order to allow complex programs to be very short and thus feasible to be flooded into the network with limited energy-costs. *Maté* code is sent through the network in small capsules of 24 instructions, each of which is a single byte long (thus, a single capsule fits in a *TinyOS* packet). *Maté*'s high-level abstraction provides an efficient way to frequently reprogram a sensor network, reducing code transmission costs and thus saving precious energy resources.

MagnetOS. Energy-saving issues have also been the guidelines that led to the design of *MagnetOS* [12], a power-aware adaptive operating system, specifically developed to work both on single nodes as well as across a large ad hoc network. *MagnetOS* provides a unified system abstraction to applications that see the entire network as a single unified Java virtual machine. The *MagnetOS* system is able to adapt to changes in resource availability, network topology and application behavior. It also supports efficient power consumption policies and hides the network's heterogenity to the applications.

The use of virtual machine architectures is also widespread for larger COs like PDAs and mobile phones. The most common and well-known is *Java*. Some specific motivations exist in this context: in devices like mobile phones, the core services (such as voice communication) are provided by the native environment, while additional applications/services (such as games) are confined in the virtual machine. This prevents additional services to compromise the operation of the core system, which is considered as dependable, without preventing extensibility. In addition, the isolation provided by the virtual machine avoids untrusted code to access sensitive data, such as contact information.

Despite the many attractive features of the *Java* virtual machine, we must highlight some practical limitations which exist for COs: the heterogenity abstraction is somewhat limited, as platform fragmentation is important (PJava, MIDP profiles, various API scattered in various optional JSRs, etc.). Also, the native environment may still be impacted by malicious or buggy code running in the virtual machine, as it usually shares some resources with the virtual machine: the CPU, memory and energy. While CPU and memory usage can easily be controlled, energy is more difficult.

5.3.2.5. *Mobile code and mobile agents*

Mobile code is a general notion that indicates a software program transmitted from one entity to another through a network to be executed at the destination. *Remote evaluation*, *code-on-demand* and *mobile agents* are the three basic paradigms that are encompassed in the notion of *mobile code*. *Mobile agents*, in particular, represent a mobile code that autonomously migrates between entities and they are therefore well suited for the implementation of distributed applications.

The notion of *mobile agents* may be easily seen as an efficient programming strategy for sensor networks, since sensing tasks may be specified as mobile code that may spread across the network piggybacking collected sensor data. These scripts may be injected into the network at any point and are able to travel autonomously through the network and distribute themselves where and when necessary.

A possible approach to the definition, implementation and deployment of such scripts is provided by Boulis *et al.* in [19], where the design and implementation of *SensorWare*, an active sensor framework for sensor networks, is presented.

SensorWare. In *SensorWare*, programs are specified in Tcl [61], a dynamically typed, procedural programming language. The functionality specific to *SensorWare* is implemented as a set of additional procedures in the Tcl interpreter. The most notable extensions are the query, send, wait, and replicate commands. query takes a sensor name and a command as parameters. One common command is value, which is used to obtain a sensor reading. Send takes a node address and a message as parameters and sends the message to the specified sensor node. Node addresses currently consist of a unique node ID, a script name and additional identifiers to distinguish copies of the same script. The replicate command takes one or more sensor node address as parameters and spawns copies of the executing script on the specified remote sensor nodes. Node addresses are either unique node identifiers or "broadcast" (i.e. all nodes in transmission range). The replicate command first checks whether a remote sensor node is already executing the specified script. In this case, there are options to instruct the run-time system to do nothing, to let the existing remote script handle the additional user or to create another copy of the script. In *SensorWare*, the occurrence of an asynchronous activity (e.g. reception of a message, expiry of a timer) is represented by a corresponding event. The wait command expects a set of such event names as parameters and suspends the execution of the script until one of the specified events occurs. The following script is a simplified version of the *TinyDB* query and calculates the maximum volume over all rooms (i.e. over all sensor nodes in the network):

```
set children [replicate]
set num_children [length children]
set num_replies 0
set maxvolume [query volume value]
while {1} {
    wait anyRadioPck
```

```
if {maxvolume < msg_body} {
    set maxvolume msg_body }
incr num_replies
if {num_replies = num_children} {
    send parent maxvolume
    exit }
}
```

The script first replicates itself to all nodes in the communication range. No copies are created on nodes already running the script. The replicate command returns a list of newly "infected" sensor nodes (children). Then, the number of new children (num_children) is calculated, the reply counter (num_ replies) is initialized to zero, and the volume at this node is measured (maxvolume). In the loop, the wait blocks until a radio message is received. The message body is stored in the variable msg_body. Then, maxvolume is updated according to the received value and the reply counter is incremented by one. If we received a reply from every child, then maxvolume is sent to the parent script and the script exits. Due to the recursive replication of the script to all nodes in the network, the user will eventually end up with a message containing the maximum volume among all nodes of the network. Further details about the *SensorWare* approach are reported in section 5.4.1.3.

In the broader context of cooperating personal devices, a few systems use the mobile agent paradigm, for example the Sony *Shop Navi* project [59]. Beyond security issues, a major problem is the need of a common agent hosting environment, which is a difficult constraint given the heterogenity of the devices. An important advantage of mobile code is the support for the dynamic deployment of software (over the air provisioning), which allows dynamic adaptation of the same devices in new situations requiring additional software support.

5.3.2.6. Role-based abstractions

Many sensor network applications require some form of self-configuration, where sensor nodes take on specific functions or roles in the network without manual intervention. These roles may be based on varying sensor node properties (e.g. available sensors, location, network neighbors) and may be used to support applications requiring heterogenous node functionality (e.g. clustering, data aggregation).

The concept of role assignment is thus an implicit part of many networking protocols as well as a common function of middleware platforms for sensor networks. Heinzelman *et al.* proposed in [34], *MiLAN*, a middleware able to control the allocation of functions to sensor nodes in order to meet certain quality of service requirements specified by the user. In [70], another high-level role-based programming approach for sensor networks is presented. In this work, a high-level task specification is compiled into a set of node-level programs that must be properly allocated to sensor nodes taking into account the node capabilities. These approaches typically support only very specific role assignment tasks and do not offer a general solution for the role specification problem. Providing such a general solution is the primary effort of Frank *et al.* [31, 66], whose framework for *generic role assignment* is briefly sketched here.

Generic role assignment. In this approach, a developer can specify user-defined roles and rules for their assignment using a high-level configuration language. Such a *role specification* is a list of role-rule pairs. A role is simply an identifier. For each possible role, an associated rule specifies the conditions for assigning this role. Rules are Boolean expressions that may contain predicates over the local properties of a sensor node and predicates over the properties of well-defined sets of nodes in the neighborhood of a sensor node. *Properties* of individual sensor nodes are available sensors (e.g. temperature) and their characteristics (e.g. resolution), other hardware features (e.g. memory size, processing power, communication bandwidth), remaining battery power, or physical location and orientation. A *distributed role assignment algorithm* assigns roles to sensor nodes, taking into account role specifications and sensor node properties. A separate instance of the role assignment algorithm is executed on each sensor node. Triggered by property and role changes on nodes in the neighborhood, the algorithm evaluates the rules contained in the role specification. If a rule evaluates to true, the associated role is assigned.

Consider the following *coverage* example. A certain area is said to be covered if every physical spot falls within the observation range of at least one sensor node. In dense networks, each physical spot may be covered by many equivalent nodes. The lifetime of the sensor network can be extended by turning off these redundant nodes and by switching them on again when previously active nodes run out of battery power. Essentially, this requires the proper assignment of the ON and OFF roles to sensor nodes. The following role specification implements this:

```
ON :: {
    temp-sensor == true &&
    battery >= [threshold] &&
    count(2) {
        role == ON &&
        dist(super.pos, pos) <= [sensing-range]
    } <= 1 }
OFF :: else
```

The rule in lines 1-7 specifies the conditions for a node to have ON status: it must have a temperature sensor and enough battery power (lines 2 and 3). As a third condition, we require that at most one other ON node should exist within this node's sensing range. This is specified by the count operator in line 4. It expects a hop-range as its first parameter and returns the number of nodes within this range for which the expression in curly braces evaluates to true. Here we request to evaluate nodes within 2 network hops. Note that the used property names (e.g. `role` in line 5, `pos` in line 6) in the nested expression refer to properties of the specified neighbor nodes. To access properties of the current node, the prefix `super` is used (e.g. `super.pos` in line 6). The `dist` operator used in line 6 returns the metric distance between two positions. In the example, it specifies that only nodes located within this node's sensing range should be counted.

5.3.2.7. Group-based approach

The *clustering* paradigm is well-known in the field of distributed systems and ad hoc networks ([14, 64]) and offers a suitable programming abstraction for systems collecting a manifold of complex devices, which cooperate and coordinate to reach a common goal [30].

In a sensor network, for example, nodes that share a neighborhood relationship can organize themselves in *groups* that constitute single addressable entities for the programmer, and within which nodes can efficiently communicate and collectively exploit local resources. The *Hood* and *abstract regions* paradigms are the approaches that will now be analyzed.

Hood. Whitehouse *et al.* managed to define the neighborhood concept as a proper programming primitive by designing *Hood*, an abstraction "which allows users to think about algorithms directly in terms of neighborhoods and data sharing instead of decomposing them into messaging protocols, data caches and neighbor lists" [76]. For a given network task, *Hood* defines the membership criteria and the attributes to be shared within a group, and provides an interface that shows the names of the current neighbors as well as the list of the shared attributes, hiding from the applications all the nesting details about neighbors discovering and data sharing, data caching and messaging within a single group. This group-based approach seems a suitable solution for sensor networks, since it scales well for increasing network size, it is robust to node failures and allows dynamic network reconfiguration.

Abstract regions. Another neighborhood-based approach is also followed in the design of *abstract regions*, a set of "general-purpose communication primitives for sensor networks that provide addressing, data sharing and reduction within local region of the network" [74]. Regions are just a collection of nodes and may be defined on the basis of geographical, topological or connectivity predicates, such as "*all nodes within 10 meters*" or "*all nodes in 1-hop communication distance*", and nodes within a region communicate, share variables and data and provide aggregation. *Abstract regions* provide a communication abstraction that simplifies application design by hiding local actions within regions (communication, data dissemination and aggregation), and by allowing applications to explicitly trade off resource consumption and accuracy of global operations [75].

5.3.2.8. *Spatial programming*

Accessing network resources using *spatial references*, in the same way that traditional imperative programming variables are accessed using memory references, is the underlying idea of *spatial programming*, a space-aware programming paradigm particularly suitable for distributed embedded systems. In this view, a networked embedded system is seen as a single virtual address space and applications can access network resources by defining a *spatial reference*, i.e. a pair *{space:tag}*, where *space* indicates the expected physical location and *tag* indicates a property of the demanded network resource [17].

A system that supports the *spatial programming* abstraction has been implemented by Borcea *et al.* using *Smart Messages*, a software architecture that recalls many concepts and constructs typical of mobile agents [18].

Other systems that exploit the *spatial programming* approach are *SIS* [10], *Close Encounters* [40] and *Ubibus* [11]. In these works, COs are used as data symbols that cover a given geometrical shape and the physical space is used as a way of structuring information and processing. The idea is to annotate existing interactions of physical entities with computing actions. These actions are triggered according to geometrical conditions, in particular physical proximity.

5.3.2.9. *Shared information space*

Devices that need to cooperate to accomplish a global task also need to share data and information about their internal states. In traditional centralized systems, *shared information* is usually stored in a physical central place, accessible for all entities participating in the system.

In systems like sensor networks, sensor nodes need to cooperate and coordinate and thus need to share information in order to efficiently perform high-level sensing tasks. A centralized solution is, however, not suitable for such systems, since making single nodes reporting data and state information to a central unit poses an extremely high communication overhead (thus, inefficient energy management) and provides a single point of failure. Since useful information is "spread" among a manifold of single entities, a distributed solution is needed. Koberstein *et al.* proposed assuming that a sensor network behaved like a *swarm*, where nodes cooperate with each other by sharing knowledge about swarm state and external conditions [39]. This knowledge is "stored" in a *distributed virtual Shared Information Space (dvSIS)*, an abstract entity that may be represented as a union of local instances stores on single sensor nodes. Since a single local instance may contain information that is no longer valid, or may be inconsistent with the local instance of other nodes, requirements on consistency and completeness need to be very strict. Using a broadcast, a single node can publish new acquired data in the *dvSIS*, thus other nodes can enhance their local instances just by listening to the broadcasts. Even if it is often desirable that all participating nodes

have identical views on this space (i.e. all nodes see the same data items), a more efficient implementation can be provided if nodes are allowed to have slightly different
views on the space, without affecting a correct global behavior of the network.

The shared information space paradigm is also useful to coordinate mobile computing and ubiquitous computing applications. Tuple spaces similar to Linda are used
in mobile computing with *LIME* [62], and a spatial tuple space enabling spatial programming for ubiquitous computing is proposed in the *SPREAD* system [23].

5.3.2.10. *Other approaches*

Among the other approaches presented, we would also like, even if only briefly, to
discuss the following.

Service discovery. Systems adopting this approach allow a node to discover services available in the current context, in particular, those provided by neighborhood
nodes. This abstraction is related to traditional distributed computing approaches, such
as client-server interactions or distributed objects method invocations. Typically, an
object uses the service discovery protocol prior to being able to interact with the neighboring objects. Some existing systems based on this approach are *JINI* [71], *Bluetooth
SDP* [6] and *Cooltown URL beaming* [38].

Client-server approach. In this simple approach, each CO hosts one (or more)
server(s) enabling other objects to use its services. The Cooltown system is typical
of this approach, where COs (printers, etc.) are running embedded web servers,
accessed by other objects (PDAs, etc.) running web clients.

Distributed objects. This approach is similar to the previous one, but services are provided by *objects* (in the object-oriented sense), and interactions between these objects
are supported through remote method invocation. Some examples of this approach
include JINI and CORBA's ORB [71, 72].

Some other interesting approaches presented in the research community, such as
EnviroTrack [7, 15], *MiLAN* [34], *Impala* [49] and *TinyCubus* [56, 57] are discussed
in sections 5.4.1.4 and 5.4.1.5.

5.3.3. *Summary and evaluation*

In section 5.3.1, we discussed the fundamental attributes of COs and derived from
them the requirements with which an adequate programming abstraction should be
able to comply in order to allow an end-user to make a proper and efficient use of
the system. On the basis of the listed requirements, we outlined the characteristics

that, in our opinion, should be considered as the most significant when evaluating the suitability of a programming abstraction for COs. In particular, we stated that the following aspects should be considered:

- **Ease of programming and expressiveness.** Does the abstraction provide easy-to-use programming primitives? How expressive is the provided set of programming primitives? Does this set allow access to all the functionalities the network is able to provide? Does the abstraction allow an application developer to program the network as a whole, i.e. to specify global network tasks instead of single devices roles?

- **System diversity.** Does the abstraction hide device heterogenity to the programmer?

- **System dynamic.** Is the abstraction able to deal with unknown/unstable connectivity, topology and/or network size? How do the abstractions deal with temporary or permanent device failures?

- **Environmental dynamic.** Does the abstraction provide adequate primitives to define, detect and react to real-life events?

- **Resource constraints.** Is the abstraction able to cope with hardware, computing and power constraints?

- **Scalability.** Is the set of provided programming primitives suitable for growing network size and/or density?

- **Deployment and maintenance.** Is it possible to perform resource-efficient debugging at application level?

In section 5.3.2, we provided a survey on the past and present work on programming models for COs. The surveyed approaches were sorted in different categories and for each category we presented a concrete implementation example. We thus discussed along the way the most known and successful programming abstractions specifically targeted to COs, and provided an overview on existing running systems such as *TinyDB, DSWare, Maté, MagnetOS, SensorWare, generic role assignment, Hood* and *abstract regions*, to name just a few.

Most of the proposed approaches were designed for a specific application scenario or were tailored to some specific design goals, and thus appear able to comply with only a subset of the above-listed requirements. Even if most of them (e.g. database approach, agent-based approach, event-based approach, virtual markets) are not new, they required significant adaptation to be used for COs and/or sensor networks. These approaches differ with respect to ease-of-use, expressiveness, scalability, overhead, etc., as we will now outline by means of three representative examples, namely, *TinyDB, SensorWare* and *DSWare*.

TinyDB provides the user with a declarative query system which is very easy to use. The database approach hides distribution issues from the user, and rather than programming individual objects, the network can be programmed as a single (virtual) entity. On the other hand, the expressiveness of the database approach is limited in various ways. Firstly, adding new aggregation operations is a complex task and requires modifications to the query processor on all objects. More importantly, it is questionable whether more complex tasks can be appropriately supported by a database approach. For example, the system does not explicitly support the detection of spatio-temporal relationships among events in the real world (e.g. expressing interest in a certain sequence of events in certain regions). The system also suffers from some scalability issues, since it establishes and maintains network-wide structures (e.g. spanning tree of the network, queries are sent to all devices). In contrast, many sensing tasks exhibit very local behavior (e.g. tracking a mobile target), where only very few devices are actively involved at any point in time. This suggests that *TinyDB* cannot provide optimal performance for such queries. Additionally, the tree topology used by *TinyDB* is independent of the actual sensing task. It might be more efficient to use application-specific topologies instead. A clearly different approach is the one followed in *Sensor-Ware*, where an imperative programming language is used to assign tasks to individual nodes. Even fairly simple sensing tasks result in complex scripts that have to interface with operating system functionality (e.g. querying sensors) and the network (e.g. sending, receiving, and parsing messages). On the other hand, *SensorWare*'s programming paradigm allows the implementation of almost arbitrary distributed algorithms. Typically, there is no need to change the runtime environment in order to implement particular sensing tasks. However, the low performance of interpreted scripting languages might necessitate the native implementation of complex signal processing functions (e.g. Fast Fourier Transforms, complex filters), thus requiring changes in the runtime environment in some cases. *SensorWare* allows the implementation of highly scalable applications, since the collaboration structures among sensor nodes are up to the application programmer. For example, it is possible to implement activity zones of locally cooperating groups of sensor nodes that "follow" a tracked target. The *Sensor-Ware* runtime does not maintain any global network structures. One potential problem is the address-centric nature of *SensorWare*, where specific nodes are addressed by unique identifiers, thus potentially leading to robustness issues in highly dynamic environments. Another interesting comparison can be made with respect to *DSWare*, a system that provides compound events as a basic programming abstraction. However, a complete COs application will require a number of additional components besides compound event detection. For example, code is needed to generate basic events from sensor readings, or to act on a detected compound event. *DSWare* does not provide support for this glue code, requiring the user to write low-level code that runs directly on top of the sensor node operating system. This makes the development of any application a complex task, while at the same time providing maximum flexibility. *DSWare* supports only a very basic form of compound events: the logical "and" of event occurrences enhanced by a confidence function. It might be worthwhile to consider more complex compound events, such as explicit support for spatio-temporal relationships among events (e.g. sequences of events, non-occurrence of certain events). Note that more restrictive compound event specifications can avoid the transmission of event notifications and can hence contribute to better energy-efficiency and scalability. Without re-discussing all the abstractions surveyed in section 5.3.2, we can conclude that the examined approaches exhibit a trade-off between ease-of-use and expressiveness. While *TinyDB* is easy to use, it is restricted to a few predefined aggregation functions. More complex queries either require changes in the runtime environment, are inefficient or cannot be expressed at all. While *SensorWare* and *DSWare* support the efficient implementation of almost arbitrary queries, even simple sensing tasks require significant programming efforts. Narrowing this gap between ease-of-use and expressiveness while concurrently enabling scalable and energy-efficient applications is one of the major challenges in the design of adequate programming abstractions for COs. It is not yet clear whether suitable programming models will be inspired by known paradigms as in the presented examples, or if completely new approaches need to be defined.

5.4. System architectures

As we already mentioned in the introduction, we will refer to a *system architecture* as "the structure and organization of a computing system, as a set of functional modules and their interactions". In this section, we will survey the state of the art of system architectures for COs. We divide this study into two parts:

• **Node internals.** This part presents several possible abstraction levels in a single node. Operating systems, the simplest approach, provide basic system functionality including a uniform way of accessing the hardware. A *data management middleware* hides the data sources and offers precomputed information to applications, in some cases also in collaboration with other COs. *Virtual machines* completely abstract the hardware, offering a virtual execution environment to the user. *Adaptive system software* hides the changing environment of a real-life CO from the user.

• **Interaction of nodes.** This part presents system architectures topics related to the interaction of nodes. We abstract two main sets of functionalities, namely low-level functionalities, including tasks corresponding to physical, link, routing and transport layers, as well as high-level functionalities, including coordination and support, clustering, timing and localization, addressing, lookup, collaboration, failure detection and security.

5.4.1. *System architectures: node internals*

In this section, we will survey the state of the art of node internal system architectures. Our survey discusses operating systems, virtual machines, data management middleware and adaptive system software for COs.

The most important requirements are related to the requirements discussed in section 5.3.1. The main difference to the architectures designed for traditional systems are the resource-constraints, in particular, the memory footprint and the limited energy budget of the target system. Therefore, energy-efficiency and a small memory footprint are indispensable features of these architectures. Other important requirements include the flexibility to cope with different applications and hardware, as well as adaptivity and a small learning curve. Furthermore, portability can be regarded as a desirable feature.

5.4.1.1. *Data-centric and service-centric approach*

The field of COs comprises a wide range of applications[3]. Sensor network scenarios are said to be *data-centric*, whereas pervasive or ubiquitous computing scenarios are more *service-centric*. In this section, both approaches are described.

A service is a well-defined and self-contained function that does not depend on the context or the state of other services. The service is executed on the explicit request of a caller, who has to know the interface of the service. A response is returned after the completion of the service.

In a data-centric approach, the execution is controlled by the data. For example, on the basis of the type of incoming data, the appropriate function is called which is able to handle this type. Although the user of a data-centric system may communicate with it by functions of a well-defined interface, and although the components of a system may internally exchange the data over such interfaces, in contrast to a service-centric approach, the desired functionality is not specified explicitly by the name of the service but implicitly by the passed data.

There is no obvious hierarchical ordering of service-centric and data-centric approaches. On the one hand, a data-centric system could use several services to fulfill its task. On the other hand, a service-centric approach could fall back on a lower-level data-centric query. Thus, this approach abstracts from the data.

5.4.1.2. *Operating systems*

In this section, we present different operating systems for COs. Unlike general-purpose desktop operating systems such as *Windows* or *Linux*, these operating systems run on devices that are designed for special-purpose tasks. The main tasks of these operating systems is to provide an abstract interface to the underlying hardware and to schedule system resources.

After a short discussion on scaled-down versions of *Linux* and *Windows*, we briefly present some operating systems designed for handheld devices such as PDAs and mobile phones. These devices can be regarded as part of a network of COs (the first

---

3. See also Chapter 2.

mobile phones with attached sensors are already available), or they can be used to access WSNs.

Many of the traditional embedded operating systems are designed for real-time systems with small memory footprints such as robot arms or break systems, whereas most current operating systems targeted for WSNs are not real-time systems. We first present three of these operating systems in more detail, namely *TinyOS*, *Contiki*, and *Mantis*. These systems span the whole spectrum of concurrency: *TinyOS* does not provide any multithreading, *Contiki* provides multithreading as a library for those applications that explicitly require it and *Mantis* is a layered multithreaded operating system. Next, other sensor node operating systems, *SOS*, *kOS*, *Timber* and *DCOS*, are briefly presented:

- **Scaled down versions of desktop operating systems**

  - **Windows.** Microsoft *Windows* is the most common operating system for desktop computers. There are also embedded systems that run *Windows XP*, for example ATMs, set-top boxes and ticket vending machines. *Windows XP Embedded* is a modular cut-down version of XP that allows the designer to choose the modules to be used. This way, the size of a system without networking, GUI and device drivers is limited to about four to five MB of memory.

  - **Linux.** Since *Linux* is covered under the GPL license[4], anyone can customize *Linux* to his/her PDA, Palmtop or other mobile or embedded device. Therefore, a multitude of scaled-down *Linux* versions exist. These include **RTLinux** (Real-time Linux), an extension of the *Linux* kernel that provides real-time guarantees by inserting an additional abstraction layer between the kernel and the hardware; **uClinux**, a scaled-down *Linux* version for system without a memory mapping unit, and thus no isolation between kernel and user-space processes; **Montavista Linux** with Linux distributions for ARM, MIPS and PPC; **ARM-Linux** and many others.

- **Operating systems for handheld devices**

  - **Palm OS.** The *Palm OS* is specifically designed for PDAs featuring a small screen, with less processing power than desktop PCs and limited memory. In *Palm OS*, the kernel is responsible for thread scheduling, handling hardware interrupts and other low-level management tasks. Although Palm-applications are single-threaded, the kernel itself uses multiple threads.

  - **Symbian OS.** *Symbian OS* is a robust multi-tasking operating system, designed specifically for wireless environments and the constraints of mobile phones. The core kernel's size is less than 200 KB. The OS has support for handling low memory situations and a power management model. *Symbian OS* runs on fast, low-power, low-cost CPU cores such as ARM processors.

  - **Windows CE.** In contrast to Windows XP Embedded, Windows CE has a different codebase than Windows XP. Windows CE is particularly designed for

---

4. http://www.gnu.org/copyleft/gpl.html.

small hand-held devices. Windows CE is a pre-emptive multitasking operating system allowing multiple applications or processes to run within the system simultaneously. Furthermore, Windows CE provides deterministic interrupt latencies and real-time properties. Windows CE also provides programmable power conserving mechanisms. According to Microsoft, the code size is 200 KB without graphics, but the code size increases dramatically when graphics and networking are included.

• **Embedded real-time operating systems.** There are a large number of embedded real-time operating systems. Here we present a few of them:

- **eCos.** *eCos* [1] is an open-source system designed to be highly configurable. *eCos* has extensive configuration possibilities and can be scaled up from a few hundred bytes in size to hundreds of KB. *eCos* provides features such as pre-emptable tasks with multiple priority levels, low latency-interrupt handling, multiple scheduling policies and multiple synchronization methods. The *eCos* development environment contains a set of GNU-based tools that assist in making application specific configurations of *eCos* for each particular embedded system. *eCos* has compatibility layers for *POSIX* and *uITRON*.

- **QNX.** *QNX* [4] is a Unix-like operating system with real-time properties, and is the most prominent example of a successful micro-kernel design. The micro-kernel is surrounded by cooperating processes that provide higher-level services such as inter-process and low-level networking communication, process scheduling and interrupt dispatching. *QNX* features a very small kernel of about 12 KB. *QNX* is designed for systems running x86, MIPS, PowerPC or ARM CPUs.

- **XMK.** *XMK* (eXtreme Minimal Kernel) [2] is an open-source real-time kernel designed to fit very small micro-controllers, yet be scalable up to larger systems. A minimal kernel configuration requires only 340 bytes of ROM and 18 bytes of RAM.

• *TinyOS.* *TinyOS* [35] is an operating system specially designed for the constraints and requirements of WSNs. It is currently the most widely used system for academic research in the area of sensor networks. *TinyOS* is available for several platforms, e.g. Mica, Telos, EYES, iMote. Additionally, a *TinyOS* simulator called *TOSSIM* is included. We will now briefly review the most relevant aspects of this tiny operating system:

- **System modularity.** *TinyOS* builds on a component architecture where both applications and the operating system consists of single, interlinked components. Thus, there is no strict separation between the operating system kernel and application software. The operating system is not a separate program on which separate applications can rely on. Rather, the necessary components of *TinyOS* are compiled together with the application to a single executable that contains both the operating system and the application components.

A component (see Figure 5.1) consists of a fixed-sized state and tasks. Interaction between components is provided via function call interfaces that are sets of commands and events. Commands are used to initiate an action, such as the transmission of a message. Events denote the completion of a request, such as the completion

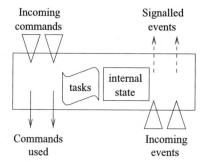

Figure 5.1. *Schematic component of* TinyOS *consisting of tasks, internal state, commands and events (after [35])*

of the transmission of a packet or an external event such as the reception of a packet. An application is composed by choosing a number of components and "wiring their interfaces together" [44].

- **Scheduling hierarchy.** *TinyOS* uses a two-level scheduling hierarchy that lets high-priority events pre-empt low priority tasks. Events are invoked because of external input such as incoming data, sensor input or a timer. They are allowed to signal other events or call commands. Events and commands must not block the processor with time-consuming operations, but have to complete in a short time. Longer calculations have to be performed in a task. Tasks are a form of deferred procedure call enabling postponed processing. This way, events can post tasks for later processing. When the calculation is done, the task can signal an event to inform the component about the result. Both events and tasks must run to completion after being invoked. This precludes the use of blocking statements. Events are implemented using hardware interrupts, and tasks are implemented using a linear FIFO dispatcher. The dispatcher has a queue of tasks, where each task is represented by a pointer to a function. In the case where the task queue is empty, the system can go into a sleep state and wait for the next interrupt. If this event posts a task, the dispatcher takes it from the task queue and runs it.

- **Concurrency.** *TinyOS* currently does not support multi-threading, blocking or spin loops. Therefore, and as a consequence of the command-event model, many operations in *TinyOS* are so-called split-phase. A request is issued as a command that immediately returns. The completion of the request is signalled by an event. While this approach enables, for example, implicit error handling, it complicates program design and development since it more or less forces the programmer to implement blocking sequencing in a state-machine style.

- **Programming.** The entire *TinyOS* system, as well as all applications running under it, is implemented in the *nesC* language [32]. *nesC* is an extension to the C language that supports event-oriented programming, i.e. the execution of a function as a reaction to certain system events which is common for sensor networks. *nesC* introduces several keywords that allow the modeling of *TinyOS* components as described above. Using `command` and `event`, a component signal can be defined. The invocation of commands is done via `call` and the triggering of events via `signal`. A set of commands and events can be encapsulated in an interface. The actual implementation is located in a module which can use and provide (keywords `uses` and `provides`) several interfaces. In a `configuration`, components are wired together, connecting interfaces used by components to interfaces provided by others. Every *nesC* application is described by a top-level configuration that wires together the components used. Additionally, the *nesC* compiler provides compile time checks for finding race conditions. Therefore, it finds all asynchronous code, i.e. code that is reachable from at least one interrupt handler. Every use of a shared variable from such asynchronous code is a potential race condition if the programmer does not use the `atomic` statement. This check can be done since the compiler processes the complete code including application and operating system components, and therefore has knowledge about the interaction between them.

The foremost feature of *TinyOS* is its small code size and memory usage, and its component model that lets the system designer specify the system dependencies at compile time. The main drawback is the event-driven concurrency model which restricts applications to be implemented as explicit state machines. *TinyOS* is an open-source software, published under a three-clause BSD license. Because of its widespread use in the wireless sensor networking research community, there is a wealth of implementations of various communication protocols for sensor networks available.

• **Contiki.** *Contiki* [27] is an operating system designed for networked and memory constrained systems. *Contiki* is, in many aspects, similar to *TinyOS*, but has additional support for threads and dynamically loadable programs. *Contiki* includes the uIP stack for TCP/IP communication. Important properties of *Contiki* include its execution models, its system architecture, dynamic reprogramming and portability.

- **Execution models.** In order to keep its memory footprint small, *Contiki* is based around an event-driven kernel. Unlike *TinyOS*, *Contiki* allows applications to be written in a multi-threaded fashion. Multi-threading is implemented as a library that is optionally linked only with those applications that specifically require a threaded model of execution. The event-driven nature of the kernel makes the system compact and responsive, whereas the multi-threading makes it possible to run programs that perform long-running computations without completely blocking the system. For example, performing user authentication on a mote requires up to 440 seconds [13]. Multi-threading allows the system to handle incoming packets while performing long computations. Additionally, *Contiki* provides a third execution model called `protothreads` [28]. In event-based systems, programs usually have to be implemented as explicit state machines, and thus are hard to debug and maintain. `Protothreads` are stack-less thread-like constructs. They allow programs to be written in a sequential fashion and, like threads, provide conditional blocking on top of

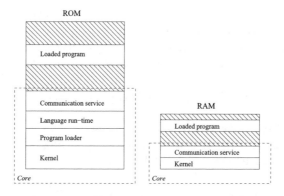

Figure 5.2. *Contiki: partitioning into core and loaded programs [27]*

the event-driven system. Unlike threads, `protothreads` are extremely lightweight, requiring only two bytes of memory per `protothread` and no additional stack.

- **System architecture and partitioning.** A *Contiki* system is partitioned into two parts: the `core` and the `loaded programs` (see Figure 5.2). The partitioning is made at compile time and is specific to the deployment in which *Contiki* is used. Typically, the core consists of the *Contiki* kernel, the program loader, the most commonly used parts of the language, run-time and support libraries, and a communication stack with device drivers for the communication hardware. The core is compiled into a single binary image that is stored in the devices prior to deployment. The core is generally not modified after deployment, even though it is possible to use a special boot loader to overwrite or patch the core. A *Contiki* system consists of the kernel, libraries, the program loader and a set of processes. A process may be either an `application program` or a `service`. A `service` is a process that implements functionality that is used by other processes such as protocol stacks and data handling algorithms. Services can be seen as shared libraries which can be replaced during run-time. The kernel consists of a lightweight event scheduler that both dispatches events to running processes and periodically calls polling handlers used to, for example, check for status updates of hardware devices. The kernel also supports two kinds of events, namely, asynchronous events which are a form of deferred procedure calls and synchronous events mainly used for interprocess communication. The kernel enqueues asynchronous events in a special event queue and dispatches the events later to the target process.

- **Dynamic reprogramming.** When developing applications for sensor networks, the ability to reprogram the sensor nodes without requiring physical access to the nodes greatly simplifies the development and reduces the development time. *Contiki* has support for loading individual programs from the network, which makes it possible to dynamically reprogram the behavior of the network. After a program has been loaded into the memory, the program's initialization function is called that may replace or start processes. A thin service layer, conceptually situated next to the kernel, provides service discovery and run-time dynamic service replacement within each sensor node. The ability to load and unload individual applications is important for WSNs, because an individual application is much smaller than the entire system binary and therefore requires less energy when transmitted through a network. Additionally, the transfer time of an application binary is less than that of an entire system image.

- **Portability.** *Contiki* is designed to be portable across a wide range of different platforms. *Contiki* runs on several platforms, including the ESB nodes from FU Berlin, Amtel AVR and the Intel x86, and the Z80 platform.

Module	Code size (AVR)	Code size (MSP430)	RAM usage
			$10 +$
Kernel	1044	810	$+ 4e + 2p$
Service layer	128	110	0
Program loader	-	658	8
Multi-threading	678	582	$8 + s$
Timer library	90	60	0
Replicator stub	182	98	4
Replicator	1752	1558	200
			$230 + 4e +$
Total	3874	3876	$+ 2p + s$

Table 5.1. *Size of compiled Contiki code, in bytes [27]*

Contiki's memory requirements of an example sensor data replicator application are shown in Table 5.1. They depend on the maximum number of processes that the system is configured to have (p), the maximum size of the asynchronous event queue (e) and, if multi-threading is used, the size of the thread stacks (s).

• **Mantis.** One of the main design goals of the *Mantis* system [8] is ease-of-use to allow for a small learning curve and rapid prototyping, while meeting the resource constraints of WSNs in terms of limited memory and power. The other key goal of *Mantis* is flexibility by providing experienced programmers sophisticated sensor networks features, including dynamic reprogramming over the radio and remote debugging of sensor nodes. The architecture of *Mantis* includes the following entities:

- **Kernel and scheduler.** The goal of the *Mantis* and its kernel is to leverage familiar, traditional OS services in the realm of resource-constrained WSNs. *Mantis'* design resembles a traditional layered multithreaded design as shown in Figure 5.3. The *Mantis* kernel is designed similar to a traditional Unix-style scheduler, providing a subset of *POSIX*-threads. In particular, the scheduler supports priority-based thread scheduling, as well as binary and counting semaphores. The main data structure of the

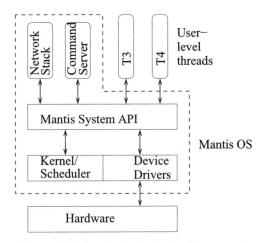

Figure 5.3. Mantis *classical OS architecture [8]*

kernel is a table holding one entry per thread. The size of the table cannot be adjusted dynamically, but is determined at the compile time (12 entries by default). Therefore, there is a fixed level of memory overhead. Each thread table entry has a size of 10 bytes including, for example, a stack pointer, priority level and the thread's starting function. The scheduler provides pre-emptive scheduling with time slicing. The hardware posts timer interrupts that cause the scheduler to initiate a context switch. Other interrupts than these are handled by the corresponding device drivers. On reception of an interrupt, device drivers post a semaphore to activate the corresponding thread. There is a special thread called the idle thread. This thread has the lowest priority and therefore runs when no other thread is runnable. The idle thread can be used to implement power-aware scheduling.

- **Network stack.** The design goal of *Mantis'* networking stack is the efficient use of the limited memory as well as flexibility and convenience. The networking stack features a traditional layered design where the different layers can be implemented in several user-level threads or in one thread. The latter solution minimizes memory usage, avoids copying data between different threads and allows for cross-layer optimizations. The advantages of the layered solution are modularization and flexibility. The *Mantis* standard network stack consists of four layers: application layer, network layer, MAC and physical layer. The latter two are implemented as one user-level thread called the base thread. If the network layer is implemented as a separate thread, the network layer thread blocks a semaphore until the base thread posts the semaphore after, for example, the reception of a packet. The threaded model also makes it possible to activate or deactivate a particular protocol, e.g. a routing protocol. *Mantis* also enables the coexistence of more than one thread for a given task, for example, several routing protocol threads that can run at the same time. In such a scenario, incoming packets are directed to the corresponding thread on a per-packet basis.

- **Dynamic reprogramming.** *Mantis* enables reprogramming of both the entire operating system and parts of the program memory by downloading a program image onto EEPROM, from where it can be burned into a flash ROM. This capability is implemented as a system call library. The entire *Mantis* kernel uses about 12 KB ROM and approximately 500 bytes RAM. Thus, the memory requirements of *Mantis* are larger than those of *TinyOS* and *Contiki*. *Mantis* also features what the *Mantis* developers call a multimodal prototyping environment, where the same code can be executed on both physical and virtual sensor nodes across X86 and Atmel platforms. In this prototype environment, both types of nodes can coexist and communicate with each other in a hybrid network. This is made possible by preserving a common C API across the platforms. With minor modifications, *Mantis* can be executed as an application running on X86 on both *Windows* and *Linux*, since both operating systems support the AVR microcontroller.

	TinyOS	**Contiki**	**Mantis**
Concurrency	events	optional threading and protothreads	threads
Code size (ROM and RAM)	smallest	medium	largest
Ease of learning	hardest	medium	ease of programming major design goal

Table 5.2. *Comparison of TinyOS, Contiki and Mantis*

Table 5.2 compares the important features of *TinyOS*, *Contiki* and *Mantis* with respect to concurrency, code size and ease of learning. *TinyOS* is highly optimized to achieve a small code size, but is hard to program, while *Mantis* is a more traditional structure of a layered multi-threaded operating system, which leads to the largest code size, but simplifies programming. *Contiki* by design combines flexibility with low memory footprint. While for non-expert programmers, *TinyOS* is tricky to program [42], *Mantis* simplifies programming. The same is true for *protothreads* available in the *Contiki* operating system. However, *Mantis'* large memory footprint makes it impossible to implement it on systems with limited memory such as some PIC micro-controllers, to which *TinyOS* has been ported [50]. Smaller memory is both more cost-efficient and energy-efficient. For example, micro-controllers featuring less RAM have also been used for sensor nodes that solely rely on energy scavenged from the environment [54].

- **Other sensor node operating systems**
- **SOS.** The *SOS* operating system [33] is very similar to *Contiki*, particularly the design emphasis on dynamically loadable modules, motivated by the need for code updates during deployment of a sensor network. Like *Contiki*, *SOS* consists of dynamically-loaded modules and a small kernel. The kernel implements messaging, and dynamic memory and module loading and unloading at runtime. In *SOS*, modules are position independent binaries implementing a specific function. Applications are

composed of one or more modules. Unlike *Contiki*, *SOS* does not (yet) provide multi-threading. The code size of the *SOS* core including a facility to distribute programs is about 20 KB, and the RAM usage is more than 2 KB. The ROM usage is comparable to *TinyOS* running *Deluge*, a reliable distribution protocol used to distribute, for example, OS images in a sensor network. The RAM usage of *TinyOS* and *Deluge* is, however, only a quarter of the RAM requirements of *SOS*.

- **kOS.** The *kOS* (*kind-of* or *kilobit*) operating system [20] is designed for iterative applications. In order to keep the duty cycle of applications below a certain threshold and thus to save power, the *kOS* scheduler adapts the periods of the applications. To keep the OS small and simple, the execution model used requires all applications to run to completion before other applications execute. *kOS* interacts with applications using a simple messaging interface. While these design decisions keep *kOS* small, they limit its usage to iterative applications and do not enable, for example, more sporadic tasks such as the dynamic loading of modules. The minimal memory footprint of *kOS* without any application is similar to the size of *Mantis*, requiring slightly more than 12 KB of ROM and about 500 bytes of RAM.

- **Timber.** *Timber* [37] is a self-contained functional language based on an extension of Haskell. Since *Timber* is self-contained, it can be run without any other run-time or operating system and can thus be regarded as a stand-alone operating system. In fact, *Timber* is used as the operating system for the Mulle sensor nodes [36]. The language semantics are the most fundamental part of the OS, and hence the run-time features such as scheduling, threading and memory management can be tailored to each individual application. In contrast to other operating systems for sensor nodes, *Timber* also supplies sufficient infrastructure for reactive concurrent programming and realizes real-time constraints.

- **DCOS.** The main objectives of the *DCOS*[5] (*Data-Centric Operating System*) [26] is to provide real-time guarantees, energy-efficient operation and online reconfigurability. *DCOS* uses an architecture which the authors call data-centric. In this architecture, data is the main abstraction of events. The scheduled entities in *DCOS* are software components called `Data Centric Entities` (DCE). These entities produce data and they can be triggered by other data. While the system is running, *DCOS* adapts the system behavior by dynamically replacing DCEs and/or reconfiguring the data flow between DCEs.

5.4.1.3. *Virtual machines*

Some systems use a virtual machine instead of an operating system running native machine code. Since the virtual machine code can be made smaller, the energy consumption of transmitting the code over the network can be reduced. One of the drawbacks is the increased energy spent in interpreting the code for long running programs, as the energy saved during the transport of the binary code is instead spent in the execution of the code:

---

5. Also discussed in Chapter 4 (section 4.7).

• **Maté.** In order to provide run-time reprogramming for *TinyOS*, Levis and Culler have developed *Maté* [42] (also called *Maté Bombilla*), a virtual machine for *TinyOS* devices. Code for the virtual machine can be downloaded into the system at run-time. The virtual machine is specifically designed for the needs of typical sensor network applications.

Maté is a bytecode interpreter running on *TinyOS*. Bytecode is broken into capsules of 24 instructions, each one byte in length. This makes the capsules fit into a single *TinyOS* packet. Four types of capsules exist: timer capsules, message receive capsules, message send capsules and subroutine capsules. With the latter, programs can be longer than 24 instructions. Capsules contain their type and a version number. *Maté* installs a received capsule if it contains a more recent version of the specified type than the currently installed one. Capsules are broadcast with a single forw instruction.

Three execution contexts are known in *Maté*: clock timers, message receptions and message send requests. Each context has its own two stacks: an operand stack and a return address stack. The first is used for all instructions handling data, while the latter is used for subroutine calls. The maximum depth of the operand stack is 16 and the maximum depth of the call stack is 8. The three types of events correspond to the execution contexts.

There are three operand types: values, sensor readings and messages. Many instructions behave differently for different operands or combinations thereof.

Maté begins execution in response to an event and starts executing the first instruction of the corresponding context until it reaches the halt instruction. Contexts can run concurrently; their execution is interleaved at instruction granularity. *Maté* does not allow asynchronous operations. For example, it waits until a message is sent successfully or until an analog-to-digital conversion is done.

Sending a message or sampling a sensor can be done as a single bytecode instruction. Messages are automatically routed to the destination; a task is automatically enqueued on arrival. Eight instructions can be defined by users, for example, to do some complex data processing. They are implemented in *TinyOS*. Thus, a specially tailored version of *Maté* is to be built.

Maté on the Atmel AVR requires almost 40 KB ROM and more than three KB RAM, and is thus larger than *SOS* and *TinyOS* using *Deluge* [33]. The energy cost of the CPU overhead of a bytecode interpreter is outweighed by the energy savings of transmitting such concise program representations for a small number of executions. Native code is preferable for a large number of executions of the code since for simple instructions (e.g. logical operations) *Maté* takes 33 times more clock cycles than the native *TinyOS* implementation. Thus, any non-trivial mathematical operation is infeasible. Even though such operations can be implemented in native code and called user instructions, they cannot be changed during runtime. An advantage is that all code runs in a sandboxed virtual environment and benefits from all of its safety guarantees.

Application Specific Virtual Machines (*ASVM*) [43] is a enhancement of *Maté* and addresses its main limitations with respect to flexibility, concurrency and propagation. *ASVM* supports a wide range of application domains, whereas *Maté* is designed for a single domain only. In *Maté*, only a single shared variable can be used. In *ASVM*, an operation component can register several shared variables, and the system ensures race-free and deadlock-free execution. The code propagation is not only done via broadcasts, but with a control algorithm based on Trickle to detect when code updates are really needed on other nodes.

Each *ASVM* consists of a template, which includes a scheduler, a concurrency manager and a capsule store, and of extensions, which are the application-specific components that define a particular *ASVM*. Handlers as an extension of *Maté*'s event contexts are code routines that run in response to system events. Operations are the units of execution functionality, divided into primitives and functions. A particular language is compiled to primitives which are, therefore, language specific. Functions are language independent and provided by the user to tailor an *ASVM* to a particular application domain. Capsules are again the units of code propagation, but can be longer than in *Maté* and are, therefore, split into fragments during propagation. Building an *ASVM* involves connecting handlers and operations to the template.

Again, an *ASVM* instruction has an overhead of 400 cycles, thus making complex mathematical code in *ASVM* infeasible. However, compared to Maté, it is easier in *ASVM* to push expensive bytecode operations to native code using functions to minimize the amount of interpretation.

- **MagnetOS.** *MagnetOS* [12, 48] is a distributed operating system that simplifies the programming of ad hoc networking applications by making the entire network appear as a single virtual machine. It provides adaptation to the resource constraints and changes in the network (e.g. topology, application behavior, available resources), increases the system longevity through good power utilization, supports nodes with heterogenous resources and capabilities, and is highly scalable. Unlike *Maté*, *MagnetOS* targets larger platforms such as x86 laptops, Transmeta tablets and StrongArm PocketPC devices.

Applications consist of a set of event handlers that are executed in response to a system, sensor or application-initiated occurrence. An event handler stores the instance variables and is free to move across nodes in the network. Execution consists of a set of event invocations that may be performed concurrently.

The *MagnetOS* system provides the image of a virtual Java machine. Regular Java applications are partitioned into distributable components that communicate via events by a static partitioning service. Therefore, applications are rewritten at bytecode level. For example, object creations are replaced by calls to the *MagnetOS* run-time, which selects an appropriate target node and constructs a new event handler at that location. Remote data accesses, lock aquisitions and releases, type-checking and synchronization instructions are converted as well.

For object creation, *MagnetOS* provides at-most-once semantics. The system performs health checks using keep-alive messages only for long-running synchronous event invocations.

Several algorithms in the core of the operating system decide when and where to move application components. All of them try to shorten the mean path length of data packets sent between components of an application by moving communicating objects to topologically closer nodes. *LinkPull* (formerly *NetPull*) operates at a physical link level and migrates components one hop at a time in the direction of greatest communication. *PeerPull* (formerly *NetCenter*) operates at a network level and is, therefore, able to migrate a component multiple hops at a time directly to the host with which a given object communicates most. *NetCluster* migrates a component to a randomly chosen node within the cluster it communicates with most. Finally, *TopoCenter* migrates components to a node such that the sum of migration cost and estimated future communication costs are minimized. Therefore, a partial view of the network is needed, which is gathered along a packet's path by each node attaching to its single-hop neighborhood.

When *MagnetOS* decides to move an event handler, it sets a flag. The rewritten code detects the flag and checkpoints its current state. *MagnetOS* transports this state and resumes the computation at the destination. Application writers can also manually control the placement of components. A component can be strictly bound to a node, or a starting node can be defined where the component is migrated to first and from where it can migrate further using the mechanisms described above.

• **SensorWare.** In *SensorWare* [19], lightweight and mobile control scripts based on Tcl can be defined. Their replication and migration allows the dynamic deployment of distributed algorithms on sensor nodes, thus making them easily (re)programmable. *SensorWare* targets richer platforms than *Maté*, such as iPAQs, since the framework is almost 180 KB in size. Although energy-efficiency is also a main consideration of the project, it is not clear how such a script-based solution can be implemented efficiently in sensor nodes with high resource limitations. On the other hand, since code is more compact in *SensorWare*, it provides built-in multi-user support and it is portable to other platforms.

SensorWare is built upon the operating system of the sensor nodes and uses its standard functions and services. The operating system provides hardware abstraction. Control scripts rely completely on the *SensorWare* layer, while other static applications and services can use the standard functions and services of both *SensorWare* and the operating system.

Tcl is used as the basic scripting language in *SensorWare*. All *SensorWare* functions are defined as new Tcl commands, thus fully integrating into Tcl. Such commands abstract specific tasks, such as communication with other nodes, or data sensing and filtering. Special commands allow the forwarding of the current program to other nodes, while trying to avoid unnecessary code transfers by transmitting the code only if the script is not already running on the neighboring nodes.

SensorWare is an event-based language. However, as *SensorWare* supports multithreading, control scripts can use blocking waits until an event occurs. Examples for such events are the reception of a message, the expiration of a timer, or the availability of one or more items of sensor data.

There are two different types of task classes in the run-time environment of *SensorWare*: fixed tasks and platform-specific tasks. The former are always included in every *SensorWare* implementation and handle system functions such as the spawning of new scripts, surveillance of resource contracts, radio transmission and reception, etc. The latter depend on the hardware configuration and have to do with the specific types of sensors available at a given sensor node.

Threads in *SensorWare* are coupled with queues. Queues of scripts are receiving events. Queues for radio, sensors, times, etc. receive events of the device they are connected to as well as messages that declare interest in this event type. System messages can be exchanged between system threads.

To address the problem of heterogenity, any module or service (e.g. radio, sensing device, timer service) in *SensorWare* is represented as a virtual device. Every device implements a common interface with four commands to communicate with the device. Using these commands, it is possible to ask for information, to trigger an action, and to create and dispose of event IDs. To facilitate porting the framework to other platforms, wrapper functions for several operating systems functions and hardware accesses are used.

5.4.1.4. *Data management middleware*

Operating systems usually provide the only means to access the hardware – especially sensors – in a uniform way, but no means to manage the data flow. Thus, the actual data management functionality is located in the application layer. The *Data management middleware* is an abstraction layer between the sensor network and the application layer that provides access to information on a higher semantic level, including the storage, distribution and querying of this information. Thus, it also allows the transparent addition of new sensors and new sensor types.

• **EnviroTrack.** The main purpose of the *EnviroTrack* project [7, 15] is to provide an efficient implementation for the class of applications that need to track the locations of entities in the environment. *EnviroTrack* distinguishes itself from traditional localization systems in the assumption that cooperative users can wear beaconing devices that interact with location services in the infrastructure for the purposes of localization and tracking.

Therefore, *EnviroTrack* provides a new abstraction based on context labels and tracking objects so that as the tracked entity moves, the identity and location of the sensor nodes in its neighborhood change. The programmer interacts with a changing group of sensors through a simple object interface.

EnviroTrack works by providing the user with programming abstractions that allow him to specify context types. This is done by activating a condition called `sense` which is used by sensors to join and leave the group of sensors in charge of tracking a given object. The second part of the context type definition is composed of a series of context variables that define an aggregate state. An aggregation is composed of readings taken by the various sensors of the group about a specific object. For example, this variable could be the average location for cars or a maximum temperature in a fire-detection scenario. For each context variable, applications define an aggregation function and two constants: a critical mass and freshness, which modify the aggregation process.

The architecture of *EnviroTrack* is composed of the following elements:

- **Pre-processor.** This component emits code that initializes the structures to track context labels and periodically calls the sense functions to allow entity discovery. It also translates the definition of context types.

- **Group management.** This protocol manages joins and leaves of sensors for a specific entity. It also ensures that there is always at least one leader for the group.

- **Aggregate state computation.** Group members send their data to the leader periodically and the leader collects them, aggregates them and forwards them again to the appropriate location.

- **Directory services.** Context types are hashed to a location which serves as a directory for that type. Whenever new context labels are created, they register their position on the directory service which, if they move, is updated accordingly.

EnviroTrack is able to track entities using an energy-efficient protocol and is able to provide fault tolerance to message loss, leader failures and aggregate loss.

- **DSWare.** *Data Service Middleware* (*DSWare*) [45, 46] is a specialized layer that performs the integration of various real-time data services for sensor networks. It provides a database-like abstraction to sensor networks similar to *TinyDB* or *Cougar*.

The distinguishing characteristics of *DSWare* are its support for group-based decision making and reliable storage to improve real-time system performance, reliability of aggregated results and reduction in communication overhead. In order to provide its functionality, *DSWare* is composed of the following components:

- **Data storage.** This component takes care of storing data in the network according to the semantics associated with the data. It also provides some primitives to work with spatially correlated data, which has two advantages: it allows for easy data aggregation and also makes it possible for the system to perform in-network processing.

- **Data caching.** The purpose of this component is to provide multiple copies of popular data in the regions that need it, so that its availability is high and query execution can be performed in a faster way. There is a feedback mechanism that allows *DSWare* to decide on-the-fly whether or not copies of data should reside in frequent queries nodes. The control scheme uses the proportion of periodic queries relative to the total number of queries, average response time, etc. to make its decisions.

- **Group management.** This allows for the cooperation between various nodes in order to perform distributed computations, value comparison, etc. This component also supports the implementation of energy-saving actions such as putting some nodes to sleep. For this, the formation and management of groups of sensor nodes is a crucial function of this component.

- **Data subscription.** This component allows for the definition of continuous queries in the network. It defines the characteristics of the data feeding paths and uses stable traffic nodes to select the optimal routes. When many base stations make subscriptions for data from the same sensor node, the data subscription service puts copies of the data at intermediate nodes in order to save on communication costs. It is also able to merge several feeding paths into one if this saves communication costs.

- **Scheduling.** This component allows for the scheduling of services to all *DSWare* components. It provides two options: energy-aware and real-time scheduling.

• **TinyDB.** TinyDB [77, 52] is a project developed at the University of Berkeley in cooperation with Intel Research that aims at providing efficient data acquisition primitives for sensor networks. For the *TinyDB* project, the most important type of query to be supported in sensor networks are continuous queries, since all other types can be mapped to a continuous one.

TinyDB, which has been implemented on top of *TinyOS* and runs on the MICA family of sensor nodes (also developed at UC Berkeley), defines an *Acquisitional Query Language (ACQL)*, which is very similar in its structure to traditional SQL, that allows for the efficient retrieval of data within the network. Where possible, *TinyDB* performs in-network processing of data in order to reduce the size of transmissions. For example, the developers of *TinyDB* have developed *TAG*, a *Tiny AGgregation* engine that supports arbitrary decomposable aggregation functions using a generic and extensible framework.

For *ACQL*, all queries create a continuous data stream that can be mapped to each sensor node. *TinyDB* assumes the presence of a sink that is able to process the

ACQL statement and translate it into an efficient binary representation that can be processed by each sensor. In *TinyDB*, the entire sensor network is a single table where the columns contain all the attributes in the network, and the rows specify the individual sensor data. Using this data model and special language capabilities explicitly designed for sensor networks, *TinyDB* is able to process the following types of queries:

> - **event-based queries** have a precondition usually given by an event that triggers the execution of the query;

> - **storage-based queries** are able to perform caching and intermediate storage of data at specific locations in the network;

> - **lifetime-based queries** run for a specified amount of time in the network, collecting the necessary data.

Although *TinyDB* provides a nice abstraction to retrieve data from a sensor network, it is only able to support applications that obtain data from the sensors and process it outside the sensor network. For any other type of processing, it is necessary to create *TinyOS* components that provide the required functionality.

• **Cougar.** *Cougar* [78] is a project developed at Cornell University whose research dictates that monitoring is best described in a declarative manner. For the Cougar project, the sensor network is the database itself, thus they provide abstractions to represent the different devices in the sensor network as a database.

The distinguishing feature of *Cougar* is the use of *Abstract Data Types* (ADTs) and virtual relations, which are a tabular representation of the functions that define the type of data available at different sensors. Using this information, the sensor network is seen as one large table that contains the data to be queried by the user. *Cougar* assumes the presence of a front-node that implements a full-fledged database server, and translates the queries issued by the user into a format that can be understood by each sensor node to answer the query. Each sensor contains an instance of a mini-server that is able to understand these messages and return the appropriate answer.

The mini-server supports synchronous queries whose results need to be returned immediately and on-demand, and asynchronous queries, used to monitor threshold events. Using this distinction, *Cougar* is able to answer the following types of queries:

> - **Historical queries** which usually refer to aggregate queries over historical data, such as: *"For each rainfall sensor, display the average level of rainfall for 1999"*.

> - **Snapshot queries** which refer to the values of data at a given point in time, such as: *"Retrieve the current rainfall level for all sensors in Tompkins County"*.

> - **Long-running queries** which refer to the values of data over a certain time interval, such as: *"For the next 5 hours, retrieve every 30 seconds the rainfall level for all sensors in Tompkins County"*.

Although the idea of providing a well-known abstraction to represent the nodes in a sensor network is interesting, there are certain abstractions, such as events or publish/subscribe mechanisms, which cannot be mapped to the classic view. However, *Cougar* is able to provide distribution transparency for queries issued to the sensor network.

5.4.1.5. *Adaptive system software*

Since the requirements to COs or the system environment can change significantly during the lifetime, and a constant manual adjustment is too costly, several systems have been developed that perform automatic adaptation:

• **MiLAN.** *MiLAN* [34] is a "proactive" middleware that aims at providing a bridge between the capabilities of current middleware platforms and the need for proactive rules that have an effect on the network and the sensors themselves. It achieves its goal by allowing sensor network applications to specify their quality needs and subsequently making adjustments on specific properties of the sensor network to meet these needs.

MiLAN supports data-driven applications that collect and perform an analysis of the data from the environment. The quality of data is affected by noise, redundancy and the capabilities of the sensors to detect this information. Furthermore, *MiLAN* also supports state-based applications which, being of dynamic nature, change the specific needs on the quality of acquired data over time.

The three types of information used by *MiLAN* to perform adaptation are:

- data about the QoS level defined by the application;

- data about the overall performance of the system and about the user;

- data about the sensor network, such as available resources, state of sensors, energy level and channel bandwidth.

Using this information, *MiLAN* is able to adapt the configuration of the sensor network to optimize the functionality of the system by proactively specifying which sensors need to send data and the role each sensor should play in the overall scheme.

• **Impala.** *Impala* is a middleware system that was designed as part of the *ZebraNet* mobile sensor network, and its architecture allows for application modularity, adaptivity and reparability in WSNs [49].

Impala supports multiple applications by adopting an event-based modular programming model and providing a friendly programming interface. It also features a lightweight system layer that performs on-the-fly application adaptation based on parameters and device failures, which makes it possible to improve the performance, reliability and energy-efficiency of the system. The modular application structure is used to perform application updates in small, modular pieces over the radio, which is similar to *Contiki* and *SOS*. However, the system is implemented on devices similar to PDAs rather than simple sensor nodes.

Impala adopts a layered approach in which the upper layer comprises the application protocols and programs. The lower layers are composed of the following middleware agents:

- the *application adapter*, which performs adaptation on the application protocols at runtime to adapt to the changing environmental conditions;

- the *application updates*, which receive and transmit updates using wireless technology in order to install new code versions on a node;

- the *event filter*, which is responsible for capturing and dispatching the appropriate messages to the other two layers.

Although *Impala* supports a certain degree of adaptation, it does not address the issue of heterogenity, which, in other middleware platforms, plays a very important role.

• **TinyCubus.** The overall architecture of *TinyCubus* [56, 57] mirrors the requirements imposed by the heterogenity of applications and the hardware they operate on. *TinyCubus* has been developed with the goal of creating a generic reconfigurable framework for sensor networks.

TinyCubus is implemented on top of *TinyOS* [35] using the nesC programming language [32], which allows for the definition of components that contain functionality and algorithms. *TinyOS* is primarily used as a hardware abstraction layer. For *TinyOS*, *TinyCubus* is the only application running in the system. All other applications register their requirements and components with *TinyCubus* and are executed by the framework.

TinyCubus itself consists of three parts: the *Tiny Cross-Layer Framework*, the *Tiny Configuration Engine* and the *Tiny Data Management Framework*.

- **Tiny Data Management Framework.** The *Tiny Data Management Framework* provides a set of data management and system components. For each type of standard data management component, such as replication/caching, prefetching/hoarding, aggregation, as well as each type of system component, such as time synchronization and broadcast strategies, it is expected that several implementations exist. The *Tiny Data Management Framework* is then responsible for the selection of the appropriate implementation based on the current information contained in the system.

The *Tiny Data Management Framework* contains a Cubus which combines optimization parameters, such as energy, communication latency or bandwidth; application requirements, such as reliability or consistency level; and system parameters, such as mobility or network density. For each component type, algorithms are classified according to these three dimensions. For example, a tree-based routing algorithm is energy-efficient, but cannot be used in highly mobile scenarios with high reliability requirements. The *Tiny Data Management Framework* selects the best suited set of components based on the current system parameters, application requirements and optimization parameters. This adaptation has to be performed throughout the lifetime of the system and is a crucial part of the optimization process.

      – **Tiny Cross-Layer Framework.** The *Tiny Cross-Layer Framework* provides a generic interface to support the parametrization of components that use cross-layer interactions. Strict layering is not practical for WSNs because it might not be possible to then apply certain desirable optimizations. For example, if some of the application components as well as the link layer components need information about the network neighborhood, this information can be gathered by one of the components in the system and provided to all the others. On the other hand, cross-layer interactions can negatively influence the desirable properties of the software architecture, such as modularity. For example, if cross-layer interactions are not applied in a controlled way, it might not be possible to exchange a module without major changes to others. Therefore, in the *Tiny Cross-Layer Framework*, a state repository is used to store the cross-layer data of all the components, i.e. the components do not interact directly with each other. Thus, architectural properties are better preserved than with the unbridled use of cross-layer interactions.

Other examples of cross-layer interactions are callbacks to higher-level functions, such as the ones provided by the application developer. The *Tiny Cross-Layer Framework* also supports this form of interaction. To deal with callbacks and dynamically loaded code, *TinyCubus* extends the functionality provided by *TinyOS* to allow for the dereferencing and resolution of interfaces and components.

      – **Tiny Configuration Engine.** In some cases, parametrization, as provided by the *Tiny Cross-Layer Framework*, is not enough. Installing new components, or swapping certain functions is necessary, for example, when a new functionality, such as a new processing or aggregation function for the sensed data, is required by the application. The Tiny Configuration Engine addresses this problem by distributing and installing code in the network. Its goal is to support the efficient configuration of both system and application components with the assistance of the topology manager.

The topology manager is responsible for the self-configuration of the network and the assignment of specific roles to each node. A role defines the function of a node based on properties such as hardware capabilities, network neighborhood, location, etc. The topology manager uses a generic specification language and a distributed role assignment algorithm to assign roles to the nodes.

Since in most cases the network is heterogenous, the assignment of roles to nodes is extremely important: only those nodes that actually need a component can receive and install it. This information can be used by the configuration engine, for example, to distribute code efficiently in the network.

5.4.1.6. *Summary and evaluation*

This section has presented the state of the art of system architectures for single nodes comprised of operating systems, virtual machines, data management middleware and adaptive system software. From the operating system point of view, the dominant operating systems *TinyOS*, *Contiki* and *Mantis* include all required functionalities except for real-time support, which has been addressed in recent operating systems such as *Timber* and *DCOS*. While we believe that most functionalities are available in today's operating systems, the main issue seems to be the programmability of these systems, which relates to the issues discussed in section 5.3.

Virtual machine code is more compact than native code, which reduces the energy consumption when sending code updates through the network, even when there is support for loadable modules as in *Contiki* or *SOS*. However, code interpretation is more expensive using virtual machines. This trade-off (which is apparently application-dependent) has not been evaluated thoroughly.

The view of a sensor network as a database might be good for an external user querying the network; however, an internal application, this approach seems to be costly. Other *data management middleware* schemes are too application-specific, even if they include several generic parts that are useful for many other scenarios as well.

The need for adaptive system software is obvious and several approaches exist. While *MiLan* focuses on the quality of the sensor data, *Impala* and *TinyCubus* deal with the optimization of the application itself. Unfortunately, it is hard to model *MiLan*'s functionality in the other two systems.

5.4.2. *System architecture: interaction of nodes*

5.4.2.1. *Introduction*

In this section we, focus on system architecture and topics related to cooperation among nodes. We can roughly abstract two main sets of functionalities defining how nodes interact:

1. *low-level functionalities*, including tasks corresponding to physical, link, routing and transport layers;

2. *high-level functionalities*, including coordination and support, clustering, timing and localization, addressing, lookup, collaboration, failure detection and security.

An approximate mapping of the above functionalities onto the traditional ISO/OSI protocol layers is shown in Figure 5.4.

This mapping shows a peculiarity of the sensor network, namely that most of the functionalities extend over and depend on several traditional protocol layers [9]. This is because sensor networks have to provide functionalities that are not present in traditional networks. Furthermore, the efficiency constraints imposed by sensor networks' limited resources imply that any strictly layered approach, while possible, is likely to

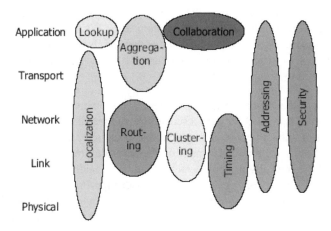

Figure 5.4. *Approximate mapping of the interaction functionalities to ISO/OSI protocol layers*

produce suboptimal solutions. On the other hand, a complete monolithic approach is unlikely to be manageable in complexity. Hence, a suitable level of integration has to be found.

5.4.2.2. *Communication models*

Conceptually, communication within a sensor network can be classified into two categories [69]: application and infrastructure. The network protocol must support both these types of communication. Application communication relates to the transfer of sensed data with the goal of informing the observer about the phenomena. Within application communication, there are two models: cooperative and non-cooperative. Under the cooperative sensor model, sensors communicate with other sensors to realize the observer's interest. This communication is beyond the relay function needed for routing. In-network data processing is an example of cooperative sensors. Non-cooperative sensors do not cooperate for information dissemination; the cooperation is strictly limited to packet relay in multi-hop networks. Infrastructure communication refers to the communication needed to configure, maintain and optimize operation. More specifically, because of the ad hoc nature of sensor networks, sensors must be able to discover paths to other sensors of interest to them, and have to be able to discover paths to the observer regardless of sensor mobility or failure. Thus, infrastructure communication is needed to keep the network functional, ensure robust operation in dynamic environments, as well as to optimize its overall performance. We note that such infrastructure communication is highly influenced by the application interests, since the network must reconfigure itself to best satisfy these interests. As infrastructure communication represents the overhead of the network operation, it is important to minimize this communication while ensuring that the network can support efficient application communication. In sensor networks, an initial phase of infrastructure communication is needed to set up the network. Furthermore, if the sensors are energy-constrained, there will be additional communication for reconfiguration. Indeed, the role of each node will be re-negotiated on the basis of its remaining energy. For example, the role of a cluster head in clustered networks is typically rotated in order to balance the energy consumption.

Similarly, if the sensors are mobile or the observer interests are dynamic, additional communication is needed for path discovery/reconfiguration. For example, in a clustering protocol, infrastructure communication is required for the formation of clusters and cluster head selection; under mobility or sensor failure, this communication must be repeated (periodically or upon detecting failure). Finally, infrastructure communication is used for network optimization. Consider the Frisbee model, where the set of active sensors follows a moving phenomenon to optimize energy efficiency. In this case, the sensors wake up other sensors in the network using infrastructure communication. Sensor networks require both application and infrastructure communication. The amount of required communication is highly influenced by the networking protocol used. Application communication is optimized by reporting measurements at the minimal rate that will satisfy the accuracy and delay requirements given known sensor capabilities, and the quality of the paths between the sensors and the observer. The infrastructure communication is generated by the networking protocol in response to application requests or events in the network. Investing in infrastructure communication can reduce application traffic and optimize overall network operation.

5.4.2.3. *Network dynamics*

A sensor network forms a path between the phenomenon and the observer. The goal of the sensor network protocol is to create and maintain this path (or multiple paths) under dynamic conditions by means of interaction of nodes, while meeting the application requirements of low energy, low latency, high accuracy and fault tolerance.

The way in which these communication paths are established and maintained, and thus the way in which nodes interact, strongly depends on the network dynamics. Network dynamics can be roughly classified as static sensor networks and mobile sensor networks:

• **Static networks.** In static sensor networks, the communicating sensors, the observer and the phenomenon are all static. An example is a group of sensors spread for temperature sensing. In this type of network, sensor nodes require an initial set-up infrastructure communication to create the path between the observer and the sensors, with the remaining traffic mainly being available for application communication. Reconfiguration can still occur for task reassignment or failures due to energy consumption.

• **Dynamic networks.** In dynamic sensor networks, either the sensors themselves, the observer, or the phenomenon are mobile. Whenever any of the sensors associated with the current path from the observer to the phenomenon moves, the path may fail. In this case, either the observer or the concerned sensor must take the initiative to rebuild a new path. During initial set-up, the observer can build multiple paths between itself and the phenomenon and cache them, choosing the one that is the most beneficial at that time as the current path. If the path fails, another of the cached paths can be used. If all the cached paths are invalid, then the observer must rebuild new paths. This observer-initiated approach is a reactive approach, where path recovery action is only taken after observing a broken path. Another model for rebuilding new paths from the observer to the phenomenon is a sensor-initiated approach. In a sensor-initiated path recovery procedure, path recovery is initiated by a sensor that is currently part of the logical path between the observer and the phenomenon, and is planning to move out of range. Dynamic sensor networks can be further classified by considering the motion of the components. This motion is important from the communications perspective since the degree and type of communication is dependent on network dynamics:

- **Mobile observer.** In this case, the observer is mobile with respect to the sensors and phenomena. For example, a plane might fly over a field periodically to collect information from a sensor network. A model that is well-suited to this case is the Data MULES model [63]. The MULE architecture provides wide-area connectivity for a sparse sensor network by exploiting mobile agents such as people, animals or vehicles moving in the environment. The system architecture is comprised of a three-tier layered abstraction.The top tier is composed of access points/central repositories, which can be set up at convenient locations where network connectivity and power are present. These devices communicate with a central data warehouse that allows them to synchronize the data that they collect, detect duplicates and return acknowledgments to the MULEs. The intermediate layer of mobile MULE nodes provides the system with scalability and flexibility at a relatively low cost. The key traits of a MULE are large storage capacities (relative to sensors), renewable power and the ability to communicate with the sensors and networked access points. MULEs are assumed to be serendipitous agents whose movements cannot be predicted in advance. However, as a result of their motion, they collect and store data from the sensors, as well as deliver ACKs back to the sensor nodes. In addition, MULEs can communicate with each other to improve system performance. The bottom tier of the network consists of randomly distributed wireless sensors. Work performed by these sensor nodes should be minimized as they contain the most constrained resources of any of the tiers. Depending on the application and situation, a number of the tiers in our three-tier abstraction could be collapsed onto one device. As data MULEs perform the collection of information to and from the sensor nodes when they are in the sensor's radio range, sensor nodes can be very simple (they are only required to sense data and communicate them to the data MULE when in range). This, in turn, may reduce complexity, and thus the cost, of sensor nodes, enabling their adoption in very large scale systems. The price to pay in the case where a data MULE-like solution is adopted is an energy-latency trade-off. Complexity is shifted from the sensor nodes to the MULEs, and the energy consumption is reduced as nodes only have to communicate the data they generate, but communication may experience high delays as sensors have to wait for a MULE to pass by before the sensed data is communicated.

- **Mobile sensors.** In this case, the sensors are moving with respect to each other and the observer. For example, consider traffic monitoring, implemented by attaching sensors to vehicles [68]. If the sensors are cooperative, the communication paradigm imposes additional constraints, such as detecting the link layer addresses of the neighbors, and constructing localization and information dissemination structures. As sensor nodes are energy-constrained devices, their mobility can be foreseen as the (often uncontrollable) mobility of the mobile devices they are attached to (users, cars, etc.).

- **Mobile phenomena.** In this case, the phenomenon itself is moving. A typical example of this paradigm is sensors deployed for animal detection. In this case, the infrastructure-level communication should be event-driven. Depending on the density of the phenomena, it will be inefficient if all the sensor nodes are active all the time. Only the sensors in the vicinity of the mobile phenomenon need to be active. The number of active sensors in the vicinity of the phenomenon can be determined by application specific goals, such as accuracy, latency and energy-efficiency. A model that is well-suited to this case is the Frisbee model [22].

5.4.3. *Architectures and functionalities summary*

In this section, we sketch a general architecture design based on the considerations made in the above sections. Furthermore, we briefly discuss a set of functionalities to allow flexible, but efficient interaction among nodes to reach a common goal.

We can identify two main layers of abstraction: the sensor and networking layer, and the distributed services layer:

• The sensor and networking layer is made of sensor nodes and network protocols. Ad hoc routing protocols allow messages to be forwarded through multiple sensor nodes, taking into account the mobility of nodes and the dynamic change of topology. Communication protocols must be energy-efficient because of the limited energy and computational resources.

• The distributed services layer is made of distributed services for the mobile sensor applications. Distributed services co-ordinate with each other to perform decentralized services. Resources might be replicated for higher availability, efficiency and robustness. A lookup service supports mobility, instantiation and reconfiguration. Finally, the information service deals with aspects of collecting data. This service allows vast quantities of data to be easily and reliably accessed, manipulated, disseminated and used by applications.

Applications run on the top of this architecture and exploit the functionalities provided by the sensor and networking layer, and distributed services layer (see Figure 5.5).

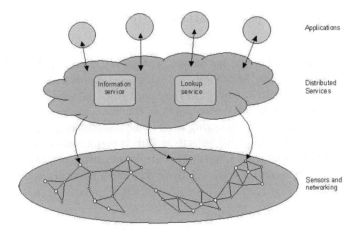

Figure 5.5. *Architecture overview*

We can identify two distinct phases which bring the sensor network to a fully operational mode: the initialization phase and the operation phase:

● **Initialization phase.** When new sensor nodes are added to the sensor network, they should learn about the capabilities and functions of other sensors, and work together to perform cooperative tasks and networking functionalities.

The initialization phase is further divided in the following phases:

- **Discovery phase.** In this phase, nodes explore their environment and establish contact with neighbors using ad hoc networking techniques. The nodes perform a distributed coordination protocol in order to find a suitable networking topology that allows all nodes to communicate.

- **Synchronization phase.** In this phase, nodes try to synchronize with neighboring nodes in order to have a common notion of time. This is required, for example, to allow dynamic power management strategies in networking and data processing protocols to use TDMA-like MAC protocols, and to be able to identify the time at which events occur according to a common reference time.

- **Positioning phase.** In this phase, nodes try to find out their relative, and possibly, geographic positions.

- **Registration phase.** In this phase, nodes can communicate, and there is a common agreement on time and position/topology. A node connects to the lookup service to register its presence, its position and its capabilities.

- **Configuration phase.** In this phase, the lookup service configures a sensor node, or the node configures itself to perform a specific task.

Once these phases are completed, the sensor network becomes operational. Sensors may move and may be required to perform different tasks. This implies that some of the above-mentioned phases would be performed again in the reconfiguration of tasks.

• **Operation phase.** Once the sensor network has been initialized, it will perform the tasks as specified in the initialization phase. Sensor nodes are organized in ad hoc and highly dynamic networks. They must react to mobility, changes in task assignment, and network and device failures. Therefore, each sensor node must be autonomous and capable of organizing itself in the community of sensors to perform coordinated activities with global objectives. This is performed via self-organizing in possibly hierarchical structures, topology control protocols, protocols addressing and exploiting mobility in sensor networks, dynamic routing protocols, etc.[6]

5.5. Conclusions and future work

This section presents summaries of the different parts of this chapter, namely for programming models and system architectures in the form of node internals and interaction of nodes. An overview of the corresponding part is given, followed by a discussion on identified trends and open issues.

5.5.1. *Programming models*

In order to support the development, maintenance, deployment and execution of applications for COs, an appropriate abstraction sitting between the operating system and the application itself needs to be provided to the programmer [67]. In the last couple of years, different approaches for the design and development of such abstractions were presented in the research community, and some of them evolved to running systems, most of which were surveyed in section 5.3.2.

Most of the discussed approaches were designed for a specific application scenario or were tailored to some specific design goals, and thus appear able to comply with only a subset of the requirements listed in section 5.3.1. We can also conclude that for complex systems like networks of COs, the design of a "unique" suitable programming abstraction, able to comply with the requirements and constraints of the many envisioned application scenarios, appears as an extremely challenging task.

Trends

Most of the surveyed programming paradigms (e.g. database approach, agent-based approach, event-based approach) are not new, but require significant adaptation to be used for COs. These approaches differ with respect to ease-of-use, expressiveness, scalability, overhead, etc., as outlined in section 5.3.3. Some paradigms,

---

6. An extensive survey of the different protocols proposed in the literature for performing such tasks has been reported in Chapter 3.

like the database or the group-approaches, allow an easy way to program the network as a single (virtual) entity, but the definition and completion of more complex tasks than just simple sensing, is not yet appropriately supported by such approaches. Virtual machines, like *Maté* or *MagnetOS*, give priority to energy-saving issues and are well-suited for hiding device heterogenity to the programmer, but imply cumbersome local code interpretation. The *generic role assignment* and the *virtual market* approaches are tailored to provide a general and easy-to-use framework for supporting task assignment and device coordination in networks of COs and are still objects of active research.

Some open issues

Since most of the investigated programming abstractions are only partially able to comply with the several requirements posed by complex systems such as a network of COs, we believe that the research community should try to better understand which models and abstractions are the most appropriate for the different classes of devices covered by the notion of COs.

More effort should also be put on the provision of easy-to-use programming primitives, since most systems still require experienced programmers. Since future applications envisage cooperation among devices which may strongly differ in terms of dimension, power and computing resources, mobility, etc., we also recognize the need for further research on paradigms that allow for an efficient programming of networks that incorporate very heterogenous devices. Scalability issues should also gain more attention, since it is still unclear if the existing approach will perform satisfactorily for very large networks of COs. Finally, we believe that the design and development of adequate tools for debugging at an application level should also be considered as a fundamental issue for the success of a programming abstraction for COs.

5.5.2. Node internals

The main tasks of operating systems for COs is to provide an abstract interface to the underlying hardware and to schedule system resources. The main challenges are the small memory footprint and limited energy budget of the nodes. Therefore, scaled-down versions of traditional operating systems have been successful for larger devices such as IPAQs, but not for extremely resource-constrained sensor network nodes. For these nodes, operating systems have been developed from scratch and are in use.

Other abstractions for COs have been developed: *virtual machines* provide a high-level programming language that allows users to write complex programs with only a few commands. *Data management middleware* offers a uniform view to the data gathered in a CO and its processing, storage and distribution. *Adaptive system software* frees the user from adjusting the application to every possible system condition.

Trends

TinyOS is the dominant operating system for sensor networks. Its most outstanding feature is its small memory footprint, but it does not have support for loadable modules, its concurrency model is based on events only and *TinyOS* is hard to program. Therefore, newer operating systems such as *Contiki, Mantis* and *SOS* have support for loadable modules, and offer more sophisticated concurrency models which slightly increases the memory footprint of the systems, but simplifies OS and application development. Fairly recent operating systems such as *DCOS* and *Timber* also address the aspect of real-time support not found in the earlier, dominating operating systems.

Virtual machines also allow for efficient code updates since virtual machine code is more compact than native code, which reduces the energy consumption when sending code updates through the network compared to native code even when there is support for loadable modules as in *Contiki* or *SOS*. However, code interpretation is more expensive using virtual machines.

Virtual machines, data management middleware, and *adaptive system software* provide several different abstractions and functionality that help to simplify the programming and operation of COs. This is still a hot topic. *Data management middleware* first concentrated on database abstractions for the whole network, but they could only be used from the outside. Currently, middleware for applications with common characteristics (e.g. location tracking, real-time data services) are being developed. Adaptive systems also initially focused on only one aspect, for example, to ensure the quality of the sensor readings. This is currently extended to other algorithms, and finally the whole system is to be adapted.

Some open issues

Real-time aspects will become more important in sensor networks in the future. While there have been some developments, real-time operating systems for sensor networks can still be seen as an open issue.

While we believe that most functionalities are available in today's operating system, the main issue seems to be the programmability of these systems, particularly for highly optimized systems such as *TinyOS* and *Contiki*. Abstractions such as *protothreads* found in the *Contiki* operating sytems are a step in the right direction, but more work needs to be done, as also pointed out above in the discussion on programming models.

The support by *data management middleware* is another step, but either the approaches are not usable inside the network of objects, or they are too application-specific and, thus, not interoperable. Therefore, a general concept for the basic functionality of such a middleware is clearly needed.

Since code updates are essential for COs with long-term installations, every system will support updates in the future. For *virtual machines*, updates are most simple since the script or bytecode to interpret can be loaded from every place. For this reason, all of the virtual machines mentioned have this capability. Updates are more difficult for native code, since function calls and variable access have to be performed at a low-level. Either the user is willing to accept overhead for indirections, or the new code has to be fully integrated into the existing code. No general solutions exist here.

With *virtual machines, data management middleware*, and *adaptive system software*, several approaches exist that abstract from different parts of a CO. It has not been studied yet if and how these approaches can be combined to a single overall framework for COs.

5.6. Bibliography

[1] *eCos free open source runtime system*, Web page: http://www.ecoscentric.com/, Visited 2005-06-27.

[2] *eXtreme Minimal Kernel*, Web page: http://www.shift-right.com/xmk.htm, Visited 2005-06-27.

[3] *IrisNet: Internet-scale Resource-Intensive Sensor Network Service*.

[4] *QNX RTOS*, Web page: http://www.qnx.org, Visited 2005-06-27.

[5] *TinyDB: A Declarative Database for Sensor Networks*.

[6] *Bluetooth SIG, Specifications of the Bluetooth System – core, version 1.1*, February 2001.

[7] T. Abdelzaher, B. Blum, Q. Cao, Y. Chen, D. Evans, J. George, S. George, L. Gu, T. He, S. Krishnamurthy, L. Luo, S. Son, J. Stankovic, R. Stoleru and A. Wood, EnviroTrack: Towards an Environmental Computing Paradigm for Distributed Sensor Networks, in *Proceedings of the 24th International Conference on Distributed Computing Systems (ICDCS 2004)*, Washington, DC, USA, IEEE Computer Society, 2004, pp. 582–589.

[8] H. Abrach, S. Bhatti, J. Carlson, H. Dai, J. Rose, A. Sheth, B. Shucker, J. Deng and R. Han, MANTIS: system support for multimodAl NeTworks of in-situ sensors, in *Proceedings of the 2nd ACM International Conference on Wireless Sensor Networks and Applications*, 2003, pp. 50–59.

[9] I.F. Akyildiz, W. Su, Y. Sankarasubramaniam and E. Cayirci, Wireless Sensor Networks: A Survey, *Computer Networks*, vol. 38, no. 4, pp. 393–422, 2002.

[10] M. Banâtre and F. Weis, SIS: a new paradigm for mobile computer systems, in *Information Society Technologies Conference*, Helsinki, Finland, November 1999.

[11] Michel Banâtre, Paul Couderc, Julien Pauty and Mathieu Becus, Ubibus: Ubiquitous Computing to Help Blind People in Public Transport, in *6th International Symposium Mobile Human-Computer Interaction (Mobile HCI 2004)*, Glasgow, UK, September 2004, pp. 310–314.

[12] R. Barr, J.C. Bicket, D.S. Dantas, B. Du, T.W.D. Kim, B. Zhou and E.G. Sirer, On the Need for System-Level Support for Ad Hoc and Sensor Networks, *Operating System Review*, vol. 36, no. 2, pp. 1–5, 2002.

[13] Z. Benenson, N. Gedicke and O. Raivio, Realizing Robust User Authentication in Sensor Networks, in *First Workshop on Real-World Wireless Sensor Networks*, Stockholm, Sweden, June 2005.

[14] K. P. Birman, The process group approach to reliable distributed computing, *Communications of the ACM*, vol. 36, no. 12, pp. 37–53, 1993.

[15] Brian M. Blum, Prashant Nagaraddi, Anthony D. Wood, Tarek F. Abdelzaher, Sang Hyuk Son and Jack Stankovic, An Entity Maintenance and Connection Service for Sensor Networks, in *Proceedings of the First International Conference on Mobile Systems, Applications, and Services (MobiSys 2003)*, San Francisco, CA, USA, USENIX, May 2003, pp. 201–214.

[16] P. Bonnet, J.E. Gehrke and P. Seshadri, Querying the Physical World, *IEEE Journal of Selected Areas in Communications*, vol. 7, no. 5, pp. 10–15, 2000.

[17] C. Borcea, C. Intanagonwiwat, P. Kang, U. Kremer and L. Iftode, Spatial Programming Using Smart Messages: Design and Implementation, in *24th International Conference on Distributed Computing Systems (ICDCS 2004)*, Hachioji, Tokyo, Japan, March 2004, pp. 690–699.

[18] C. Borcea, D. Iyer, P. Kang, A. Saxena and L. Iftode, Cooperative Computing for Distributed Embedded Systems, in *22nd International Conference on Distributed Computing Systems (ICDCS 2002)*, July 2002, pp. 227–236.

[19] A. Boulis, C.C. Han and M.B. Srivastava, Design and Implementation of a Framework for Programmable and Efficient Sensor Network, in *MobiSys 2003*, San Franscisco, USA, May 2003.

[20] M. Britton, L.L. Shum, L. Sacks and H. Haddadi, A Biologically-Inspired Approach to Designing Wireless Sensor Networks, in *2nd European Workshop on Wireless Sensor Networks*, Istanbul, Turkey, January 2005.

[21] E. Bugnion, S. Devine and M. Rosenblum, Disco: Running Commodity Operating Systems on Scalable Multiprocessors, in *16th ACM Symposium on Operating Systems Principles*, Saint Malo, France, October 1997.

[22] A. Cerpa, J. Elson, D. Estrin, L. Girod, M. Hamilton and J. Zhao, Habitat Monitoring: Application Driver for Wireless Communications Technology, in *Proc. ACM SIGCOMM Workshop on Data Communications in Latin America and the Caribbean*, April 2001.

[23] P. Couderc and M. Banâtre, Ambient computing applications: an experience with the SPREAD approach, in *36th Annual Hawaii International Conference on System Sciences (HICSS '03)*, Big Island, Hawaii, January 2003.

[24] D. Culler, A. Sah, K. Schauser, T. von Eicken and J. Wawrzynek, Fine-grain Parallelism with Minimal Hardware Support: A Compiler-Controlled Threaded Abstract Machine, in *4th International Conference on Architectural Support for Programming Languages and Operating Systems*, April 1991.

[25] L.I. Dickman, Small Virtual Machines: A Survey, in *Workshop on Virtual Computer Systems*, Cambridge, MA, USA, June 1973, pp. 191–202.

[26] Stefan Dulman, Tjerk Hofmeijer and Paul Havinga, Wireless sensor networks dynamic runtime configuration, in *Proceedings of ProRISC 2004*, 2004.

[27] A. Dunkels, B. Grönvall and T. Voigt, Contiki – a Lightweight and Flexible Operating System for Tiny Networked Sensors, in *First IEEE Workshop on Embedded Networked Sensors*, 2004.

[28] A. Dunkels, O. Schmid and T. Voigt, Using Protothreads for Sensor Network Programming, in *First Workshop on Real-World Wireless Sensor Networks*, Stockholm, Sweden, June 2005.

[29] D. Estrin, D. Culler, K. Pister and G. Sukhatme, Connecting the Physical World with Pervasive Networks, *IEEE Pervasive Computing*, vol. 1, no. 1, pp. 59–69, 2002.

[30] D. Estrin, R. Govindan, J. Heidemann and S. Kumar, Next Century Challenges: Scalable Coordination in Sensor Networks, in *International Conference on Mobile Computing and Networks (MobiCOM '99)*, 1999.

[31] Christian Frank and Kay Römer, Algorithms for Generic Role Assignment in Wireless Sensor Networks, in *Proceedings of the 3rd ACM Conference on Embedded Networked Sensor Systems (SenSys)*, San Diego, CA, USA, November 2005.

[32] D. Gay, P. Levis, R. von Behren, M. Welsh, E. Brewer and D. Culler, The nesC language: A holistic approach to networked embedded systems, in *Proceedings of the ACM SIGPLAN 2003 Conference on Programming Language Design and Implementation*, San Diego, CA, USA, 2003, pp. 1–11.

[33] C. Han, R. Rengaswamy, R. Shea, E. Kohler and M. Srivastava, SOS: A dynamic operating system for sensor networks, in *MobiSys 2005*, Seattle, USA, June 2005.

[34] W.B. Heinzelman, A.L. Murphy, H.S. Carvalho and M.A. Perillo, Middleware to Support Sensor Network Applications, *IEEE Network*, vol. 18, pp. 6–14, 2004.

[35] J. Hill, R. Szewczyk, A. Woo, S. Hollar, D. Culler and K. Pister, System Architecture Directions for Networked Sensors, in *ASPLOS 2000*, Cambridge, USA, November 2000.

[36] J. Johansson, M. Völker, J. Eliasson, Å. Östmark, P. Lindgren and J. Delsing, Mulle: A minimal sensor networking device – implementation and manufacturing challenges, in *IMAPS Nordic 2004*, 2004.

[37] M. Kero, P. Lindgren and J. Nordlander, Timber as an RTOS for Small Embedded Devices, in *First Workshop on Real-World Wireless Sensor Networks*, Stockholm, Sweden, June 2005.

[38] T. Kindberg, J. Barton, J. Morgan, G. Becker, D. Caswell, P. Debaty, G. Gopal, M. Frid, V.Krishnan, H. Morris, J. Schettino and B. Serra, *People, Places, Things: Web Presence for the Real World*, Tech. Report TR HPL-2000-16, Internet and Mobile Systems Laboratory, HP Laboratories Palo Alto, February 2000.

[39] J. Koberstein, N. Luttenberger, C. Buschmann and S. Fischer, Shared Information Spaces for Small Devices: The SWARMS Software Concept, in *Workshop on Sensor Networks at Informatik 2004*, Ulm, Germany, September 2004.

[40] G. Kortuem, Z. Segall and T. G. Cowan Thompson, Close Encounters: Supporting Mobile Collaboration through Interchange of User Profiles, in *1st International Symposium on Handheld and Ubiquitous Computing (HUC '99)*, Karlsruhe, Germany, November 1999.

[41] M. Langheinrich, F. Mattern, K. Römer and H. Vogt, First Steps Towards an Event-Based Infrastructure for Smart Things, in *Ubiquitous Computing Workshop (PACT 2000)*, Philadelphia, USA, October 2000.

[42] P. Levis and D. Culler, Maté: A Tiny Virtual Machine for Sensor Networks, in *ASPLOS X*, San Jose, USA, October 2002.

[43] Philip Levis, David Gay and David Culler, Active Sensor Networks, in *Proceedings of the Second USENIX/ACM Symposium on Networked Systems Design and Implementation (NSDI 2005)*, Boston, USA, May 2005.

[44] Philip Levis, Sam Madden, David Gay, Joe Polastre, Robert Szewczyk, Alec Woo, Eric Brewer and David Culler, The emergence of networking abstractions and techniques in tinyos, in *1st USENIX/ACM Symposium on Networked Systems Design and Implementation (NSDI '04)*, San José, USA, March 2004, pp. 29–42.

[45] S. Li, S.H. Son and J.A. Stankovic, Event Detection Services Using Data Service Middleware in Distributed Sensor Networks, in *IPSN 2003*, Palo Alto, USA, April 2003.

[46] Shuoqi Li, Ying Lin, Sang H. Son, John A. Stankovic and Yuan Wei, Event Detection Services Using Data Service Middleware in Distributed Sensor Networks, *Telecommunication Systems*, vol. 26, no. 2-4, pp. 351–368, 2004.

[47] T. Lindholm and F. Yellin, *The Java Virtual Machine Specification*, 2nd ed., Addison-Wesley, 1999.

[48] Hongzhou Liu, Tom Roeder, Kevin Walsh, Rimon Barr and Emin Gün Sirer, Design and implementation of a single system image operating system for ad hoc networks, in *MobiSys '05: Proceedings of the 3rd International Conference on Mobile Systems, Applications, and Services*, New York, NY, USA, ACM Press, 2005, pp. 149–162.

[49] Ting Liu and Margaret Martonosi, Impala: A Middleware System for Managing Autonomic, Parallel Sensor Systems, in *ACM SIGPLAN Symposium on Principles and Practice of Parallel Programming (PPoPP '03)*, June 2003.

[50] Ciaran Lynch and Fergus O'Reilly, Pic-based tinyos implementation, in *Second European Workshop on Wireless Sensor Networks (EWSN 2005)*, January 2005, pp. 93–107.

[51] S.R. Madden, *Query Processing for Streaming Sensor Data (PhD Qualifying Exam Proposal)*, May 2002.

[52] S.R. Madden, M.J. Franklin, J.M. Hellerstein and W. Hong, TAG: A Tiny Aggregation Service for Ad hoc Sensor Networks, in *OSDI*, Boston, USA, December 2002.

[53] S.R. Madden, M.J. Franklin, J.M. Hellerstein and W. Hong, The Design of an Acquisitional Query Processor For Sensor Networks, in *SIGMOD*, San Diego, CA, USA, June 2003.

[54] S. Mahlknecht and M. Roetzer, Energy supply considerations for self-sustaining wireless sensor networks, in *Second European Workshop on Wireless Sensor Networks (EWSN 2005)*, January 2005, pp. 93–107.

[55] G. Mainland, L. Kang, S. Lahaie, D. Parkes and M. Welsh, Using Virtual Markets to Program Global Behavior in Sensor Networks, in *11th ACM SIGOPS European Workshop*, Leuven, Belgiun, September 2004.

[56] Pedro José Marrón, Andreas Lachenmann, Daniel Minder, Jörg Hähner, Robert Sauter and Kurt Rothermel, TinyCubus: A Flexible and Adaptive Framework for Sensor Netwo rks, in Erdal Çayırcı Şebnem Baydere and Paul Havinga, editors, *Proceedings of the 2nd European Workshop on Wireless Sensor Networks*, Istanbul, Turkey, January 2005, pp. 278–289.

[57] Pedro José Marrón, Daniel Minder, Andreas Lachenmann and Kurt Rothermel, Tiny-Cubus: An Adaptive Cross-Layer Framework for Sensor Networks, *Information Technology*, vol. 47, no. 2, pp. 87–97, 2005.

[58] T. Mullen and M.P. Wellman, Some Issues In The Design of Market-Oriented Agents, in M. Wooldridge, J. Mueller and M. Tambe, editors, *Intelligent Agents II: Agent Theories, Architectures and Languages*, Springer-Verlag, 1996.

[59] K. Nagao and J. Rekimoto, Agent Augmented Reality: A Software Agent Meets the Real World, in *2nd International Conference on Multi-Agent Systems (ICMAS-96)*, San José, USA, 1996, pp. 228–235.

[60] S. Nath, Y. Ke, P. B. Gibbons, B. Karp and S. Seshan, *IrisNet: An architecture for enabling sensor-enriched Internet service*, Tech. report, Intel Research Pittsburgh, December 2002.

[61] J. K. Ousterhout, *Tcl and the TK toolkit*, Addison-Wesley, 1994.

[62] G. P. Picco, A. L. Murphy and G.-C. Roman, LIME: Linda Meets Mobility, in *21st International Conference on Software Engineering*, Los Angeles, USA, May 1999, pp. 368–377.

[63] Sushant Jain Rahul C. Shah, Sumit Roy and Waylon Brunette, Data mules: modeling a three-tier architecture for sparse sensor networks, in *IEEE Workshop on Sensor Network Protocols and Applications (SNPA)*, May 2003.

[64] G.-C. Roman, Q. Huang and A. Hazemi, Consistent Group Membership in Ad Hoc Networks, in *23rd International Conference in Software Engineering (ICSE)*, Toronto, Canada, May 2001.

[65] K. Römer, Programming Paradigms and Middleware for Sensor Networks, in *GI/ITG Workshop on Sensor Networks*, 2004, pp. 49–54.

[66] K. Römer, C. Frank, P.J. Marron and C. Becker, Generic Role Assignment for Wireless Sensor Networks, in *11th ACM SIGOPS European Workshop*, Leuven, Belgium, September 2004, pp. 7–12.

[67] Kay Römer, Oliver Kasten and Friedemann Mattern, Middleware Challenges for Wireless Sensor Networks, *ACM Mobile Computing and Communication Review* vol. 6, no. 4, pp. 59–61, 2002.

[68] Cory Sharp, Shawn Schaffert, Alec Woo, Naveen Sastry, Chris Karlof, Shankar Sastry and David Culler, Design and implementation of a sensor network system for vehicle tracking and autonomous interception, in *Second European Workshop on Wireless Sensor Networks (EWSN 2005)*, January 2005, pp. 93–107.

[69] S. Tilak, N. Abu-Ghazaleh and W. Heinzelman, *A Taxonomy of Wireless Microsensor Network Models*, 2002.

[70] A. Ulbrich, T. Weis, G. Mühl and K. Geihs, Application Development for Actuator and Sensor Networks, in *GI Workshop on Sensor Networks*, ETH Zurich, Switzerland, March 2005.

[71] J. Waldo, The Jini Architecture for Network-Centric Computing, *Communication ACM*, vol. 42, no. 7, pp. 76–82, 1999.

[72] J. Weatherall, A Ubiquitous Control Architecture for Low Power Systems, in *ARCS 2002 International Conference on Architecture of Computer Systems*, Springer-Verlag, April 2002.

[73] M.P. Wellman, *Market-oriented Programming: Some Early Lessons*, (1996).

[74] M. Welsh, Exposing Resource Trade-offs in Region-Based Communication Abstractions for Sensor Networks, in *2nd ACM Workshop on Hot Topics in Networks (HotNets-II)*, San José, USA, November 2003.

[75] M. Welsh and G. Mainland, Programming Sensor Networks Using Abstract Regions, in *1st USENIX/ACM Symposium on Networked Systems Design and Implementation (NSDI '04)*, San José, USA, March 2004, pp. 29–42.

[76] K. Whitehouse, C. Sharp, D. Culler and E. Brewer, Hood: A Neighborhood Abstraction for Sensor Networks, in *MobiSys 2004*, Boston, USA, June 2004.

[77] Alec Woo, Sam Madden and Ramesh Govindan, Networking support for query processing in sensor networks, *Communications of the ACM*, vol. 47, no. 6, pp. 47–52, 2004.

[78] Y. Yao and J.E. Gehrke, The Cougar approach to in-network query processing in sensor networks, *ACM Sigmod Record*, vol. 31, no. 3, pp. 9–18, 2002.

Chapter 6

Cooperating Objects Roadmap and Conclusions

Taking into account the information contained in the previous chapters that summarizes the trends and gaps of current research directions in the field of COs, it is possible to organize this chapter in the form of a roadmap. In order to do this properly, we started a compilation and consolidation process with the following data:

• the individual *expertise* and *practical experiences* of each of the partners involved in the Embedded WiSeNts project presented in Chapter 1;

• the analysis of the current technologies and *current trends* that show future research directions, as presented in previous chapters;

• a *market analysis* of COs performed with the input from industrial partners and other research institutes;

• *visionary applications* obtained from a wider audience;

• and the *identification of gaps* and research agendas in the different areas that comprise the field of COs.

Given the research background of all partners involved in the Embedded WiSeNts project and the fact that the field of COs is advancing rapidly, this chapter does not pretend to present a roadmap that can be used "as is" to drive development for any industrial sector involved in this field. It can, however, be used as further input for development and innovation departments that wish to benefit from information about the possible direction and the timeframe for CO research.

Chapter written by Pedro José MARRÓN, Daniel MINDER and the Embedded WiSeNts Consortium.

The final sections contain a series of recommendations that can be used by financing institutions and organizations in order to drive research in the direction shown. It is worth mentioning that a full version of the contents of this chapter is available from Logos Verlag, Berlin and from the Embedded WiSeNts webpage (http://www.embedded-wisents.org/).

6.1. Intended audience

The Embedded WiSeNts research roadmap has been written with three different audiences in mind, as follows.

Researchers: those that work or intend to work in the field of COs and would like to understand the current state of the art, current trends and possible gaps for future research.

Industry: those that would like to understand the current state of the art and possible market developments to be used as an additional source of information for the definition of specific strategies and business opportunities related to COs.

Funding institutions: to achieve a better understanding of the field of COs and its potential as a topic that can be included in upcoming framework programs or other financing instruments.

6.2. Methodology and structure

Figure 6.1 gives an overview of the different parts of the roadmap and of the relationship between its components. The boxes represent main topics discussed in the roadmap that can usually be found in different chapters or sections within a chapter. The arrows represent inputs and preliminary information that has been used in the process of writing the corresponding sections.

The figure has been coded using the following convention: boxes with a thick dashed border represent chapters that have been obtained from other work packages during the execution of the Embedded WiSeNts project. The detailed information can be obtained from the project website (http://www.embedded-wisents.org/). Boxes with a thick solid border represent documents that have been obtained from external sources for the project, such as companies that provide market information used in this roadmap to identify market trends and predictions. Boxes with a thin solid border show intermediate results derived from other chapters and additional documents. These intermediate results are in themselves interesting for certain audiences that would like to understand the process used to get to the final results. Finally, boxes with a faint plotted border represent the actual output of this roadmap (final results) in terms of recommendations for further work for the three types of audiences described above.

Following a parallel route, the results of the studies have been evaluated and validated in a survey among the participants of the "From RFID to the Internet of Things"

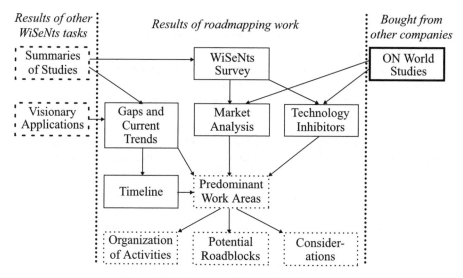

Results of other   *Results of roadmapping work*   *Bought from*
WiSeNts tasks   *other companies*

Figure 6.1. *Structure of the roadmap document*

workshop organized by the EU. Using the information obtained there and data from the market analysis we have bought from ON World Inc., we were able to provide a market analysis study. These same inputs were also used to determine the most prominent inhibitors for COs described later on.

These four intermediate results (gaps and current trends, timeline, market analysis and inhibitors) were used to determine the predominant work areas that served as a driving force and input for the final results of the roadmap, namely, the organization of activities, potential roadblocks or major inhibitors that hinder the acceptance of CO technologies in society, and the final considerations for research programs.

6.3. Executive summary

According to a market study performed by ON World Inc. on wireless sensor networks called "Wireless Sensor Networks – Growing Markets, Accelerating Demands" in July 2005, 127 million wireless sensor network nodes are expected to be deployed in 2010 and the growth of this market later on is expected to increase in certain application domains.

Wireless sensor networks are a canonical example of a wider field dealing with COs that attempts to create the necessary technologies to make Weiser's vision of the disappearing computer a reality. COs are, in the most general case, small computing devices equipped with wireless communication capabilities that are able to cooperate and organize themselves autonomously into networks to achieve a common task.

In order to validate some of the results detailed in previous chapters, the members of the Embedded WiSeNts consortium have performed a survey among the participants of the EU Workshop "From RFID to the Internet of Things" held in Brussels on March 6th/7th, 2006. In it, we prepared a series of questions that were posed to volunteers who answered in an anonymous way. We obtained answers from 51 people working in both university and industrial research, and product development in the industry.

Over 80% of all participants estimate that the breakthrough for the technology will happen in the next 5 to 10 years and it is safe to assume that they assume that most technological inhibitors and research issues will be solved by then.

As expected, there are a certain number of optimists that believe CO technology is almost ready for general release. A slightly larger number of pessimists (or realists?) do not expect COs to be widely used in the industry until at least 15 years from now.

6.4. Research gaps and timeline

Looking at the state of the art described in the previous chapters of this book, we have been able to identify a series of gaps that we believe to be interesting from the point of view of research for the coming years. These gaps have been grouped into 5 different categories as follows.

Hardware: these are gaps that have to do with the development of the devices that physically constitute networks of COs. The gaps that fall into this category are: *sensor calibration, power efficiency, energy harvesting, new sensor and low-cost devices* and *miniaturization*.

Algorithms: these are gaps that deal with functional properties of COs, that is, specific protocols, types of procedures, etc. The gaps that fall into this category are: *localization, context-aware MAC and routing, clustering techniques, data storage and search* and *motion planning*.

Non-functional properties: these are gaps that deal with Quality-of-Service-type characteristics. Properties that do not affect the functionality of the network, but its quality. The gaps that fall into this category are: *multiple sinks, scalability, quality of service, robustness, mobility, security, heterogenity* and *real-time*.

Systems: these are gaps that have to do with the specific architecture or support for the rest of the system. Normally, systems work at the individual CO level, but have to provide support for networking. The gaps that fall into this category are: *adaptive systems, operating systems, programming models* and *system integration*.

Others: this category collects the gaps that do not fit anywhere else or that might be hard to classify within other categories because they do not really fall into the computer science umbrella. The ones we have selected for description here are the following: *modeling and analysis, experimentation* and *social issues*.

In general, we assume the largest number of breakthroughs in the areas mentioned above to happen in roughly between 5 and 10 years, which coincides with the estimation obtained from our survey regarding the point in time where COs will start to be used widely in industry.

Most of these are already being worked on in some form or another. Only two of the gaps are expected to start 5 years from now: *real-time* and *social issues*, the first one because of the nature of the problem and the second one because social issues will only arise as soon as the early adopters (especially from industry) start introducing COs more aggressively into our everyday lives. In general, gaps that are not being investigated yet or that need investigation throughout the predicted period seem to be the most promising lines of research.

6.5. Potential roadblocks

Working with members from the industry and with the participants of the Embedded WiSeNts survey, it seems clear that, although the opinion of all experts indicates that CO technology has clear chances of success, there is always the possibility of failure if certain issues are not solved properly or in a timely manner.

The following are the potential roadblocks that have appeared during these conversations and interactions.

No clear business models: one of the main potential roadblocks for the adoption of any kind of technology is the lack of a business model that supports it. For the case of COs, it is probably too early to determine whether or not strong enough business models will appear and, as far as early adopters are concerned, there are enough examples of companies that survive by selling technology that can and will be used in this field. However, it might be necessary to work tightly with end-users to identify the real needs and, therefore, business models with high potential.

Lack of standards: with the creation of a new field, it is obvious that early adopters need to provide a pragmatic solution in order to "show something that works", but after a certain time, the industry needs to come together and agree on a common ground for future developments. There are already some attempts to standardise ZigBee and UWB, which will possibly play an important role as communication protocols that bring together networks of COs. However, this is just the beginning and further action needs to be taken in order for CO technology to take off.

Confidence in technology: for more sophisticated applications, currently available CO technology is, for the most part, unable to deliver the desired characteristics, such as lifetime or robustness. Therefore, the immaturity of current solutions in individual fields hinders the adoption of COs in more general application domains. However, in some cases this reluctance is based on prejudices, e.g.

against wireless communication. It is hard to convince people that a new technology which is generally considered as more error-prone can deliver almost the same Quality of Service as the old, wired technology when designed carefully.

Social issues: even if the technological issues are solved and the industry is able to pull together a set of standards that support CO technology, the end-users are still the ones that decide whether or not they will want to make use of this technology. The main question is whether or not the vast majority of people are willing to accept tiny computing devices "interfering" with their lives. People are not willing to have "Big Brother" watching them unless they see a benefit to it. In general, people are reluctant to provide private information that might give an insight on their daily activities or habits and, therefore, if COs can be misused for this purpose, finding a killer-application might take longer than expected, if at all. An increase in the awareness of security and privacy issues is surely needed for the proper understanding of the capabilities of COs, so that the end-users can put this new technology into perspective and not feel threatened.

6.6. Recommendations

Several issues have been identified by the research community, the industry and end-users as important work items. We have identified three different types of actions: actions related to research, actions related to educational activities and other actions.

Research actions: regarding research, there are six major topics that should be addressed and that summarize some of the major gaps listed above:

• Research on algorithms, such as data search and storage, aggregation, consistency and the integration of COs into existing software.

• Research on hardware, such as new small, cheap components, the miniaturization of existing components and energy harvesting techniques for their powering.

• Research on non-functional properties, such as Quality of Service, power consumption, scalability, mobility and security.

• Research on adaptive systems that are easy to use, easy to combine with existing software, adapt to changes in the environment and are fully integrated in operating system and programming abstraction solutions.

• Research on system integration and combination of heterogenous techniques into one working system.

• Research on social issues of COs in order to improve the understanding of such systems both for research and for possible commercialization.

Research on all topics should be application-driven to ensure the convergence of research and applications. Real-life experiments should be carried out in order to understand and solve the underlying scientific challenges of providing practical and efficient solutions to real world problems.

Educational actions: there is a need for education of industry and end-users in order to promote the new technology and to allow potential users and customers to understand the benefits and risks associated with the use of COs. At the same time, companies that would like to enter this market need to be taught about the potential and the power of this new technology so that they can properly address the needs of their customers.

Other actions: the most prominent action that needs support from all major players is the process of standardizing the hardware and software available for COs. There will also be a need to support the creation of regulations and legislative actions that create a legal framework that supports the correct use of COs and hinders possible misuse of this technology.

6.7. Summary and final conclusions

The field of COs is a very dynamic one, with such potential that it could really revolutionize our lives even more profoundly than the Internet has done in recent years. In this book, we have tried to provide a thorough overview of the direction we expect research to take in the next 10 to 15 years, and in the way researchers, industrial partners, end-users and financing institutions should work together to make COs possible in the near future.

For this purpose, we have given a thorough summary of the state of the art regarding CO technology, especially in the field of wireless sensor networks, and have identified the trends that guide current research in the field. Knowing what is available allows us to have a glimpse as to what should be done next and what gaps and missing technologies should be addressed from the point of view of research in order to push this technology even further.

We have also given information about a probable timeline for the development of research and have attempted to characterize the points in time where major breakthroughs will allow for CO technology to become mainstream. Additionally, we have tried to pinpoint major inhibitors and potential roadblocks and given concrete suggestions to avoid them.

Finally, we have provided a list of recommendations for future research and have tried to organize the proposed activities in such a way that the major players (the research community, the industry, the end-users and financing institutions) can collaborate and cooperate in such as way as to complement their efforts and work together.

For the researchers already working in this field, it is obvious that we are in the process of creating a technology that has the potential to change our lives for the better. However, we cannot work in this exciting endeavor alone. Only coordinated actions from all interested parties will make the vision of COs a reality.

Feel free to jump in!

List of Authors

Michel BANÂTRE
INRIA, Institut National de Recherche en Informatique et Automatique
France

Şebnem BAYDERE
Yeditepe University
Istanbul
Turkey

Philippe BONNET
DIKU, University of Copenhagen
Denmark

Erdal ÇAYIRCI
Yeditepe University
Istanbul
Turkey

Paul COUDERC
INRIA, Institut National de Recherche en Informatique et Automatique
France

George COULOURIS
University of Cambridge
UK

Onur ERGIN
Yeditepe University
Istanbul
Turkey

Elena FASOLO
DEI, University of Padova
Italy

İsa HACIOĞLU
Yeditepe University
Istanbul
Turkey

Paul HAVINGA
University of Twente
Netherlands

Maria LIJDING
University of Twente
Netherlands

Pedro José MARRÓN
University of Bonn
Germany

Ivan MAZA
AICIA, Asociacion de Investigacion y Cooperacion Industrial de Andalucia
Spain

Nirvana MERATNIA
University of Twente
Netherlands

Daniel MINDER
University of Stuttgart
Germany

Anibal OLLERO
AICIA, Asociacion de Investigacion y Cooperacion Industrial de Andalucia
Spain

Chiara PETRIOLI
University of Rome
La Sapienza
Italy

Marcelo PIAS
University of Cambridge
UK

Kay ROEMER
ETH, Eidgenoessische Technische Hochshule
Zurich
Switzerland

Silvia SANTINI
ETH, Eidgenoessische Technische Hochshule
Zurich
Switzerland

Antidio VIGURIA
AICIA, Asociacion de Investigacion y Cooperacion Industrial de Andalucia
Spain

A. VITALETTI
CINI, National Inter-University Consortium for Computer Science
Italy

Thiemo VOIGT
SICS, Swedish Institute of Computer Science
Sweden

Adam WOLISZ
TUB, Technische Universitaet Berlin
Germany

Andrea ZANELLA
DEI, University of Padova
Italy

Michele ZORZI
DEI, University of Padova
Italy

Index